The Anthropocene: Politik—Economics—Society—Science

Volume 11

Series Editor

Hans Günter Brauch, Peace Research and European Security Studies (AFES-PRESS), Mosbach, Baden-Württemberg, Germany

More information about this series at http://www.springer.com/series/15232
http://www.afes-press-books.de/html/APESS.htm
http://www.afes-press-books.de/html/APESS_11.htm

Zerrin Savaşan

Paris Climate Agreement: A Deal for Better Compliance?

Lessons Learned from the Compliance Mechanisms of the Kyoto and Montreal Protocols

Zerrin Savaşan
Department of International Environmental
Law and Politics
Selçuk University
Konya, Turkey

The cover photo of the final ceremony after the adoption of the Climate Change Agreement in December 2015 (© UNFCCC 2015) was made by the Secretariat of the United Nations Framework Commission in Bonn, which granted permission for it to be used here.

More on this book is at: http://www.afes-press-books.de/html/APESS_11.htm

ISSN 2367-4024   ISSN 2367-4032   (electronic)
The Anthropocene: Politik—Economics—Society—Science
ISBN 978-3-030-14312-1   ISBN 978-3-030-14313-8   (eBook)
https://doi.org/10.1007/978-3-030-14313-8

Library of Congress Control Number: 2019932614

Copy-editing: PD Dr. Hans Günter Brauch, AFES-PRESS e.V., Mosbach, Germany
English Language Editor: Dr. Vanessa Greatorex, Chester, England

This Springer imprint is published by the registered company Springer Nature Switzerland AG
The registered company address is: Gewerbestrasse 11, 6330 Cham, Switzerland

To all those
who understand their responsibility
to the environment
and the future

Preface

"However much you know, what you say is as much as what
is understood of it."

Mevlana Jalaluddin Rumi (1207–1273)

With the inspiration of Mevlana, the main aim of this book has been to make it as easy to understand as possible. Therefore, from the beginning, simplification and clarification have been the fundamental targets during the writing process. Yet, due to the fact that its main research subject, 'compliance/compliance mechanisms', with their increasing number of legal documents and institutions, has a very complicated structure, achieving these targets has not been easy in some phases. In fact, during the writing process, while trying, on the one hand, to simplify the issues in order to make every aspect of the book more understandable, I have tried, on the other hand, to avoid restricting its scope and meaning.

Despite the difficulties, undertaking detailed research on innovative compliance mechanisms (CMs) has enabled me to gain extensive knowledge of compliance issues in general and compliance mechanisms in particular, and also gain experience in researching this kind of highly complex issue. In addition, as the CMs have been created in very recent decades in most of the MEAs, and most of those are still in the process of being developed, but have not been completely developed yet, the book's attempt to be one of the first to ask how to enhance compliance with CMs under MEAs, and to contribute to discussions on them and also to the recent debate on the options for an improved compliance system under the Paris Climate Agreement (PCA), has made both the research and the writing process more challenging but, at the same time, more exciting for me.

While coming to the end of this research adventure—full of challenges but also full of joy and motivation resulting from new avenues endowing the author with new areas for exploration—in line with Mevlana's words, I just hope now that I have succeeded in explaining everything clearly enough to be understood correctly by the book's readers.

Konya, Turkey

Zerrin Savaşan

Acknowledgements

This work is based on my unpublished Ph.D. thesis (entitled 'CMs under Multilateral Environmental Agreements (MEAs): A Comparative Analysis of CMs under Kyoto and Montreal Protocols'), which was submitted to the Graduate School of Social Sciences, Middle East Technical University (METU), on 10 May 2013. It was updated/revised in line with the recent developments, and its focus was shifted towards fostering discussion on whether the Paris Climate Agreement (PCA) is a deal for better compliance, what it promises for improving compliance and how to enhance compliance under the PCA.

It was supported by scholarships granted by TÜBİTAK for participating in the '2012 Summer Session on Environmental Law' held 29 May–22 June 2012, by Washington College of Law within American University in Washington, DC, USA; by The Council of Higher Education for a six-month research visit (from 1 October 2010 to 31 March 2011) to Max Planck Comparative Public Law and International Law Institute, Heidelberg University, Germany; by the Academic Staff Training Programme for a six-month research visit (from 1 February 2010 to 31 July 2010) to the Center for Environmental Studies, Vrije University, Amsterdam, Holland; and finally by the opportunity granted me in the context of the Project (No. 1288) conducted under State Planning Organization-Scientific Research Projects to visit the World Trade Organization (WTO) (Geneva, Switzerland) from 4 to 28 December 2009 to interview international experts and officials from the WTO. I am very thankful to all these institutions for providing me with financial support and thereby giving me the chance to conduct my research with the aid of better facilities.

Furthermore, I greatly appreciate my thesis advisor, Assoc. Prof. Dr. Şule Güneş, for her academic guidance, insights, criticism and attention throughout the research. I also would like to thank each member of my Dissertation Committee, Prof. Dr. Meliha Altunışık, Assoc. Prof. Dr. Funda Keskin, Assoc. Prof. Dr. Oktay Tanrısever and Assist. Prof. Dr. Osman Balaban, for their valuable comments, suggestions and contributions.

I wish to express my deepest gratitude to my family and friends for their support and inexhaustible patience throughout the research. They always motivated me throughout the most difficult times, and never stopped encouraging me, even though it was a very long and difficult journey for them as well. I have always been proud of my family, whose existence has always been the most valuable thing to me.

I am also very appreciative of Dr. Vanessa Greatorex's extra effort in getting the book ready to be published. I'm so glad to have her as an English language editor of the book. She has been right there with her willingness to help out wherever needed. She has really helped to make things run smoothly taking the time out of her busy schedule. I couldn't have done it without her assistance.

Last but not least, I am especially indebted to PD Dr. Hans Günter Brauch, the editor of my book, who is also the chairman of Peace Research and European Security Studies (AFES-PRESS) and the editor of the Hexagon Book Series on Human and Environmental Security and Peace, the Springer Briefs in Environment, Security, Development and Peace (ESDP), and the Springer Briefs in Pioneers in Science and Practice (PSP). I would like to express my deepest appreciation to him for his support of young authors/scholars of different disciplines, countries, religions and cultures, and of their career goals, giving them time, energy and assistance to pursue those goals. I am especially grateful to him for his endless humane support and patience with me throughout the whole process. This work would not have been possible without his support; it would certainly have been more challenging to complete it successfully.

Contents

Acronyms and Abbreviations

AAUs	Assigned amount units
AF	Adaptation Fund
AG-13	Ad Hoc Group on article 13, UNFCCC
ARR	Annual report review
Art./Arts.	Article/Articles
CBDR	Common but differentiated responsibility
CBDR-RCs	Common but differentiated responsibility and respective capabilities
CDM	Clean development mechanism
CEIT	Countries with economies in transition
CERs	Certified emission reductions
CFI	Court of first instance
CH_4	Methane
CISs	Commonwealth of Independent States
CITES	Convention on International Trade in Endangered Species of Wild Fauna and Flora
CM	Compliance mechanism
CMA	Conference of the Parties serving as the meeting of the Parties to the Paris Agreement
CO_2	Carbon dioxide
ComplCom	Compliance Committee
COP	Conference of the Parties
CSD	Commission on Sustainable Development
DSP	Dispute settlement procedure
DSU	Understanding on rules and procedures governing the settlement of disputes
EC	European Court of Justice
ECHRs	European Convention on Human Rights
ECtHRs	European Court of Human Rights
EEAP	Environmental Effects Assessment Panel
EEC Treaty	European Economic Community Treaty

EGTT	Expert Group on Technology Transfer
EMEP	Cooperative programme for the monitoring and evaluation of the long-range transmission of air pollution in Europe
ERR	Expedited review report
ERT	Expert review team
ERUs	Emission reduction units
ET	Emission trading
GC	Governing Council
GCF	Green Climate Fund
GEF	Global environment facility
GEG	Global environmental governance
GEOMAP	Global Environmental Observing, Monitoring and Assessment Programme
GHGs	Greenhouse gases
HFCs	Hydrofluorocarbons
IAEA	International Atomic Energy Agency
ICJ	International Court of Justice
IDR	In-depth review
IEL	International Environmental Law
IEP	International Environmental Politics
IET	International Emission Trading
IGOs	Intergovernmental organizations
IL	International Law
ILO	International Labour Organization
IMF	International Monetary Fund
ImplCom	Implementation Committee
INDCs	Intended nationally determined contributions
IO	International organization
IPCC	Intergovernmental Panel on Climate Change
IRs	International relations
IRD	International Regimes Database
IRR	Initial review report
ITLOS	International Tribunal on the Law of the Sea
ITPGRFA	International Treaty on Plant Genetic Resources
ITTA	International Tropical Timber Agreement
IUCN	International Union for the Conservation of Nature
JI	Joint implementation
JWG	Joint Working Group
KP	Kyoto Protocol
LDCF	Least Developed Countries Fund
LoA	Logic of appropriateness
LoC	Logic of consequences
LoTs	Law of Treaties
LRTAP	Convention on Long-range Transboundary Air Pollution
LULUCF	Land use, land-use change and forestry

MARPOL	International convention for the prevention of maritime pollution from ships
MCP	Multilateral consultative process
MEA	Multilateral Environmental Agreement
MF	Multilateral Fund
MOPs	Meeting of the Parties
MP	Montreal Protocol on substances that deplete the ozone layer
N₂O	Nitrous oxide
NAPAs	National Adaptation Programmes of Action
NCM	Non-compliance mechanism
NCP	Non-compliance procedure
NDC	Nationally determined contributions
NGOs	Non-governmental organizations
ODSs	Ozone-depleting substances
OECD	Organisation for Economic Co-operation and Development
OEWG	Open-ended Working Group
Para./Paras.	Paragraph/Paragraphs
PCA	Paris Climate Agreement
PCA-CPA	Permanent Court of Arbitration
PFCs	Perfluorocarbons
PPFC	Prototype Carbon Fund
PRTR	Protocol on Pollutant Release and Transfer Registers
QoI	Question of implementation
RoP	Rules of procedure
RUs	Removal Units
SAP	Scientific Assessment Panel
SBI	Subsidiary Body for Implementation
SBSTA	Subsidiary Body for Scientific and Technological Advice
SCCF	Special Climate Change Fund
SF₆	Sulphur hexafluoride
TEAP	Technology and Economic Assessment Panel
TF	Trust Fund
UN	United Nations
UNCLOS	United Nations Convention on the Law of the Sea
UNCTAD	UN Conference on Trade and Development
UNDP	UN Development Programme
UNEP	United Nations Environment Programme
UNESCO	United Nations Educational, Scientific and Cultural Organization
UNFCCC	United Nations Framework Convention on Climate Change
UNIDO	UN Industrial Development Organization
UNITAR	UN Institute for Training and Research
VC	Vienna Convention for the Protection of the Ozone Layer
VCLT	Vienna Convention on the Law of Treaties
WCO	World Customs Organisation
WEC	World Environment Court

WG   Working Group
WHO   World Health Organization
WMO   World Meteorological Organization
WSSD   World Summit on Sustainable Development
WTO   World Trade Organization

Chapter 1
Introduction

Intellectuals solve problems, geniuses prevent them.

Albert Einstein (1879–1955)

1.1 The Development of a Global Environmental Governance and the Problem of Effective Compliance with Multilateral Environmental Agreements

There have been always environmental problems in human history.[1] Of these, local and regional problems were first recognized as a major cause of concern. Global environmental problems, on the other hand, were recognized just four decades ago.

Indeed, human beings realized that environmental problems constituted a major cause of global concern in the late 1960s. In this period, it was realized that global environmental problems, like global economic politics, cut across state borders and give rise to troubles and conflicts arising from interdependence; and these troubles can only be dealt with through an internationally coordinated global environmental policy that can be achieved merely through effective international cooperation.

International cooperation has actually been taking place through participation in international organizations at global, regional and subregional levels since the early nineteenth century. To illustrate, the Congress of Viennna in 1815 and the series of conferences, such as the Hague Peace Conferences, which followed it, were the forerunners of the international cooperation that takes place today in the United Nations (UN).

When the UN was established in 1945, immediately after the Second World War (WW II), there were a large number of independent states participating in the

---

[1]In the final analysis, environmental problems result not just in the destruction of nature, but also in the destruction of human nature and so in the loss of moral values as well (Kalpaklı 2017). As the positive sciences are not enough to raise environmental consciousness among 'humans', who are the primary cause of the destruction, social sciences emerge as fundamental tools for creating environmentally-friendly generations (Kalpaklı 2014).

© Springer Nature Switzerland AG 2019
Z. Savaşan, *Paris Climate Agreement: A Deal for Better Compliance?*
The Anthropocene: Politik—Economics—Society—Science 11,
https://doi.org/10.1007/978-3-030-14313-8_1

international arena. Such a large number of states with diverse interests had made the task of achieving international cooperation complicated (Soroos/Marvin 1999: 27).

Despite this, the UN was able to promote, create and operate international cooperation over time. Particularly, with regard to addressing global environmental problems, it has played a fundamental role at the international level. Thus, greatly changing the existing system, it has become an important actor in the entire international legal order, and also in the development of globally coordinated environmental law and policy (Birnie/Boyle 1992: 33).

Indeed, the UN, with its crucial organs, significant conferences, specialized agencies and semi-autonomous bodies, has greatly contributed to the development of international environmental law and environmental policy-making.

The five global conferences, namely,

- The United Nations Conference on the Human Environment (UNCHE), commonly referred to as the Stockholm Conference convened in Stockholm, Sweden, in 1972
- The United Nations Conference on Environment and Development (UNCED), also known as the Rio Conference or the Earth Summit 1992, held at Rio de Janeiro, Brazil, in 1992
- The World Summit on Sustainable Development (WSSD), commonly referred to as the Rio+10 or the Earth Summit 2002, which took place in Johannesburg, South Africa, in 2002
- The United Nations Conference on Sustainable Development (UNCSD), also known as the Rio+20 or the Earth Summit 2012, convened in Rio de Janeiro, Brazil, in 2012
- The United Nations Sustainable Development Summit (UNSDS) held for the adoption of the post-2015 development agenda and convened in New York, United States of America (USA), in 2015

Have represented a massive effort to strengthen institution-building on environmental issues and to reduce various problems (Andresen 2007a: 318; Andresen 2007b: 457, 458; DeSombre 2006: 37).

It is here also necessary to mention the existence of numerous UN specialized agencies, such as the International Labour Organization (ILO), the World Health Organization (WHO), the International Monetary Fund (IMF), the United Nations Educational, Scientific and Cultural Organization (UNESCO) and semi-autonomous bodies like the UN Development Programme (UNDP), the UN Institute for Training and Research (UNITAR), the UN Conference on Trade and Development (UNCTAD) and the UN Industrial Development Organization (UNIDO), which have greatly contributed to intergovernmental forums for different aspects of the global environmental agenda.

In brief, the UN system has mostly shaped and framed the environmental agenda to date. It has fostered the development of international environmental law and tried to cope with the challenges of global environmental problems which range from climate change to pollution, from extinction of species to deforestation. However, at the same time, it has promoted the establishment of numerous structures, and thus the proliferation of international environmental institutions having mandate over different aspects of environmental problems. This is because, in the UN system, there has been no single institution which controls the management of global environmental issues. There have been a great number of other official and semi-official organizations and agencies also working in fields concerning global environmental protection (Hunter 2002: 217). So there have always been problems of coherence and compliance in the system, despite the strengthening – more or less – of the legal framework negotiated under the UN.

In addition to the continuing growth of the UN system, the number of MEAs has also rapidly increased, particularly in the period from Stockholm to Rio (Sand 2007), since the MEA, namely, the Convention on the Rhine, adopted in 1868 (Dodds 2001). This activity had led to almost 700 MEAs and 1,000 bilateral environmental agreements (BEAs) by the early twenty-first century (Mitchell 2003).[2] This increase in the number of MEAs, while creating 'treaty congestion' on the one hand, has also caused the decentralization of the institutional structure, on the other hand (Hunter et al. 2002; Sand 2007; Weiss 1995).[3] This is because all MEAs have established their own institutional bodies, such as the Conference of the Parties (COPs) and Meeting of the Parties (MOPs), which provide their operation in practice.

In recent decades, the international system has also changed in many aspects. In fact, even though states have remained as the main actors of the system, particularly in some areas, like global environmental issues, non-state actors – such as inter-governmental organizations (IGOs), non-governmental organizations (NGOs) and transnational corporations – playing progressive roles that directly or indirectly affect the global environment policies have also begun to be effective in global environmental law and policy.

Overall, to date, a great number of massive efforts-agreements, organizations and mechanisms have been created and developed for international cooperation for global environmental protection. This proliferation of institutions, agreements and actors in environmental issues has led to a great number of growing challenges which wait to be resolved; suffice here to mention fragmentation of global environmental governance (GEG),[4] lack of cooperation and coordination among

---

[2]It should be noted that as there is no consensus on their number among the researchers, it is possible to find different numbers in different researches. See also Kanie (2007: 68).

[3]Hunter et al. (2002: 455) also distinguish two categories of treaty congestion: substantive treaty congestion, implying the overlap of treaties' requirements, and procedural treaty congestion, implying the over-extension of the limited time and resources necessary to ensure compliance and to negotiate new agreements.

[4]Biermann et al. (2009) argues that there have been various degrees of fragmentation and that three criteria can be employed to differentiate them: degree of institutional integration; degree of overlap between decision-making systems; existence and degree of norm conflicts and type of actor constellations. Based on these criteria, he proposes to differentiate between three types of

international organizations, and lack of implementation, enforcement, effectiveness and compliance (Andresen 2007a, b; Arts/Leroy 2006; Gupta 2002; Mehta et al. 2001; Najam et al. 2006).

This situation has also led to an immense debate among scholars on the possible need and potential directions of a reform in the GEG to deal with these challenges. Within this debate, many proposed schemes – ranging from the reform of UNEP (Meyer-Ohlendorf/Knigge 2007), or the UN (Chambers 2005) to the creation of a World or Global Environmental Organization (Biermann 2007; Charnovitz 2002, 2005; Montini 2011; Rechkemmer 2005; Simonis 2002a, b), or a sort of permanent G-20 meeting on environment and sustainable development, based on the present G-20 mode (Montini 2011), or a World Environment Court (Pauwelyn 2005; Rest 2000) – have been discussed among scholars to eliminate the weaknesses of the system.

Yet, despite efforts and attempts to strengthen and reform it, it seems that the immense growth of the system seems to continue making it increasingly more incoherent and less effective than it should be, with the problems that it has created. So, recently, in the Earth Summit 2012, it has been particularly stressed that a strengthened institutional framework, which can enhance coherence, coordination and cooperation, can reduce fragmentation, duplication and overlap, and can increase effectiveness, efficiency and transparency within the operation of the system (Rio+20 Report 2012: 14), is highly necessary for "respond[ing] coherently and effectively to current and future challenges and efficiently bridg[ing] gaps in the implementation of the sustainable development agenda" (Rio+20 Report 2012: 14).

Among those challenges associated with environmental issues, the need to strengthen 'compliance' as an escalating problem in GEG has been specifically stressed in Agenda 21 of the Earth Summit 1992, as a strategy for the effective governance of environmental problems.

In fact, Agenda 21 provides that:

> [e]ach country should develop integrated strategies [including "mechanisms for promoting compliance"] to maximize compliance with its laws and regulations relating to sustainable development (section I, chapter 8, para. 21).

It is also explicitly underlined that:

> [s]tates should further study and consider methods to broaden and make more effective the range of techniques available at present, [which] may include mechanisms and procedures for the exchange of data and information, notification and consultation regarding situations that might lead to disputes with other States in the field of sustainable development and for

---

fragmentation: (1) synergistic fragmentation, (2) co-operative fragmentation, and (3) conflictive fragmentation. He also reviews the claims in favour and against more integrated or more fragmented governance organized around the questions of (1) the relative speed of reaching agreements (2) the level of regulatory ambition that can be realized (3) the level of potential participation of actors and sectors and (4) the equity concerns involved. For details see, Biermann et al. (2009: 14–40).

effective peaceful means of dispute settlement in accordance with the Charter of the United
Nations, including, where appropriate, recourse to the International Court of Justice, and
their inclusion in treaties relating to sustainable development (Section IV, chapter 39,
para. 9).

Although Agenda 21 is a soft-law document, when the efforts for creating an
all-encompassing UN Treaty (such as, for instance, the 'Draft International
Covenant on Environment and Development', which contains provisions for
reporting, compliance mechanisms and dispute-settlement mechanisms) are taken
into account, its soft-law recommendations become more important. This is particularly because the draft Covenant aims not only to consolidate existing legal
principles related to the environment and development, but also to convert Agenda
21's soft-law recommendations into legally binding hard international law (IUCN
2010).

Through Agenda 21, UNEP and other organizations have also been empowered
to support compliance activities. The UNEP Special Session of the Governing
Council (GC) held in 2000 also recognizes the central importance of environment,
compliance, enforcement and liability as well as capacity-building. UNEP, particularly with the adoption of the 'Programme for the Development and Periodic
Review of Environmental Law for the First Decade of the 21st Century
(Montivideo III Programme)' in 2001, has started to make 'compliance issue' a core
concern, stressing the ineffectiveness of environmental law (United Nations
Environment Program Governing Council 2001).

As a result of these efforts, via the Earth Summit 2002, the need to promote
compliance and so the establishment and operation of mechanisms that can improve
and maintain compliance has emerged as a core theme of GEG. In the same year,
UNEP adopted Guidelines on Compliance with and Enforcement of Multilateral
Environmental Agreements (MEAs) whose Chapter I specifically addresses compliance with MEAs (UNEP 2006).

In the Earth Summit 2012, compliance issues were not specifically emphasized,
most probably due to the fact that the Conference's particular focus was on
establishing the concept of green economy and building an institutional framework
for materializing sustainable development goals. However, through the
Conference's non-binding document, 'The Future We Want,' it is also reaffirmed
that:

the means of implementation identified in Agenda 21, the Programme for the Further
Implementation of Agenda 21, the Johannesburg Plan of Implementation (...) are indispensible for achieving the full and effective translation of sustainable development commitments into tangible sustainable development outcomes (Rio+20 Report 2012: 48).

Thus, through this gradually growing recognition that 'compliance should be
taken as a priority in the coming near future,' 'compliance issue' has turned out to
be one of the major concerns for GEG in both IEL and IEP in the current decade.
Indeed, it has recently been realized that, to cope successfully with environmental
problems, it is not sufficient to adopt legally binding commitments on environmental issues; the adoption of environmental agreements is only the beginning of

the process. It is also vital to provide full compliance to ensure the effectiveness of both these agreements and environmental governance, due to the fact that:

> [c]ompliance remains an important objective not only because it often correlates with better levels of environmental protection, but also because it improves environmental governance by maintaining the credibility of environmental regimes... (Stephens 2009: 63–64).

In fact, for effective global governance, besides the other areas of concern, it is also necessary to provide the complying actors with the agreements that they have participated. If all the actors adhere to their agreement commitments, even if the problem cannot be solved completely, it forms an important step towards incrementally better compliance, better governance and ultimately better sustainable development.

This awareness of the importance of compliance has led to a serious debate among IEP and IEL scholars, and there have begun to be discussions about which ways best enable states to meet their environmental agreement obligations.

Consequently, to improve compliance with MEAs, the search has begun to find new methods which will not merely solve the problems but prevent them before they occur, in accordance with Einstein's words at the start of this chapter. That is, these new methods should be based on a preventative approach to identify compliance problems and find solutions (through assistance or any other measures) before non-compliance actually occurs. They should also be complex, detailed and technical, but at the same time more flexible and dynamic so that they can adapt easily to the frequent changes in global environmental issues.

Thus, the gradually increased attention which is being given to the issue of compliance and intense studies by scholars have generated new procedures called 'compliance mechanisms' in the field of environmental protection which supplement the methods previously established under the rules of international law.

Currently, it is possible to see these new elaborated and flexible mechanisms – institutions and procedures – established within some MEAs, whose mandate would be to strengthen compliance with obligations deriving from related MEAs. Though the number of those that have already been developed is not high, and most of them are in the process of being developed, they are expected to strengthen compliance – and, as a consequence, environmental governance – or at least to play a significant role.

However, their effective operation and ability to induce better compliance remain important questions both in academic studies and in practice. This is because, even though they have characteristics which make them stronger than traditional mechanisms at dealing with the compliance issue more effectively (as highlighted particularly in Chap. 4), they also have some weaknesses that can undermine their effectiveness in the final analysis (Table 1.1).

In order to figure out these weaknesses and find ways to improve compliance under MEAs, it is necessary to examine their compliance mechanisms and discuss options for their improvement. Therefore, in the following chapters, the focus will be on methods of fostering compliance under MEAs, starting with an analysis of

Table 1.1 Development of compliance mechanisms as a matter of special concern

The recognition of environmental problems as a global concern

⬇

Increase in the number of international
environmental institutions-agreements-actors
tor resolve these problems

⬇

A great number of growing challenges,
including lack of compliance

⬇

Greater need to ensure
compliance with environmental commitments;
gradual recognition of the compliance issue

⬇

Need for preventative approach to identify compliance problems.
Need to establish new flexible mechanisms

⬇

The establishment of compliance mechanisms

Source The author

practices in two specific environmental areas: ozone layer protection and climate change prevention.

In particular, the research is based on two cases, the CM of the Protocol to the 1985 Vienna Convention (VC) on Substances that Deplete the Ozone Layer (Montreal, Canada, 16 September 1987) and the CM of the Protocol to the 1992 United Nations Framework Convention on Climate Change (UNFCCC) (Kyoto, Japan, 11 December 1997) and their outcomes in practice. Using the evidence from these case studies, it seeks to understand the possible advantages/disadvantages and strengths/weaknesses of different options and thus gauge the likely compliance status of the parties with regard to their commitments under the Climate Agreement recently adopted in COP 21, held in Paris from 30 November to 13 December 2015.

This Climate Agreement, known as the Paris Agreement on Climate Change (Paris, France, 12 December 2015), contains a specific provision on compliance,

Article 15, titled "Facilitating implementation and compliance." However, it gives no details on the modalities and procedures for the effective operation of the compliance system under the Agreement and leaves elaboration on them to the Conference of the Parties (COP) serving as the meeting of the Parties (MOP) to the Paris Agreement at its first session (CMA1) (Art. 15.3, Paris Agreement; COP 21, Decision 1, para. 103). Nonetheless, there is still no tangible formulation to date on compliance issues or the Compliance Committee referred to under Art. 15, PCA. So there are several topics requiring further clarification in the next CMAs, and considering/illustrating possible options on these topics in advance, in the light of the experience gained from the compliance mechanisms of other MEAs to date, can be quite useful for designing an effective compliance mechanism under the PCA.

In this respect, before illustrating the CMs under the Kyoto and the Montreal Protocols (Chaps. 5 and 6), some more general issues will be investigated and discussed, e.g. the meaning of the concepts of compliance, CM and MEA (Chap. 2); the related theories and two basic explanatory models on compliance (Chap. 3); the development of CMs and the limitations of traditional means (Chap. 4). Based on the findings, a debate will be provided on the options for an improved compliance system under the Paris Climate Agreement (PCA) (Chap. 6).

1.2   The Purpose of the Book

As stated earlier, the focus of this study is to analyse the ways of improving compliance under MEAs, one of the areas which has become a central concern for global environmental governance in recent times.

Furthermore, in order to figure out the weaknesses of the current compliance mechanisms under MEAs and to discuss ways to create a better system for improving compliance under the PCA, the lessons learned by the CMs of the Montreal Protocol and the Kyoto Protocol will be taken into account.

The aim of the following chapters is not to question the effectiveness of these mechanisms – whether, how and under what conditions they are influential in altering the behaviours of actors, and gradually the compliance and environmental quality. The main intention is just to provide an in-depth understanding of the existing mechanisms and elaborate their present features on the basis of two case studies, which are generally considered to be the most successful ones.

This is because of the following reasons:

- It remains incredibly difficult to measure the direct effectiveness of CMs on compliance. Indeed, to analyse them and reach a certain judgement on their effectiveness, even their "likely effectiveness," is a very complex issue because of various relevant processes and dynamics incorporated into the mechanisms that deserve attention to give a sufficient response to the question (Hovi et al. 2005: 4).

- Almost all mechanisms developed under different MEAs are still in the stage of establishment or development; therefore, their effectiveness cannot be regarded as fully tested. For instance, the CM developed under the Kyoto Protocol was adopted on the decision of the COP in 2001 (COP 7, Decision 24). Many issues necessary to bring the Protocol into operation – except the legal status of enforcement consequences – have been resolved in 2005 through the confirmation of COP 7, Decision 24 in MOP 1, Decision 27. Individual cases brought before the Committee, on the other hand, only began to be considered in 2006. In addition, the CM under the Montreal Protocol continues to develop and has not yet reached an end. It has been further improved through the adoption of a draft NCP in 1990 and the adoption of the current NCP in 1992. Finally, the modifications in 1998 have signalled the possibility of the need for further modifications in the future.
- Compliance is a very complex issue. A great range of factors can affect it, like the object and scope of MEA, the characteristics of the accord, the characteristics of the country, or party to the accord, and other factors in the international environment (the role of NGOs, actions of other states and the role of IGOs) (Jacobson/Weiss 2001). Therefore, it also requires a great number of efforts to be enhanced, like: strengthening its empirical and theoretical foundations, the role of civil society, the norms complementing and supporting compliance and rule of law; building capacity of regulators and those they regulate and political will; expanding funding, applying new analytical tools; diagnosing specific problems and understanding and empowering key actors (Zaelke et al. 2005).

In this respect, offering a starting point on which further research related to the compliance issue under the PCA and improving compliance under the PCA at first hand, and under MEAs in general, the main purpose of the present inquiry is:

- To lay out a research agenda on the compliance issue and CMs relying on three main components: gathering information, procedures/institutional structure, measures;
- To analyse the role of CMs in ensuring better compliance and the ways of enhancing compliance under the current systems of CMs, pursuing an approach which includes the lessons from two case studies;
- To question the options for an improved compliance system under the PCA;
- To foster discussion on whether the PCA is a deal for better compliance and what it promises for improving compliance with a greater focus among academics;
- To find out how the PCA contributes to the improvement of the compliance issue.

In brief, the present analysis intends to be one of the first attempts to ask how to enhance compliance under the PCA. At the very least, it is designed to contribute to the related discussion.

1.3   The Book's Structure

In order to figure out the weaknesses under the current systems of compliance mechanisms of MEAs and to search for ways to improve compliance under MEAs in general and under the PCA in particular, the following chapters seek responses to some basic questions:

1. *Drawing a conceptual framework for compliance and compliance mechanisms.* What do the concepts of compliance, CM and MEA mean? What does compliance mean? What are compliance's differences from related concepts such as implementation, effectiveness and enforcement? What should be understood from compliance mechanisms (CM) under multilateral environmental agreements (MEAs)? And what do the concepts of MEA and regime refer to? What do they mean in general, and in the context of this study in particular?
   In this part, the main goals are to clarify what these concepts mean, and to constrain them to the frames of these definitions, thus, to enable their correct application in practice, to show the direction of the perspective pursued in the study and to facilitate the discussion on compliance and CMs in a more comprehensible manner.
2. *Drawing a theoretical framework for compliance and compliance mechanisms.* What are the competing theoretical perspectives from both international law and international relations on compliance debate? What are their main characteristics? What are the features distinguishing them from each other? What are the main theoretical models of compliance? To what extent and how do these models provide insights into the compliance debate? What are the main features of these models? What are the similarities and differences between them? Which one can be used as a theoretical framework in the context of the book itself to ensure better compliance?
   Through all these questions, this part will try to give a theoretical explanation of these mechanisms and put forth their theoretical basis in order to make the way they function more understandable in practice.
3. *Development of CMs and their main components.* What are the reasons, the preparatory and triggering impacts behind their development? What are the limitations stemming from the traditional means of international law, i.e. law of treaties, responsibility of states, dispute settlement procedures (DSPs) , including diplomatic and judicial means? What are the features making CMs more attractive than these traditional means? What are the distinct characteristics between the CMs and DSPs? What are the institutions created under an MEA, and their functions? What are the main components of CMs? And what are the fundamental features of these components? What is the legal basis of the NCPs? What are the main powers and functions of the committees created under the NCPs? What are the procedural phases and safeguards in NCPs? How can the binding effect of the response measures be demonstrated?

The main objective here is to draw a general framework based on the main characteristics of CMs, and thus explain the current system briefly and plainly, without reference to its specific details, in order to provide background information on the system's general characteristics.

4. *Case study of the compliance mechanism regarding ozone depletion.* How does the CM function in the context of the Montreal Protocol? How did it develop? What are the basic characteristics of its components (gathering information, procedures/institutional structure, measures)? To what extent can its provisions on compliance mechanisms be materialized in practice? How well does it function in practice?

5. *Case study of the compliance mechanism regarding climate change.* While doing this analysis, there will be two main sections, one of which is on the CM of the Kyoto Protocol and the other on the Paris Climate Agreement (PCA).

In the first section, pursuing the same method followed in the previous section, the chief research question will be 'How does the CM function in the context of the Kyoto Protocol?' In this regard, the development of the CM up to the present will be studied and then the components of the mechanism (gathering information, procedures, institutional structure, measures) will be evaluated to provide a full understanding of the process. In order to illustrate and demonstrate the issues the study is concerned with, what actually happens, how well they function in practice, and to what extent their provisions manage to be influential in practice will also be explored.

Based on the analysis of the mechanisms, in both case studies – CMs under both the Montreal and the Kyoto Protocols – the lessons learned will also be discussed on the basis of the following questions: Are CMs capable of influencing compliance in ways that other traditional mechanisms cannot or in ways that are more effective? Are they capable of ensuring better compliance with the environmental commitments brought by MEAs? If they are not, what are their weaknesses, how can they be improved and can they be made more effective in shaping actors' behaviour? Can this improvement in CMs deliver better compliance? If not, which ways should also be considered to strengthen compliance? Also, what should be done in the short and long terms to strengthen compliance?

In the second section, the PCA will be illustrated in detail, and background information on the Agreement, involving both its pre-Paris period and the Paris negotiations period, will be provided. Thus, it seeks answers to the following questions: What are the implications of the PCA for compliance issues? What happens next? What are the future options for a compliance mechanism under the PCA? What is its potential for better compliance? Benefiting from the lessons learned by the CMs of the Montreal Protocol and Kyoto Protocol, the weaknesses of the current compliance mechanisms under MEAs are thereby figured out and the options and proposals for a better system for improving compliance under the PCA will be discussed.

As it would be an overly ambitious task to illustrate and analyse in detail all aspects of all CMs and compliance issues, the study will be restricted to the above-mentioned aspects which have been indentified as the major focus areas of the study.

Nevertheless, additional information will be provided on some relevant and interesting aspects which contribute to the study of compliance mechanisms and compliance issues under MEAs in general and under the PCA in particular. In order not to exceed the scope of the study, these will not be examined in detail, but will only be touched on slightly.

1.4 Methodology

The method adopted for assessing the implications of the PCA on compliance issues and discussing the ways for better compliance under both the PCA and MEAs in general consists of a comparative case analysis based on interdisciplinary research benefiting from the literatures of two disciplines, both international relations and international environmental law.

Therefore, this intensive research is based on a comprehensive literature review and data collection (from dozens of books, a variety of journals and official documents) conducted in the libraries of the following Universities: Middle East Technical University (METU); Institute for Environmental Studies (IVM), Vrije University (VU); Max Planck Institute For Comparative Public Law and International Law (MPI), University of Heidelberg; Washington College of Law, American University (AU) (from 2009 to 2012); Selçuk University; and the Institute for Transnational Legal Research (METRO), Faculty of Law, Maastricht University (from 2015 to 2016).

The book's structure has been drawn on the basis of a conceptual framework (involving clarifications of the concepts of compliance, MEAs and CMs) and a theoretical framework (involving the most prominent approaches – from both international law and international relations – to the compliance debate and basic explanatory models of compliance: the management vs. the enforcement model). To support this main structure, two cases in particular – the compliance mechanism under the Montreal Protocol relating to the 1985 Vienna Convention on Substances that Deplete the Ozone Layer and the compliance mechanism under the Kyoto Protocol relating to the 1992 United Nations Framework Convention on Climate Change (UNFCCC) – have been selected to demonstrate and illustrate the issues with which the study is concerned.

Case Studies Selection: In this book the two mechanisms mentioned above are chosen as case studies, because:

- They are both established under globally-wide ratified MEAs, including "result-oriented obligation"[5];
- They are the most developed ones, and so it is considered that a great number of examples of practice and beneficial lessons for the themes under discussion can be drawn from them;
- Lessons learnt from them can be integrated into the analysis of the book to map out the possible ways of enhancing compliance under the PCA.

In this case selection, the following parameters have been adopted:

- *Frame of Reference*: CMs created under global MEAs.
- *Grounds for Comparison*: Both Protocols of Kyoto and Montreal are within the same cluster, that is:

  - Each deals with atmospheric problems;
  - The ozone-depleting substances (ODSs) are also greenhouse gases (GHGs);
  - They are interrelated and affect each other. For example, ultraviolet radiation resulting from ozone depletion reduces the capacity of plants and marine species to sequester atmospheric carbon dioxide (CO_2) and can heighten climate change (Oberthür 2001); so, in fact, each contributes to the other's success or failure.

With respect to the field study selection, the PCA was chosen because it provides a different system from the UNFCCC system, and it is necessary to discuss what kind of model should be adopted to enhance compliance under this new system. So, the question of how to ensure the compliance of the Agreement's parties gives rise to the need for an examination of the Agreement in terms of its implications for compliance issues and compliance mechanisms.

References

Agenda 21 (1992). Available at: https://sustainabledevelopment.un.org/content/documents/Agenda21.pdf.

Andresen, S. (2007a). The Effectiveness of UN Environmental Institutions, *International Environmental Agreements: Politics, Law and Economics*, 7, 4, 317–336.

Andresen, S. (2007b). Key Actors in UN Environmental Governance: Influence, Reform and Leadership, *International Environmental Agreements: Politics, Law and Economics*, 7, 4, 457–468.

---

[5]There are two main categories of treaty obligations, namely the "result-oriented" obligations and the "action-oriented" ones. Action-oriented treaties, such as the CITES, Espoo Convention, the Basel Convention, the Biological Diversity Convention, the Cartagena Protocol on Biosafety, and the International Convention for the Regulation of Whaling, do not set concrete environmental objectives to be achieved within a precise time limit. On the other hand, result-oriented ones, such as the Montreal-Kyoto Protocols and the Stockholm Convention, set precisely defined objectives with time limits for the actions to be taken (Beyerlin, Stoll and Wolfrum 2006).

Arts, B. and Leroy, P. (2006). Institutional Processes in Environmental Governance: Lots of Dynamics, Not Much Change? Arts and Leroy (Eds.), *Institutional Dynamics in Environmental Governance* (267–282). Dordrecht: Springer.

Beyerlin, U., Stoll, P.T. and Wolfrum, R. (2006). Conclusions from MEA Compliance. Beyerlin, U., Stoll, P.T., Wolfrum, R. (Eds.), *Ensuring Compliance with Multilateral Environmental Agreements: Academic Analysis and Views from Practice* (359–369). Leiden: Koninklijke Brill NV.

Biermann, F. (2007). 'Earth System Governance' as a Crosscutting Theme of Global Change Research. *Global Environmental Change,* 17, 3–4, 326–337.

Biermann, F., Pattberg, P., Asselt H. and Zelli, F. (2009). The Fragmentation of Global Governance Architectures: A Framework for Analysis. *Global Environmental Politics,* 9, 4, 14–40.

Birnie, Patricia W. and Boyle, Alan E. (1992). *International Law and the Environment.* New York, USA: Oxford University Press.

Bonn Draft Agreement (2015). Draft Agreement and draft decision on workstreams 1 and 2 of the Ad Hoc Working Group on the Durban Platform for Enhanced Action. Ad Hoc Working Group on the Durban Platform for Enhanced Action, Second session, part eleven, 19–23 October 2015, Bonn, Germany. ADP 2-11. Available at: http://unfccc.int/files/bodies/application/pdf/ws1and2@2330.pdf.

Chambers, B.W. (2005). From Environmental to Sustainable Development Governance: Thirty Years of Coordination within the United Nations. Chambers, B.W. and Gren, J.F. (Eds.), *Reforming International Environmental Governance: From Institutional Limits to Innovative Reforms* (13–39). New York: United Nations University Press.

Charnovitz, S. (2002). A World Environment Organization. Available at: http://www.unu.edu/inter-linkages/docs/IEG/Charnovitz.pdf.

Charnovitz, S. (2005). A World Environment Organization. Chambers, W.B. and Green, J.F. (Eds.), *Reforming International Environmental Governance* (93–123). Tokyo: United Nations University Press.

Decision 24/COP 7 (2001). Procedures and Mechanisms relating to Compliance under the Kyoto Protocol, Report of The Conference of The Parties on Its Seventh Session. Addendum Part Two: Action Taken By The Conference of The Parties. Marrakesh, 29 October–10 November 2001. FCCC/CP/2001/13/Add.3. Available at: http://unfccc.int/resource/docs/cop7/13a03.pdf.

Decision 27/MOP 1 (2005). Procedures and Mechanisms relating to Compliance under the Kyoto Protocol. Report of the COP serving as the MOP to the Kyoto Protocol on its First Session. Part Two: Action taken by the COP serving as the MOP at its First Session. Montreal, 28 November–10 December 2005. FCCC/KP/CMP/2005/8/Add.3. Available at: http://unfccc.int/resource/docs/2005/cmp1/eng/08a03.pdf#page=92.

Decision 1/MOP 8 (2012). Amendment to the Kyoto Protocol pursuant to its Article 3, paragraph 9 (the Doha Amendment), Report of the COP serving as the MOP to the Kyoto Protocol on its Eighth Session. Part Two: Action taken by the COP serving as the MOP at its Eighth Session. Doha, 26 November–8 December 2012. FCCC/KP/CMP/2012/13/Add.1. Available at: http://unfccc.int/resource/docs/2012/cmp8/eng/13a01.pdf.

Decision 1/COP 21 (2015). Adoption of the Paris Agreement. Report of the Conference of the Parties on its twenty-first session, held in Paris from 30 November to 13 December 2015. Addendum Part two: Action taken by the Conference of the Parties at its twenty-first session. FCCC/CP/2015/10/Add.1. Available at: http://unfccc.int/resource/docs/2015/cop21/eng/10a01.pdf.

DeSombre, Elizabeth R. (2006). *Global Environmental Politics.* New York and London: Routledge.

Dodds, F. (2001). Inter-linkages Among Multilateral Environmental Agreements. United Nations University Centre. 3–4 September 2001. Available at: http://archive.unu.edu/inter-linkages/eminent/papers/WG2/Dodds.pdf.

Elliot, Lorraine M. (1998). *The Global Politics of the Environment.* Houndmills, Basingstoke, Hampshire: Macmillan Press.

Geneva Negotiating Text (2015). Ad Hoc Working Group on the Durban Platform for Enhanced Action Second session, part eight Geneva, 8–13 February 2015. Agenda item 3 Implementation of all the elements of decision 1/COP.17. FCCC/ADP/2015/1. Available at: http://unfccc.int/resource/docs/2015/adp2/eng/01.pdf.

Gupta, J. (2002). Global Sustainable Development Governance: Institutional Challenges from a Theoretical Perspective. *International Environmental Agreements: Politics, Law and Economics,* 2, 361–388.

Hovi J., Stokke, O.S. and Ulfstein, G. (2005). Introduction and Main Findings. O.S., Hovi J. and Ulfstein, G. (Eds.), *Implementing the Climate Regime, International Compliance* (1–16). USA: The Fridtjof Nansen Institute.

Hunter, D., Salzman, J. and Zaelke D. (2002). *International Environmental Law and Policy.* New York: Foundation Press.

International Union for Conservation of Nature and Natural Resources (IUCN) Environmental Law Programme (2010). *Draft International Covenant on Environment and Development.* Environmental Policy and Law Paper No. 31 Rev. 3 (4th edn.). Prepared in cooperation with the International Council of Environmental Law.Gland, Switzerland: IUCN. Available at: http://www.uncsd2012.org/content/documents/IUCN%20Intl%20Covenant%20on%20Env%20and%20Dev%20EPLP-031-rev3.pdf.

Jacobson, H.K. and Weiss, E.B. (2001). Strengthening Compliance with International Environmental Accords. Diehl, P.F. (Ed.), *The Politics of Global Governance, International Organizations in an Interdependent World* (2nd edn.) (406–435). Boulder, CO: Lynne Rienner Publishers.

Kalpaklı, F. (2017). The Relationship Between Nature and Human Psyche. In *The Kite Runner, The Journal of International Social Research*, 10, 48, 82–85.

Kalpaklı, F. (2014). Exploitation of Women and Nature in *Surfacing*, Journal of Selçuk University Natural and Applied Science, Digital Proceedings of the ICOEST 2014, Side, Turkey, 14–17 May 2014, 788–802.

Kanie, N. (2007). Governance with Multilateral Environmental Agreements: A Healthy or Ill-equipped Fragmentation? Swart, L. and Estelle, P. (Eds.), *Global Environmental Governance: Perspectives on the Current Debate* (67–86). New York: Center for UN Reform Education.

Kiss, Alexandre and Shelton, Dinah (1994). *International Environmental Law.* New York: Transnational Publishers.

Kyoto Protocol to the United Nations Framework Convention on Climate Change (1998). United Nations. Available at: http://unfccc.int/resource/docs/convkp/kpeng.pdf.

Lima Call for Climate Action (2014). Decision 1/COP.20, Report of the Conference of the Parties on its twentieth session. Addendum Part Two: Action taken by the Conference of the Parties at its twentieth session. Lima, 1–14 December 2014. FCCC/CP/2014/10/Add.1. Available at: http://unfccc.int/resource/docs/2014/cop20/eng/10a01.pdf.

Mehta, L., Leach, M. and Scoones, I. (2001). Editorial: Environmental Governance in an Uncertain World. *IDS Bulletin,* 32, 4, 1–14. Available at: http://www.ids.ac.uk/files/dmfile/mehtaetal32.4.pdf.

Meyer-Ohlendorf, N. and Knigge, M. (2007). A United Nations Environment Organization. 124–142. Available at: http://www.centerforunreform.org/sites/default/files/GEG_Meyer-Ohlendorf_Knigge.pdf.

Mitchell, R.B. (2003). International Environmental Agreements: A Survey of Their Features, Formation, and Effects. *Annual Review of Environment and Resources*, 28, 429–461.

Montini, M. (2011). Reshaping Climate Governance for Post-2012. *European Journal of Legal Studies*, 4, 1 (Summer 2011), 7–24. Available at: https://ejls.eui.eu/wp-content/uploads/sites/32/pdfs/Spring_Summer2011/COMMENT_%20RE-SHAPING_CLIMATE_GOVERNANCE_FOR_POST-2012_.pdf.

Montreal Protocol on Substances that Deplete the Ozone Layer (1987). Available at: https://treaties.un.org/doc/Treaties/1989/01/19890101%2003-25%20AM/Ch_XXVII_02_ap.pdf.

Najam, A., Papa, M. and Taiyab, N. (2006). Global Environmental Governance: A Reform Agenda. Available at: http://www.iisd.org/pdf/2006/geg.pdf.

Oberthür S. (2001). Linkages between the Montreal and Kyoto Protocols, Enhancing Synergies between Protecting the Ozone Layer and the Global Climate. *International Environmental Agreements: Politics, Law and Economics*, 1, 3, 357–377.

Paris Agreement (2016). Available at: http://unfccc.int/files/home/application/pdf/paris_agreement.pdf.

Pauwelyn, J. (2005). Judicial Mechanisms: Is there a Need for a World Environment Court? Chambers, W.B. and Green, J.F. (Eds.), *Reforming International Environmental Governance: From Institutional Limits to Innovative Reforms* (150–177). Tokyo: United Nations University Press.

Porter, Gareth, Brown, Janet W. and Chasek, Pamela S. (2000). *Global Environmental Politics* (3rd edn.). Boulder, CO: Westview Press.

Rechkemmer, A. (Ed.) (2005). *UNEO – Towards an International Environment Organisation*, Baden-Baden: Nomos.

Rest, A. (2000). The Role of an International Court for the Environment. *Working Paper for the Conference Giornata Ambiente 2000*, 34–58. Available at: http://www.biotechnology.uni-koeln.de/inco2-dev/common/contribs/06_resta.pdf.

Rio+20 Report (2012). Report of the United Nations Conference on Sustainable Development. Rio de Janeiro, Brazil, 20–22 June 2012. A/CONF.216/16. Available at: http://www.or2d.org/or2d/ressources_files/rapport%20rio%20+%2020.pdf.

Sand, P.H. (2007). The Evolution of International Environmental Law. Bodansky, B., Brunnée, J. and Hey, E. (Eds.), *The Oxford Handbook of International Environmental Law* (29–43). New York: Oxford University Press.

Simonis, U.E. (2002a). Advancing the Debate on a World Environment Organization. *The Environmentalist*, 22, 29–42.

Simonis, U.E. (2002b). Global Environmental Governance: Speeding up the Debate on a World Environment Organization. Available at: https://www.econstor.eu/bitstream/10419/49568/1/351004734.pdf.

Soroos, Marvin S. (1999). 'Global Institutions and the Environment: An Evolutionary Perspective,' in Vig, Norman J. and Axelrod Regina S. (Eds.), *The Global Environment, Institutions, Law and Policy* (27–51). Washington, USA: Congressional Quarterly Press.

Stephens, T. (2009). *International Courts and Environmental Protection*. Cambridge, New York: Cambridge University Press.

UNEP (2006). *Manual on Compliance with and Enforcement of Multilateral Environmental Agreements*. UNEP: Nairobi. Available at: https://wedocs.unep.org/bitstream/handle/20.500.11822/7458/-Manual%20on%20Compliance%20with%20and%20Enforcement%20of%20Multilateral%20Environmental%20Agreements-2006743.pdf?sequence=3&isAllowed=y.

United Nations Environment Program Governing Council (2001). Decision 21/23. The Programme for the Development and Periodic Review of Environmental Law for the First Decade of the Twenty-first Century. Available at: http://wedocs.unep.org/bitstream/handle/20.500.11822/9938/GC22_MontevideoIII.pdf?sequence=1&isAllowed=y.

Weiss, E.B. (1995). International Environmental Law: Contemporary Issues and the Emergence of a New World Order. *Georgetown Law Journal*, 81, 1, 675–693.

Zaelke, D. Stilwell, M. and Young, O. (2005). Compliance, Rule of Law, and Good Governance, What Reason Demands: Making Law Work for Sustainable Development.

Chapter 2
Conceptual Framework Compliance and Compliance Mechanism

If names be not correct, language is not in accordance with the truth of things. If language be not in accordance with the truth of things, affairs cannot be carried on to success. When affairs cannot be carried on to success, proprieties and music will not flourish. When proprieties and music do not flourish, punishments will not be properly awarded. When punishments are not properly awarded, the people do not know how to move hand or foot. Therefore a superior man considers it necessary that the names he uses may be spoken appropriately, and also that what he speaks may be carried out appropriately. What the superior man requires, is just that in his words there may be nothing incorrect.

Confucius, The Analects-13
(cited by Lao-Tse 1901)

Thoughts without content are empty, intuitions without concepts are blind.

Immanuel Kant
(cited by Dicker 2004: 17)

2.1 Compliance: Just Conforming to Rules or More?

As indicated above by Kant and Confucius's words, first of all, it is necessary to clarify what is really said. Unless the meaning is understood correctly, a regulation might be applied wrongly or incompletely in practice due to substantial divergence in the ways in which it has been understood. Therefore, due to the fact that, "thinking, judging and carrying knowledge require both concepts and intuitions to which the concepts are applied" (Dicker 2004: 17), as stressed in Kant's words, in order "to relate concepts to practice," it is necessary to give the concepts practical meaning, content and direction (Schachter 1991: 2). Doing this also makes it possible to show the direction of the perspective pursued in the study and constrain it to the frames of these definitions.

For the purposes of this book about the ways of improving compliance under MEAs based on the lessons learned from two case studies, the CMs of the Montreal

© Springer Nature Switzerland AG 2019
Z. Savaşan, *Paris Climate Agreement: A Deal for Better Compliance?*
The Anthropocene: Politik—Economics—Society—Science 11,
https://doi.org/10.1007/978-3-030-14313-8_2

Protocol and the Kyoto Protocol and their outcomes in practice, on the basis of the Paris Climate Agreement (PCA) as the research field, three key questions should be posed at the outset:

- What does 'compliance' mean?
- What should be understood by the term 'compliance mechanism' (CM) in relation to multilateral environmental agreements (MEAs)?
- What do the concepts of MEA and regime refer to?

In fact, in order to enable and facilitate the discussion on compliance and compliance mechanisms under MEAs, it is substantially necessary to clarify what these concepts mean in general, and in the context of this study in particular.

Regarding compliance, to better understand this notion, it will be studied emphasizing the concept's differences from related concepts such as implementation, effectiveness and enforcement.

2.1.1 Related, but Different Concepts: Enforcement, Implementation and Effectiveness

In order to clarify the meaning of 'compliance', it is first necessary to distinguish it from the related concepts of enforcement, implementation and effectiveness, since these concepts can be used interchangeably in the literature. This can result in conceptual confusion, and so in incorrect analyses and diagnoses on the issues discussed under related subjects (Faure/Lefevere 1999; Özer 2009; Najamet al. 2006).

Before entering into an academic debate on these concepts, it is helpful to be aware of their standard dictionary definitions.

The first concept, enforcement, is generally defined as "the act of enforcing, compulsion, force applied".[1]

Implementation, on the other hand, means "providing practical means for accomplishing something, carrying into effect".[2]

The other related concept, effectiveness, is simply identified as "the quality of being able to bring about an effect".[3]

Finally, the concept of compliance is described, among other things, as "acting according to certain accepted standards"[4] and, in the dictionary of environmental economy, science and policy, as "the extent to which individuals or firms conform to environmental regulations" (Grafton 2001: 51).

---

[1]https://www.webster-dictionary.org/definition/enforcement.

[2]https://www.webster-dictionary.org/definition/implementation.

[3]https://www.webster-dictionary.org/definition/effectiveness.

[4]https://www.webster-dictionary.org/definition/compliance.

Having established their dictionary definitions, it is now easier to enter into a comprehensible academic debate on these concepts, to distinguish between them by stressing their differences, and finally to clarify the meaning of compliance.

It is widely recognized that, within the international legal order, there is no government, no hierarchical system and even no enforcement mechanisms (Raustiala/Slaughter 2002). The concept of enforcement is broadly defined as "rewards or promises of rewards as a means of strengthening incentives to comply" (Breitmeier et al. 2006: 148). Yet, this broad definition is not widely accepted in the literature. Instead, the concept is generally identified as a mechanism to force the parties to international treaties or the members of an intergovernmental organization to comply with and implement their obligations by using tools like financial penalties, the withdrawal of privileges, or trade, military and economic sanctions (Crossen 2003; Najam et al. 2006). Therefore, contrary to compliance, the concept of enforcement requires an enforcer and takes place after non-compliance occurs, its purpose being to bring the non-compliant parties back to compliance.

Secondly, implementation implies the transposition of the international legal obligations to the national legal systems to make them effective and operative in practice. The main elements of the process can be split into two phases: 1. transposition/adoption of national legal measures and thus creation of national legislation in line with international obligations; 2. adoption of legal measures and creation of necessary domestic and international institutions and other tools for operation in practice.

Therefore, implementation is "neither a necessary nor a sufficient" concept for explaining compliance (Raustiala/Slaughter 2002: 539), because it cannot occur automatically; likewise being in compliance does not always entail being compliant with certain obligations. In brief, implementation can be an important step towards compliance, but, compliance indeed conceptually "goes beyond implementation" (Chayes et al. 1998: 39).

The concept of effectiveness,[5] on the other hand, concerns the outcome of the conformity between the rule and behaviour, while compliance simply identifies this conformity, yet, "draws no causal linkage" between them (Raustiala 2000: 398).

In academic literature compliance is defined in various ways, according to the discipline or the perspective followed in this discipline and to the subject studied; as a result, it has a legal definition (compliance with law, regulations adopted to implement), a political definition (the impacts of the adopted documents on behavioural change,) and also a policy-orientated definition (attaining the goals) (Levy et al. 1995: 291–292).

In international relations, it can be characterized by a different set of questions on the basis of the perspective of the individual scholar:

---

[5]To understand the concept of effectiveness better, see also Maljean-Dubois/Richard (2004) on two tests of *efficacité* and *effectivité* (referring to De Visscher [1967]. *Les effectivités du droit international public*).

- Attaining the goals, achievement of objectives (Najam et al. 2006)[6];
- Problem-solving;
- Improvement of the situation through the actions questioned for effectiveness (e.g. the assessment of the 'counterfactual' of what would have happened if it had not adopted (Mitchell 2007: 898–899);
- Any behavioural change (if it is related to environmental issues or environmental changes) (Raustiala/Slaughter 2002).

However, neither full compliance with laws nor behavioural change seem adequate to yield the complete picture of effectiveness or measure it. This is because a variety of factors can play a role in assessing effectiveness (Raustiala 2001). To illustrate, Mitchell's work (2007) on the effectiveness of environmental agreements identifies three indicators of their effectiveness: outputs, outcomes, and impacts. In particular, Mitchell (2007: 896) puts forward the apprehension that outputs ("laws, policies and regulations that states adopt to implement an environmental agreement and transform it from international to national law") are not enough to induce behavioural changes (implying outcomes) and environmental quality change (implying impacts, improvements in environmental problems).

This indicates that compliance can be "neither necessary nor sufficient" for explaining effectiveness (Mitchell 1996: 25; Raustiala/Slaughter 2002: 539), because considerable behavioural and environmental change can still occur in some cases without compliance. In addition, the mere existence of compliance is not sufficient to lead to necessary behavioural change or environmental improvement. To illustrate, many MEAs make compliance easier, since they involve the "lowest common denominator dynamic" (Raustiala/Slaughter 2002: 539). Yet, despite this, they can still fall short of inducing behavioural change. Even when they do induce behavioural change, they may not induce environmental improvement due to unavoidable/irreversible environmental harm (Mitchell 1996).

In brief, it can be concluded that the compliance/non-compliance distinction is not always equivalent to the effective/non-effective distinction. Mitchell (2007) demonstrates this argument by analysing the influence/non-influence of environmental agreements, and concluding that a strict focus on compliance for measuring effectiveness results in analytically misleading consequences.

2.1.2   The Notion of Compliance in this Book

As outlined above, the concept of compliance is generally understood in the existing literature as behaviour which conforms with legal rules (Chayes/Chayes 1995; Chayes et al. 1995, 1998; Fisher 1981; Jacobson/Weiss 1998).

---

[6]Najam et al. (2006) also raise another concept, 'performance,' as their study's main concern, defining it as "the sum of implementation, compliance, enforcement and effectiveness."

These rules can provide for procedural requirements, like the adoption of necessary secondary regulations, the establishment of related institutions, agencies, or reporting and monitoring on the compliance, and substantive requirements which are intended to achieve the goals of the legal rules in question, such as control the activity, stop it, change the application standards of it etc.

Compliance includes two main dimensions: procedural and substantive (Breitmeier et al. 2006; Chasek et al. 2006; Crossen 2003; Jacobson/Weiss 1998; Zhao 2005). In evaluating compliance, it is therefore useful to distinguish between these two dimensions but, according to certain authors, it is also necessary to consider it within the spirit of the agreement (Weiss 1999; Weiss/Jacobson 2001). In fact, compliance can also be defined as "upholding the spirit" of the agreement (Najam et al. 2006: 46).

This study is mainly focused on compliance mechanisms created by different MEAs (case studies: the CMs under the Montreal Protocol and the Kyoto Protocol and their role for ensuring better compliance; and the field study: the Paris Climate Agreement and its potential mechanism for facilitating implementation and compliance).

So, the emphasis will be primarily on the following elements:

1. Just compliance with MEAs' substantive and procedural requirements, rather than with broader categories, such as upholding the spirit of the agreement or environmental commitments in general.

Here the question arises about what the spirit of the agreement implies; thus this notion should be precisely clarified.

According to the general rule of interpretation of international treaties, a treaty is to be interpreted in accordance with the ordinary meaning given to the terms of the treaty in the context and in the light of its object and purpose (Art. 31, para. 1, the Vienna Convention on the Law of Treaties [VCLT]). For the purpose of the interpretation of a treaty, the context comprises the preamble and annexes, in addition to the text itself (Art. 31, para. 2, VCLT). The notion of 'the spirit of agreement' corresponds to the notion of 'the object and purpose of a treaty' in addition to the context of the text, so it creates a broader normative basis. For the effective assessment and measurement of compliance, this aspect must indisputably be taken into account. However, it is extremely difficult to conduct "empirical analysis" when the concept of compliance has been extended to upholding the spirit of the agreement (Crossen 2003; Mitchell 1996).

Regarding general environmental commitments, it should be stressed that a treaty is interpreted taking into account, *inter alia*, "any relevant rules of international law applicable in the relations between the parties" (VCLT, Art. 31, para. 3. c). Therefore, to carry out an effective compliance assessment and measurement, it is again necessary to take into consideration general rules of international environmental law binding upon parties to an MEA. But again, this requires a very broad scope of analysis, and makes it complicated to evaluate compliance issues empirically.

Table 2.1 Concept of compliance

Related but different concepts		
Enforcement	Implementation	Effectiveness
• Broader than enforcement • Does not necessarily require enforcement/an enforcer • Does not mean that enforcement is required for compliance to occur • Can occur without anyone being forced to do anything • Does not aim to bring the non-compliant parties back to compliance, but to prevent non-compliance before it occurs	• Broader than implementation • Does not necessarily require implementation • Can occur with partial conformity between the rule and behaviour; does not require a causal linkage between them	• Neither necessary nor sufficient for effectiveness • Not necessarily related to effectiveness • Can occur automatically without transposition/adoption of national legal documents/measures • Even though there is enough compliance, it can still fall short of inducing behavioural change
In the Context of the Book		
1. Compliance with MEAs' substantive and procedural requirements 2. Whether States' actual behaviour conforms to the prescribed behaviour 3. Conforming to rules because of MEAs' compliance mechanisms		

Source The author

As it would be an overly ambitious task to analyse compliance in its all aspects, the study is restricted to the above-mentioned areas to avoid exceeding its scope.

2. Whether actors' actual behaviour conforms to the agreements' rules, whether they have changed their behaviour to accord with them. 'Actor' here is taken to mean the States who are party to MEAs. It is obvious that MEAs can affect other actors' behaviour (public-private); yet, "because states and supranational organizations like the EU are so far the only addresses" (Matz 2006: 303) of those instruments, here the focus will mainly be on States' compliance.

3. Treaty-induced compliance (conforming to rules because of MEAs' compliance mechanisms), so not on coincidental compliance, good faith non-compliance or intentional non-compliance (Mitchell 1994, 1996, 2007).

Overall, in this study, compliance will be used to explain exclusively whether the behaviour of the States which are party to the related MEA conforms to the agreement-based requirements (procedural-substantive) because of MEAs' compliance mechanisms, rather than conforms to the broader categories of other rules. Thus, according to this definition, non-compliance occurs when the States' behaviour departs from the related MEAs' requirements (Table 2.1).

2.2   The Notion of Compliance Mechanisms Within Multilateral Environmental Agreements

Before proceeding further, it is also necessary to clarify what the concept of compliance mechanism (CM) within a multilateral environmental agreement (MEA) refers to and how it can be defined in the current study. But, before the CM, it is also essential to identify the concepts of multilateral environmental agreement (MEA) and environmental regime.

2.2.1   Multilateral Environmental Agreement

Multilateral Environmental Agreement (MEA) refers to a multilateral treaty concerning environmental matters adopted between three or more states[7] which "[is] intended to be all-inclusive in membership and geographic and substantive scope" (Kimbal 1992: 30).

The MEA's institutional arrangements can be described in three ways (Churchill/Ulfstein 2000):

- The MEA can set up an intergovernmental organization (IGO) with a legal personality; or
- An existing IGO can serve as the institutional basis for the MEA; or
- The MEA can have no institutional arrangements at all.

The IGOs created under MEAs are quite different from the traditional IGOs, due to their "less formal, more ad hoc nature" (Churchill/Ulfstein 2000: 658). Because MEAs are subject to the law of treaties and not to international institutional law, the organs created under them are seen as "self-governing bodies, and thus as informal organizations" (Churchill/Ulfstein 2000: 648).

The CMs under MEAs generally operate through the Committees (ImplCom/ComplCom) created under the NCPs and the following institutions established under MEAs which can also be called "autonomous institutional arrangements" (Churchill/Ulfstein 2000): a conference or meeting of the parties (COP/MOP), a secretariat and one or more subsidiary-advisory bodies (technical or scientific bodies) and a financial mechanism.[8]

The MEAs have been concluded due to the emergence of the need for international standards in order to deal with environmental problems of worldwide concern, to promote cooperation and to prevent conflicts between states which could arise from these problems. By agreeing to negotiate or to join to them, the states voluntarily opt to be bound by their substantive and procedural obligations.

---

[7]For a detailed information on MEAs, see Özer (2009: 49–88).

[8]See *infra* Chap. 4 for elaboration.

MEAs have been creating separate international environmental regimes for over four decades, particularly since the Stockholm Conference on the Human Environment (1972). Therefore, it is also necessary to understand what the concept of 'environmental regime' refers to.

2.2.2 Environmental Regime

Regime is defined by Krasner (1983: 2) "as a set of explicit or implicit principles, norms, rules and decision-making procedures around which actors' expectations converge in a given area of international relations."

List/Rittberger (1992) add a further behavioural component to the four normative-institutional elements (principles, norms, rules and decision-making procedures) proposed by Krasner. In this usage, a regime can exist "if a certain density of rules and durability of norm- and rule-guided behaviour can be ascertained" (List/Rittberger 1992: 89, 90). On the other hand, a broader understanding of regimes implying "international social institutions consisting of agreed upon principles, norms, rules, procedures and programs that govern the interactions of actors in specific issue-areas" is also presented (Levy et al. 1995: 274). Regime is also regarded as "conglomerates of hard law rules (treaties), soft law instruments (declarations and recommendations) and institutions that are indispensible for the effective working of hard and soft law" (Lang 1992: 110).

This way of defining regimes as institutional mechanisms specified by a multilateral agreement for promoting the agreement's objectives predominates in the literature (Chhotray/Stoker 2009; Porter et al. 2000). It implies that regimes combine the characteristics of MEAs and international institutions, creating a "legal framework" through MEAs (Romano 2000: 86) and "a permanent mechanism for adapting this framework to the new needs of and challenges to the international community" (Romano 2000: 86).

Young (1999) classifies regimes in four categories: regulatory (identifying prosprictions and prescriptions for parties); procedural (establishing regular collective decision-making processes); programmatic (fostering the pooling of parties' resources for joint projects); and generative (fostering the development of new social practices, a constructivist concept of the regime implying that it generates new social practices) (Mitchell 1996).

It is possible to find several global environmental regimes which have usually been created in the form of a binding agreement. In particular, on global environmental issues, the most common kind of legal instrument which create regimes has been a treaty.

There are different kind of treaties in GEG such as framework conventions, self-contained agreements and umbrella agreements. Framework conventions (such as the UN Climate Change Convention and the Convention on Biological Diversity (Nairobi, Kenya, 22 May 1992) are negotiated before the adoption of one or more additional, more detailed legal instruments (Protocols) which elaborate on their norms and rules. Thus, the "framework convention-protocol approach" (Lang 1995: 288–289; Hempel 1996: 124) can overcome "the initial reluctance of some states to make firm commitments under conditions of scientific uncertainty" (Lang 1995: 288–289). Self-contained agreements, such as the UN Convention to Combat Desertification(UNCCD) (Paris, France, 17 June 1994) and the Convention on International Trade in Endangered Species of Wild Fauna and Flora (CITES) (Washington, USA, 3 March 1973), do not require additional protocols for implementation. Umbrella agreements (such as the Convention on Migratory Species of Wild Animals) allow the conclusion of other related agreements in the wider mandates of the convention.

Some of the agreements, like the Ramsar Convention, recognise a group of NGOs as International Organization Partners (IOPs). Ramsar Convention indeed grants an "additional participation status" to these organizations (Prideaux et al. 2015: 10).

Obviously, just states may still become parties to the Ramsar Convention itself, but NGOs may not. Nevertheless, formal recognition of NGOs (and intergovernmental organizations) by the Convention may be provided by conferring the status of International Organization Partners (IOPs) on them, in accordance with Resolution VII.3 of 1999 of the Conference of the Contracting Parties (COP 7, Resolution VII.3, 1999).[9] Furthermore, IOPs are requested to sign a Memorandum of Cooperation with the Bureau of the Convention, where the partnership agreement should be spelt out fully (COP 7, Resolution VII.3, para. 47, 1999). As a result, several Memoranda of Understanding/Cooperation with diverse IOs/NGOs have been concluded under the Ramsar Convention.

A wide variety of different issue areas, such as whales, maritime oil pollution, fishing, endangered species protection, etc. involving massive environmental concerns such as transboundary air and ocean pollution, ozone depletion, and global warming, are addressed legally through the formation of these regimes in which the basic principle is not to cause harm to other states beyond the limits of national jurisdiction, reflecting the notion of international environmental law found in Principle 21 of the Declaration of the Stockholm Conference (1972).

A few examples can be given by the conventions which have established the following global environmental regimes: International Convention for the Regulation of Whaling (Washington, USA, 2 December 1946), International

---

[9]The formal status of IOPs was established through Resolution VII.3 of 1999, and four organizations (namely, Bird Life International, IUCN – The World Conservation Union, Wetlands International and the World Wide Fund for Nature) were granted the status of IOP (COP 7, Resolution VII.3, 1999). Subsequently, through Resolution IX.16, the International Water Management Institute (IWMI) also became an IOP of the Convention (COP 9, Resolution IX.16, 2005). In line with Resolution VII.3, it was also decided that other relevant and interested international non-governmental and intergovernmental organizations meeting the criteria for formal recognition as Partners of the Convention should present an application to the Ramsar Secretariat for its consideration (for Memoranda of Understanding/Cooperation concluded with diverse IOs/NGOs under the Ramsar Convention, see:

http://archive.ramsar.org/cda/es/ramsar-documents-mous/main/ramsar/1-31-115_4000_2__).

Convention for the Prevention of Pollution of the Sea by Oil, replaced by the International Convention for the Prevention of Maritime Pollution from Ships (MARPOL) (London, UK, 2 November 1973), the Convention of Fishing and the Conservation of Living Resources of High Seas (Geneva, Swtizerland, 29 April 1958), the Convention on International Trade in Endangered Species of Wild Fauna and Flora (CITES) (Washington, USA, 3 March 1973), the Convention on Long-Range Transboundary Air Pollution (CLRTAP) (Geneva, Swtizerland, 13 November 1979), the UN Convention on the Law of the Sea (UNCLOS) (Montego Bay, Jamaica, 10 December 1982), the Basel Convention on the Control of Transboundary Movements of Hazardous Wastes and Their Disposal (Basel, Switzerland, 2 March 1999), the Protocol on Environmental Protection to the Antarctica Treaty (Madrid, Spain, 4 October 1991), the Convention on Biological Diversity (Nairobi, Kenya, 22 May 1992, the International Convention to Combat Desertification (Paris, France, 17 June 1994), the Stockholm Convention on Persistent Organic Pollutants (Stockholm, Sweden, 22 May 2001), the Vienna Convention for the Protection of the Ozone Layer (Vienna, Austria, 22 March 1985), supplemented by the Montreal Protocol (Montreal, Canada, 16 September 1987), the Framework Convention on Climate Change (New York, USA, 9 May 1992), and its Kyoto Protocol (Kyoto, Japan, 11 December 1997).[10]

2.2.3 Compliance Mechanisms Within Multilateral Environmental Agreements

Based on this information on the concepts of MEA and environmental regime, attention can now be directed to what the concept of compliance mechanism (CM) within a multilateral environmental agreement (MEA) refers to and how it can be defined in the current study.

First of all, it is possible to speak of mainly three types of mechanisms which can ensure compliance (Epiney 2006):

- Mechanisms ensuring compliance by confrontational means, like counter measures on the basis of the LoTs, withdrawal of membership's privileges, trade restrictions, responsibility-liability and DSPs (Wolfrum 1999);
- Mechanisms ensuring compliance by non-confrontational means, including providing economic benefits to balance environmental commitments, compliance assistance, capacity-building (Wolfrum 1999);
- Some procedural safeguards which are not defined in the scope of confrontational or non-confrontational means, "such as information rights or accord standing in internal judicial review procedures" (Epiney 2006: 325).

---

[10]See https://www.thinglink.com/scene/846020646906363905 for visual storytelling of key multilateral environmental agreements with interactive media.

Different categorizations can be applied to define mechanisms which ensure compliance. To illustrate:

- Mitchell (1994, 1996) and Faure/Lefevere (1999) mention three parts of any compliance system: a primary rule system, a compliance information system, and a non-compliance response system.
- Similarly, Chayes and Chayes consider a mechanism involving dispute resolution procedures, capacity-building, a transparent information system, treaty adaptation and response systems (Chayes/Chayes 1995; Chayes et al. 1998).
- Beyerlin et al. (2006) articulate six categories of main elements involved in a CM: 1. reporting; 2. assessment; 3. supplementary means of information-gathering 4. non-compliance procedure; 5. options for responding to verified cases of non-compliance; 6. institutional setting and procedural safeguards.

UNEP has also suggested a definition, according to which, CMs created under MEAs are designed to encourage compliance by their four components and so have to be identified under these four main components (UNEP 2005).

These components are:

1. Gathering information about the national performance of MEAs: this involves mechanisms for reviewing and assessing the performance of the parties in order to identify compliance problems beforehand;
2. Institutionalised multilateral non-compliance procedures (NCPs): these involve institutional structure (ComplCom/ImplCom), powers and functions and procedural phases/guarantees;
3. Multilateral non-compliance response measures: these involve appropriate responses (positive and negative) to non-compliance to produce and maintain an acceptable level of compliance by the parties;
4. Dispute settlement procedures (DSPs): these involve traditional means of settling disputes (diplomatic and judicial means).

As stated above, UNEP's definition explicitly identifies two separate components (information-gathering and response measures) of the mechanism rather than merely mentioning the non-compliance procedures as components of the mechanism. This is in accord with the recent tendency to define the CM as "an extension of the defining characteristics of a non-compliance procedure" (Raustiala 2001: 14). Yet, in some NCPs, setting up provisions relating to two components of this definition (information-gathering, response measures), can already involve these two components in themselves; as a result, in some cases, the concept of CM can be employed to substitute NCPs as well.

In addition, in UNEP's definition, the DSPs are counted as one of the components of a CM under the MEAs together with NCPs, response measures and gathering information. This is because the DSPs are 'external' mechanisms relying on third parties", while others are 'internal' mechanisms which are "controlled by the parties to the MEA themselves" (Handl 1997: 32, footnote 14).

Table 2.2 Compliance
mechanism as a concept
employed in the Book

• Broader than NCPs (also involve information-gathering,
  response measures)
• Narrower than UNEP's definition (UNEP's definition also
  involves DSPs)
• Implying international mechanism built into MEAs, not
  covering national ones
• Requiring a broader perspective than the concept of NCM
So, it is defined as an international-internal compliance
mechanism (not NCM) of MEAs involving three components:
information-gathering, NCPs, response measures

Source The author

For the purposes of this study, the term "compliance mechanism" (CM) signifies
the internal mechanism involving three components of UNEP's definition –
information-gathering about the parties' performance, NCPs, and response measures. As part of the process, it is possible to move back and forth between these
components as required. This interpretation of "compliance mechanism" excludes
DSPs because it is intended to present a more far-reaching meaning which promises
to show better the internal process in which compliance with MEAs' obligations
can be promoted.

It should also be underlined that the concept of compliance mechanism (CM) is
deliberately being used rather than the concept of non-compliance mechanism
(NCM). This is particularly because, in a CM, the fundamental purpose is to ensure
and promote compliance, thus, to prevent non-compliance before it occurs, rather
than to redress the consequences of non-compliance or to punish the non-compliant
parties. So, a compliance system is "intended to be broader" than a non-compliance
system (Raustiala 2001: 13), since it does not just concern the process after
non-compliance occurs; the previous period before it occurs is also a remarkably
important aspect of it, and even if compliance is ensured, a compliance system
seeks ways for further improvement. That is, CM involves the processes of: prevention of potential non-compliance; if compliance is ensured, the promotion of it;
if non-compliance (breach of an MEA obligation) has occurred, the application
response measures against it, and the prevention of its repetition.

In brief, CM as a concept requires a broader perspective than the concept of
NCM. Then, it seems that CMs have been evolved under MEAs in three steps,
involving particularly three aims together:

• Reporting-monitoring-verification/preventing non-compliance;
• An institutional-procedural structure, assessment on the compliance status
  within this structure/facilitating compliance;
• Response measures/responding to non-compliance.

In this sense, this book exclusively scrutinizes CMs based on the criteria summarized in Table 2.2.

These criteria involve three types of mechanisms for ensuring compliance, which
together establish a '*sui generis*' structure for a CM. That is, as will be seen when

the general characteristics of CMs are examined in detail,[11] CMs include in themselves both confrontational means, non-confrontational means, and also some procedural safeguards which are not defined in the scope of confrontational or non-confrontational means.

References

Breitmeier, H., Young, O.R. and Zürn, M. (2006). *Analyzing International Environmental Regimes: From Case Study to Database*. Cambridge: MIT Press.

Chasek, P.S., Downie, D.L. and Brown, J.W. (2006). *Global Environmental Politics*. Boulder, CO: Westview Press.

Chayes, A. and Chayes, A. (1995). *The New Sovereignty: Compliance with International Regulatory Agreements*. Cambridge, MA: Harvard University Press.

Chayes, A., Chayes, A. and Mitchell, R.B. (1995). Active Compliance Management in Environmental Treaties. Lang, W. (Ed.), *Sustainable Development and International Law* (75–89). London; Boston: Graham and Trotman/M. Nijhoff.

Chayes, A., Chayes, A. and Mitchell, R.B. (1998). Managing Compliance: A Comparative Perspective. Weiss, E.B. and Jacobson, H.K. (Eds.), *Engaging Countries: Strengthening Compliance with International Environmental Accords* (39–62), Cambridge, Mass.: MIT Press.

Chhotray, V. and Stoker, G. (2009). *Governance Theory and Practice, A Cross-Disciplinary Approach*, Basingstoke: Palgrave Macmillan.

Churchill, R.R. and Ulfstein, G. (2000). Autonomous Institutional Arrangements in Multilateral Environmental Agreements: A Little Noticed Phenomenon in International Law. *The American Journal of International Law*, 94, 4, 623–659.

COP 7 (1999). Resolution VII.3 "People and Wetlands: The Vital Link," 7th Meeting of the Conference of the Contracting Parties to the Convention on Wetlands (Ramsar, Iran, 1971). San José, Costa Rica, 10–18 May 1999. Available at: http://www.ramsar.org/sites/default/files/documents/library/key_res_vii.03e.pdf.

COP 9 (2005). Resolution IX.16 The Convention's International Organization Partners (IOPs), "Wetlands and water: supporting life, sustaining livelihoods" 9th Meeting of the Conference of the Parties to the Convention on Wetlands (Ramsar, Iran, 1971). Kampala, Uganda, 8–15 November 2005. Available at: http://www.ramsar.org/sites/default/files/documents/pdf/res/key_res_ix_16_e.pdf.

Crossen, T.E. (2003). Multilateral Environmental Agreements and the Compliance Continuum. *The Berkeley Electronic Press Legal Series*, 36. Available at: https://www.ippc.int/sites/default/files/documents/1182330508307_Compliance_and_theory_MEAs.pdf.

Dicker, G. (2004). *Kant's Theory of Knowledge*. Oxford, New York: Oxford University Press.

Epiney, A. (2006). The Role of NGOs in the Process of Ensuring compliance with MEAs. Beyerlin, U., Stoll, P.T. and Wolfrum, R. (Eds.), *Ensuring Compliance with Multilateral Environmental Agreements, A Dialogue between Practitioners and Academia* (273–300). Leiden: Koninklijke Brill NV.

Faure, M.G. and Lefevere, J. (1999). Compliance with International Environmental Agreements. Vig, N.J. and Axelrod, R.S. (Eds.), *The Global Environment: Institutions, Law and Policy* (172–191). Washington: CQ Press.

Fisher, R. (1981). *Improving Compliance with International Law*. Charlottesville: University Press of Virginia.

---

[11]See *infra* Chap. 4.

Grafton, R.Q. (2001). *A Dictionary of Environmental Economy, Science and Policy*. Cheltenham, Northampton: Edward Elgar.

Handl, G. (1997). Compliance Control Mechanisms and International Environmental Obligations. *Tulane Journal of International and Comparative Law*, 5, 29–51.

Hempel, L.C. (1996). *Environmental Governance: The Global Challenge*. Washington, DC: Island Press.

Jacobson, H.K. and Weiss, E.B. (1998). Assessing the Record and Designing Strategies to Engage Countries. Weiss, E.B. and Jacobson, H.K. (Eds.), *Engaging Countries: Strengthening Compliance with International Environmental Accords* (511–554). Cambridge, Mass.: MIT Press.

Kimball, L. (1992). Toward Global Environmental Management the Institutional Setting. *Yearbook of International Environmental Law*, 3 (18–42). London: Graham and Trotman.

Krasner, S.D. (1983). Structural Causes and Regime Consequences: Regimes as Intervening Variables, Krasner, S.D. (Ed.), *International Regimes* (1–21). Cornell UP: London.

Lang, W. (1992). Diplomacy and International Law Making: Some Observations, *Yearbook of International Environmental Law*, 3, 108–122.

Lang, W. (1995). From Environmental Protection to Sustainable Development: Challenges for International Law. Lang, W. (Ed.), *Sustainable Development and International Law*. London: Springer.

Levy, M.A., Young O.R., Zürn M. (1995). The Study of International Regimes. *European Journal of International Relations*, 1, 3, 267–330.

List, M. and Rittberger, V. (1992). Regime Theory and International Environmental Management. Hurrel, A. and Kingsbury, B. (Eds.), *The International Politics of the Environment: Actors, Interests, and Institutions* (85–109). Oxford: Clarendon Press.

Lao-Tse (1901). *Confucius, The Analects—13, The Analects Attributed to Confucius*. Translated by James Legge, December 13, 1901. Available at: https://china.usc.edu/confucius-analects-13.

Maljean-Dubois, S. and Richard, V. (2004). Mechanisms for Monitoring and Implementation of International Environmental Protection Agreements. Available at: http://halshs.archives-ouvertes.fr/docs/00/42/64/17/PDF/id_0409bis_maljeandubois_richard_eng.pdf.

Matz, N. (2006). Financial and Other Incentives for Complying with MEA Obligations. Beyerlin, U., Stoll, P.T., Wolfrum, R. (Eds.), *Ensuring Compliance with Multilateral Environmental Agreements: Academic Analysis and Views from Practice* (301–318). The Leiden: Koninklijke Brill NV.

Mitchell, R.B. (1994). Regime Design Matters: Intentional Oil Pollution and Treaty Compliance. *International Organization*, 48, 3, 425–458.

Mitchell, R.B. (1996). Compliance Theory: An Overview. Cameron, J. Werksman, J. and Roderick, P. (Eds.), *Improving Compliance with International Environmental Law* (3–28). London: Earthscan.

Mitchell, R.B. (2007). Compliance Theory, Effectiveness, and Behaviour Change in International Environmental Law. Bodansky, B., Brunnée, J. and Hey, E. (Eds.), *The Oxford Handbook of International Environmental Law* (893–921). New York: Oxford University Press.

Najam, A., Papa, M. and Taiyab, N. (2006). Global Environmental Governance: A Reform Agenda. Available at: http://www.iisd.org/pdf/2006/geg.pdf.

Özer Kızılsümer, D. (2009). *Çok Taraflı Çevre Sözleşmeleri*. Ankara: Usak Yayınları.

Porter, G., Brown, J.W. and Pamela, C. (2000). *Global Environmental Politics: Dilemmas in World Politics* (2nd edn.), Boulder, CO: Westview Press.

Prideaux, M., Rostron, C. and Duff, L. (2015). *Ramsar and Wetland NGOs: A Report of the World Wetland Network for Ramsar CoP 12*. London: World Wetland Network.

Raustiala, K. (2000). Compliance and Effectiveness in International Regulatory Cooperation. *Case Western Reserve Journal of International Law*, 3, 2, 387–440.

Raustiala, K. (2001). Reporting and Review Institutions in 10 Selected Multilateral Environmental Agreements. Nairobi: UNEP. Available at: https://www.peacepalacelibrary.nl/ebooks/files/C08-0025-Raustiala-Reporting.pdf.

Raustiala, K. and Slaughter, A. (2002). International Law, International Relations and Compliance. Carlsnaes, W., Risse, T. and Simmonz, B.A. (Eds.), *The Handbook of International Relations* (538–558). London: Sage.

Romano, C.P.R. (2000). *The Peaceful Settlement of International Environmental Disputes*. London: Kluwer Law International.

Saunier, R.E. and Meganck, R.A. (2007). *Dictionary and Introduction to Global Environmental Governance*. London: Sterling, VA: Earthscan.

Schachter, O. (1991). *International Law in Theory and Practice*. U.S.A, Canada: Kluwer Academic Publishers.

UNEP (2005). Comparative Analysis of Compliance Mechanisms under Selected Multilateral Environmental Agreements. UNEP: Nairobi. Available at: https://elaw.org/system/files/UNEP. comp_.mea_.compliance.pdf?_ga=2.50584041.1797524369.1551612177-1091855356. 1551612177.

Wolfrum, R. (1999). *Recueil des cours: Collected Courses of the Hague Academy of International Law*, Vol. 272 (1998). The Hague, Boston, London: Martinus Nijhoff Publishers.

Young, O.R. (1999). *Governance in World Affairs*, Ithaca, N.Y.: Cornell University Press.

Zhao, J. (2005). Implementing International Environmental Treaties in Developing Countries: China's Compliance with the Montreal Protocol, *Global Environmental Politics*, 5, 1, 58–81.

Chapter 3
Theoretical Perspectives and Explanatory Models of Compliance

A social science theory is a reasoned and precise speculation about the answer to a research question, including a statement about why the proposed answer is correct.

King, Keohane, Verba (1994: 19)

3.1 Introduction

Theories can be described as the individual ways of intellectual thinking about facts, processes and relationships which could lead to a meaningful explanation of the social phenomenon under discussion.

They are necessary to understand the current situation of that phenomenon, but also to develop different views about its future.

There may be several approaches to the same phenomenon: some are more general in their terms, while others are more specific and limited, focusing only on particular issues. Some of them may be complementary to each other, but others may be conflicting.

Research on compliance encompasses a diverse range of competing theoretical perspectives with divergent implications. This is because compliance is linked to both international law (IL) and international relations (IR) in many ways.

It is apparent from the literature arising from both international law and international relations that the most relevant prominent approaches on compliance debate can be broadly grouped into two categories (INECE 2009; Raustiala 2000; Zaelke/Higdon 2006): the rationalist and normative theories which "represent opposite ends of the spectrum" (INECE 2009: 8). They are based on the distinction introduced by March/Olsen (1998) between, respectively, two different logics of behaviour, namely, the logic of consequences (LoC) and the logic of appropriateness (LoA) (INECE 2009; Mitchell 2007).

As considering this variety of perspectives offers an opportunity to gain a more holistic and deeper view of compliance, this chapter firstly gives brief explanations of these competing theoretical perspectives from both IL and IR categories. Then it focuses on two main theoretical models, management and enforcement models of

© Springer Nature Switzerland AG 2019
Z. Savaşan, *Paris Climate Agreement: A Deal for Better Compliance?*
The Anthropocene: Politik—Economics—Society—Science 11,
https://doi.org/10.1007/978-3-030-14313-8_3

compliance, which have the potential to provide complementary and beneficial insights into the compliance debate.

While surveying these perspectives, a detailed review of IR and IL literature will not be undertaken, as this is beyond the scope of this study. In addition, despite the existence of various critical approaches relevant to international environmental law, it seems that international environmental law (IEL) has, to a large extent, remained a "critique free-zone" (Mickelson 2007: 289), and, with regard to compliance in particular, there is no critical approach. Therefore critical approaches are not examined in detail in this book.

3.2 Theoretical Perspectives on Compliance

3.2.1 Rationalist Theories

Rationalist theories posit that actors follow the logic of consequence. The logic of consequence is based on the view that actors act rationally by calculating the likely consequences of the various possibilities and then choosing the option which best serves their own self-interest. That means that if, after assessing the available alternatives, the potential costs and benefits of those alternatives, and the likely actions of other actors, actors decide that it is better for their self-interest not to comply with a regulation than to comply, they choose not to comply. Taking into account these views, rationalists particularly suggest that enforcement, deterrence and sanctions are the main means to prevent and punish non-compliance.

Within rationalist theories, states are regarded as rational-choice actors. That means, more clearly, that states are assumed to be unitary, rational, self-interested actors, acting in an anarchic world order, according to their calculations of the costs and benefits of different actions.

The principal rationalist theory is realism, which regards international law as having little or no effect. Rejecting the "legalist-moralist" approach of Woodrow Wilson – or "Wilsonian liberal internationalism", according to Slaughter (1993: 207) – realism treats states as unitary actors rationally pursuing their self-interests in the international arena (Danish 2007: 207; Slaughter 1993: 208). For realists, the purpose of law is to benefit states and their interests, so legal norms are derived from policies which are formed by states and in line with their interests (Carr 2001).[1]

According to its reconceptualization as neorealism by Waltz (1979), the international system has three components (the ordering principle of the system,

---

[1] In contrast to Carr, Ku (2001) argues that Carr fell short of understanding the framework for political discourse that law provides; international law has been an important element for adaptation and political change in international relations, evolving from a passive reflection of power in the voluntarist period to the interventionist effort to manage power in the institutionalist period, and to the civil society period of diffused power.

anarchy; the character of the units in the system; the distribution of the capabilities of the units in the system) and no causality between law and politics. However, like classical realism, neorealism also explains the states as key actors in the anarchic international system.

So, it can be argued that realist theory in general assumes that non-compliance with law is usually motivated by the interests of states, which are viewed as the owners of economic power and security. For the scholars following this theory, law can be used "to maintain the stability of the system," "to solve some coordination problems" (Kingsbury 1998: 350), and, by hegemonies, "to establish and enforce standards to which weaker states are obliged to conform" (Kingsbury 1998: 351). Yet, if law is found to be inconsistent with a significant interest of a state, that state can act in a manner serving its own interest. In short, states can violate international law whenever it is in their interest to do so. Thus, law can influence the behaviour of states only with regard to unimportant issues, particularly issues on non-security. That means that international law, existing as only an 'epiphenomenon', can have little or no independent impact on the behaviour of states (Guzman 2001: 14; Kingsbury 1998: 350). Compliance with international law, on the other hand, can be determined by considerations of power and states' interests rather than of law in all important cases. Morgenthau (1960: 296) explicitly puts forward this view by stating that "considerations of power rather than of law determine compliance and enforcement," giving some role for international law, as he accepts that international law is generally observed, but as the result of either power relations or convergent interests. Similarly, Scott (1994) also accepts the ideology of international law as a must for membership of the international system, yet, characterizes the ideology as a source of power rather than legitimacy, in contrast to the legal process scholars, such as Chayes and Chayes, Koh, Henkin and Franck.

Like realism, compliance-based theory, advanced by Andrew Guzman, is founded on a model of rational self-interested states. For Guzman, when the benefits of disobeying outweigh its costs, states prefer not to comply with the rules of international law.

This theory criticizes the theories of Henkin, Franck and Koh as well as Chayes and Chayes (which will be analysed in the subsequent section). Regarding consent theory advanced by Henkin, Guzman argues that even if consent is an important factor which forces states to comply with the legal obligations that they have accepted, compliance cannot be explained by consent alone; some other factors, which can ensure compliance, should also exist (Guzman 2001). He also finds the legitimacy theory of Franck inadequate to explain why states care about legitimacy and how legitimacy can lead to compliance (Guzman 2001). He asserts that Koh's theory of transnational legal process also fails to explain why domestic actors internalize certain legal norms, how this internalization occurs, and how it leads to compliance. Finally, he also challenges the management theory of Chayes and Chayes, as it "falls short of a general theory of compliance or a complete description of international legal agreements [as it is designed for a specific category of agreements, for regulatory agreements]" (Guzman 2001: 7) since it cannot be applied to contexts other than those involving coordination games. In brief, he criticizes all these theories, as they simply assume that the norms are obeyed, and so compliance exists, yet fail to present a proper theoretical structure to explain why and how this occurs (Guzman 2001).

After presenting these criticisms on these theories, Guzman then focuses on developing a compliance-based theory which is largely grounded in IL theory, but also partly in IR theory. His theory is founded on the institutionalist tradition, as it shows that international law can influence the behaviour of a state. Here it should be underlined that, in some areas where the stakes are high, such as state security, the effects of international law and the reputation which depends on IL are weaker. On the other hand, in the field of trade and environment, international law can have a real impact. Indeed, for Guzman (2001: 16):

> [a]n institutionalist approach to international law not only reveals that international law matters more than realists claim, it also reveals that it matters less than many international law scholars seem to assume. It is shown that international law represents a force in state behavior, but one that is of limited power, and that is much more likely to affect outcomes in some cases than in others.

Like Chayes and Chayes, Guzman believes that the most important factor affecting compliance with international law is a state's concern for its reputation. However, in contrast to Chayes and Chayes, he argues that states do not obey the rules of international law when the benefits of disobeying outweigh the costs (sanctions). Chayes and Chayes say the reasons for non-compliance are ambiguity/indeterminacy of treaty language, incapacity and time lags between commitment and performance.

Liberal theory can also be perceived as a largely rationalist theory, as it too considers the state to be the main actor in the international area. Nevertheless, in terms of compliance problems, it indicates at least three issues which are not given importance by other 'standard' rationalist theories (Kingsbury 1998: 356). Firstly, it focuses on the domestic characteristics of the states and the role of other actors, such as IOs, NGOs, multinational corporations etc. Secondly, it highlights the fact that compliance involves conformity with not only the "traditional model of interstate rules implemented by national measures," but also "different sets of norms made by and directed to different sets of actors" (Kingsbury 1998: 357). Finally, as it assumes that liberal states, involving the characteristics of representative government, the protection of civil and political rights and a judicial system guided by the rule of law, tend to comply with international law (Slaughter 1993), and compliance occurs through the effects of international law. It also assumes that greater democratization leads to greater compliance (Kingsbury 1998).

Jacobson/Weiss's (1998b) analysis of the compliance rates of eight states (Brazil, Cameroon, China, Hungary, India, Japan, Russian Federation, United States) and the EU in connection with five treaties (the World Heritage Convention, the Convention on International Trade in Endangered Species, the London Convention of 1972 – formerly London Ocean Dumping Convention, the 1983 International Tropical Timber Agreement, the Montreal Protocol on Substances that Deplete the Ozone Layer), acknowledges these assumptions of the theory, as it finds that democracies have performed their obligations more completely than non-democracies. However, despite their study, it should not be ignored that these assumptions cannot always be materialized in practice, as "democratization does not necessarily lead automatically or quickly to improved compliance" (Kingsbury 1998: 357).

3.2.2   Normative Theories

Normative theories, as distinct from rationalist theories, assume that actors follow the logic of appropriateness when deciding on complying or not complying. The logic of appropriateness is based on the view that actors act according to the identities, obligations, and conceptions of appropriate action ("involving elements of both socialization and internationalization") (Mitchell 2007: 902). This is different from the logic of consequences, in that actors do not calculate the costs and benefits of their behaviour, but evaluate "what is the right thing to do in this situation for someone like [them], or how do [they] want to see [themselves] and/or how do [they] want other actors to see [them]?" (Mitchell 2007: 902). So, in the theories adopting this logic, rules, ideals, legal obligations, shared discourses and knowledge attain more importance. As "they see state interests and even states themselves as socially-constructed, in part by norms, rather than as pre-theoretical givens," they can be defined as 'cognitive' or 'constructivist' theories as well (Raustiala 2000: 405). In addition, as these theories consider that non-compliance occurs not particularly because of lack of enforcement, but mostly because of weak tools to elicit compliance and the actor's incapacity or the actor's belief that the rule is unfair, they focus on cooperation and compliance promotion without enforcement as a means to prevent non-compliance.

Henkin's consent theory (1968), like other IL theories, assumes that compliance with international law and almost all of their obligations occurs more often than non-compliance by almost all states. It is argued that the factor providing compliance is the 'consent' given by the parties to be bound by legal rules (Henkin 1968). Henkin (1968) applies legal process arguments and a number of factors, such as reputation, reciprocity and domestic politics, to demonstrate how international law forces states to move closer to compliance. However, the factors which provide a theoretical approach to compliance are generally found to be quite wide (Raustiala/Slaughter 2002). In addition, this theory is criticized because the state can often easily withdraw its consent when it desires, and under such a circumstance, explaining the rationale for compliance becomes challenging (Guzman 2001).

In Franck's (1988, 1990, 1995) legitimacy theory, the basic factor which determines the rules' "compliance-pull" on states is put forth as legitimacy, which is defined by four elements: "textual determinacy, symbolic validation, coherence and adherence." First, textual determinacy (the linguistic component of legitimacy) is, in Franck's words, "the ability of the text to convey a clear message, to appear transparent in the sense that one can see through the language to the meaning" (Franck 1988: 713). Put simply, it refers to the clarity and transparency of the rule, or "that which makes its message clear" (Franck 1990: 30, 52). Secondly, symbolic validation (the cultural component) refers to the regularized practices, rituals or pedigrees that provide the rules with symbolic validation in order to gain compliance (Franck 1990: 34, 92–94). The third element, coherence, refers to the connection and consistency between the rule and the principles (Franck 1990: 38, 142–148). Finally, adherence refers to the connection between the rule and those secondary rules used to interpret and apply the primary rule, and to the degree to which secondary rules adhere to a primary rule (Franck 1990: 41, 184). When these four elements are present, they form "right process" (Franck 1988: 706) – which is defined as one of the components of fairness (the other is substantive fairness, referring to distributive justice, or equity) (Franck 1990: 26–29) – for fostering a perception of the rule as legitimate by those to whom it is addressed, and thus, for exerting the pull to compliance (Franck 1988: 706). That is, when right process is formed, states feel a strong pressure on themselves to comply, so fair agreements become more likely to be complied with. Thus, the theory claims "a chain (or cycle) of causation between right process and state behavior" (Raustiala/Slaughter 2002: 541).

Transnational legal process theory, developed by Koh (1997, 1999), on the other hand, draws most of its examples from the area of human rights and security. It explains the formation of voluntary obedience and internalized compliance through the incorporation and internalization of international norms into domestic systems. This theory does not consider states as unitary actors; instead, it incorporates non-state actors (including individuals, multinational corporations, NGOs, businesses, scientists, networks and others) as units of analysis. It assumes that states comply with or obey rules through a transnational legal process which involves a process of 'interaction,' 'interpretation' and 'internalization' ("social, political and legal internalization"; Koh 1999: 1413, 1414) of international norms to domestic legal structures. According to this argument, when transnational actors, including both state and non-state actors, interact, some norms or behaviours which emerge in this interaction can be internalized by actors. The internalization then leads to their incorporation into the domestic legal institutions of states, also leading to compliance.[2] In contrast to Henkin's approach, the process of interaction and internalization is stated as constructive, as these processes help to reconstitute the interests of the participants in the process. This shows that domestic institutions play an important role in this theory, as these institutions provide compliance by incorporating international obligations within the domestic legal system.

In addition to those mentioned above, constructivist theory, even though it is not an IL theory, also bears traces of normative theories, though not directly. So, like

---

[2]It should here also be emphasized that the theory posits "four kinds of relationships between rules and conduct: coincidence, conformity, compliance, and obedience," implying that a step further towards compliance leads to 'obedience' (Koh 1999: 1400, Koh 1997: 2646).

Table 3.1 The most prominent approaches (both from IL and IR) on compliance debate

	Rationalist theories	Normative theories
	Realism; compliance-based theory; liberalism	Consent theory; legitimacy theory; transnational legal process theory; constructivism
Logics of behaviour	Logic of consequence Acting to maximize self-interest Choosing the best alternatives	Logic of appropriateness Acting according to the identities, applicable norms and related expectations of other actors Choosing the right alternatives
Means for improving compliance	Deterrence and enforcement **Stick**: the threat of enforcement for non-compliance	Cooperation and compliance promotion **Carrot**: promoting compliance through supporting activities like capacity-building, technical assistance, etc.

Source The author

them, it can also be called a norm-orientated theory. This is because "from a constructivist perspective, compliance is less a matter of rational calculation or imposed constraints than of internalized identities and norms of appropriate behavior" (Raustiala/Slaughter 2002: 540). Instead of relying exclusively on states' interests and the strategic pursuit of these interests for the emergence and operation of the institutions to change state behaviour, it attracts attention to the "mutually constitutive" relationship between states and structures. In fact, it argues that states do not enter the international system with fully formed interests, as they gain their fully formed interests (and also their identities) through their participation in the international system and their interactions in the system which are framed by international norms. Thus, there should be a process of social construction of identities and norms in the system (Checkel 1999, 2000, 2001) and, in such a process, actors' behaviours can not be explained solely by the logic of consequences, but also by the logic of appropriateness (Table 3.1).

3.3 Basic Explanatory Models of Compliance

The theories explained above under two categories, rationalist and normative, all contribute to an understanding of the issue of compliance to some extent, but neither can be fully adequate to understand MEAs and their compliance mechanisms. This is because they do not always function through just one form oflogic.

In fact, when these different approaches are applied to them, it appears that MEAs can influence the actors' behaviour (actors here signifies not only states but also others, such as individuals, private actors, businesses, etc.) through both behavioural forms of logic mentioned above. No single approach can fully capture the complex nature of MEAs and their compliance mechanisms. So, understanding compliance with MEAs requires both types of logic to be analysed. Within the logic of consequence, MEAs can influence actors' behaviour by signalling that certain actions are in their interest and others are not. Actors's behavioural change occurs according to the calculations of costs and benefits of single behaviours. To prevent non-compliance and to improve compliance, this logic suggests a need to enforce, to deter and to punish. Within the logic of appropriateness, MEAs can influence actors' behaviours "by signalling that certain behaviours are 'appropriate' and others are 'inappropriate'" for that situation (Mitchell 2007: 902). To prevent non-compliance and to improve compliance: this logic suggests the need not to punish or reward the actors, but to provide cooperation and compliance promotion among them.

Which of these logics predominates varies according to the type of agreement, parties, process, time and several other factors. Sometimes, one can predominate, sometimes the other, and in some cases, the two logics can mutually reinforce each other (Mitchell 2007). That is, the logics incorporated within the rationalist and normative theories can jointly explain the behavioural change that can lead to compliance (INECE 2009). March/Olsen (1998: 952) have stressed that "the two logics are not mutually exclusive," and that any action can involve the features of both logics.

So, the analysis requires the use of a combination of both logics that can shape environmental and behavioural change differently, treating them not as substitutes, but as complements.

Moreover, as previously stated, this book's main research question is the role of compliance mechanisms adopted by parties to MEAs in strengthening compliance under MEAs in general and under the PCA in particular. Yet, neither of these explicitly aforementioned rationalist-normative theories focuses on compliance mechanisms and MEAs. Rather, they attempt to develop a general theory of compliance with international law, which is insufficient to explain CMs under MEAs.

Based on these reasons, it is here necessary to scrutinize two basic explanatory approaches to compliance issues in general and also compliance mechanisms created under MEAs.

In studies on institutionalism (which can be defined as a rationalist theory, as it presupposes that international institutions and international law facilitate cooperation between states), IR theory focuses on IR-IL interdisciplinary research. While some scholars support the cooperation between IR and IL for establishing more effective environmental laws and institutions (Abbott 1999; Brunnée/Toope 2000; Byers 1999; Chayes/Chayes 1995; Keohane 1997; Mitchell 2002, Raustiala/ Slaughter 2002, Setear 1996; Slaughter et al. 1998; Slaughter Burley 1993, Simpson 2000; Young 1979), others criticize this kind of research, as it can undermine the distinctive character of international legal rules or help dominant states like the USA evade legal obligations (Kennedy 1999; Koskenniemmi 2007). Yet, in spite of criticism, there has been a trend towards IR-IL collaborative research, particularly with regard to international environmental issues, and issues related to compliance have been an important area of this cooperation and inter-action between IR-IL scholars.

The management approach of Abram and Antonia Chayes presents one of these attempts at IR-IL collaborative research on compliance issues. As seen in the subsequent sections, their model, created for explaining compliance, has been very effective in the design of the CMs under MEAs (Kizilsümer 2009). The creation of these CMs is seen as the indication of a managerial approach to compliance.

This model and the enforcement model propagated by its critics have established two basic explanatory approaches to compliance issues and also compliance mechanisms created under MEAs (Kizilsümer 2009). In the following pages, these two basic models will be taken into account to understand the functioning of the CMs.

3.3.1 Management Model of Compliance

It can be argued that the management or 'responsiveness' (Vezirgiannidou 2009: 43) model of compliance created by Chayes/Chayes (1995), in which "the strong emphasis [is] on prevention and on management-oriented and institutionalized action" (Sachariew 1991: 32), is "a synthesizing theory" on compliance which gathers together many themes within IL literature which offer insights into IR theory (Raustiala/Slaughter 2002: 543).[3]

It is indeed a synthesizing theory, as it embraces the characteristics in parallel with IL theories, institutionalism, and also the constructivist approach.

Firstly, in accordance with IL theories, it assumes that there is a "general propensity of states to comply with international law" originating from three factors: efficiency, interests and norms (Chayes/Chayes 1995: 3; Chayes et al. 1995: 78).

1. *Efficiency*: Through the established rules and procedures (domestic bureaucracy) set out in international agreements, an explicit calculation of the costs and benefits of every decision on compliance is not required; therefore, transaction costs can be saved. Guzman (2001) here criticizes Chayes and Chayes, because transaction costs can be saved by many other strategies, such as using the available evidence or investing in information gathering.
2. *Interest*: States are free to choose whether or not to enter an international agreement after assessing whether it is in their interests or not, as such

---

[3]Keohane (1997) also tries to reconcile two divergent perspectives of compliance, naming them the instrumentalist (interest-based) and the normative (legitimacy-based) optics. He argues that neither the normative nor instrumentalist optic is sufficient to adequately explain compliance, so interests, reputations and institutions should be regarded as common to both optics when attempting to synthesize the two. See also Zaelke/Higdon (2006: 383) also for the view on "blending normative and rationalist theories."

agreements are consent-based legal instruments. In addition, the negotiation process enables actors to make interest-based calculations and to form their positions against agreements. Another opportunity which facilitates compliance is that these agreements generally make provision for their parties to participate in the regimes created under them, attending their meetings and responding to their requests (Chayes/Chayes 1995: 22); thus, they can open the ways to amendments in accordance with the shifts in parties' preferences over time. Despite all these opportunities provided by the agreements, if there is still non-compliance, this stems from deliberate non-compliance, or from treaty language which does not capture the real meanings of the issues discussed during the negotiation process, or from problems in formulating the rules "to govern future conduct" (Chayes/Chayes 1995: 10).

3. *Norms*: Norms induce a sense of obligation in states to comply with international law, and hence form the "foundation for much of the compliance with international treaty rules" (Chayes et al. 1995: 79). To exemplify, underpinned by each state's efforts to prepare, negotiate, and monitor international agreements, and the voluntariness of then signing the agreements, one of the basic rules of international law, *pacta sunt servanda* ('agreements are to be kept'), provides pressure for compliance (Chayes/Chayes 1995; Chayes et al. 1995).

To Chayes/Chayes (1995), deliberate violation of international law or international agreements is the exception, not the rule. This means, for them, in general, that states, acting in good faith, prefer not to violate their international legal obligations.

In violations other than deliberate violation, non-compliance is "due to inadvertence or incapacity" (Chayes et al. 1995: 80) and occurs mainly because of three reasons (Chayes/Chayes 1995: 9–17):[4]

---

[4]As an additional note, reasons for non-compliance have been explained differently by different scholars. To exemplify, Chasek et al. (2006) trace non-compliance to several different types of factors, including: the inadequate translation of regime rules into domestic law; insufficient capacity to implement, enforce or administer relevant domestic law; a lack of respect for rule of law; the high relative costs of compliance; inadequate financial and technical assistance; the inability to monitor compliance; poorly designed regimes; and the large number of environmental conventions and the confusing and uncoordinated web of requirements they have produced. Breitmeier et al. (2006), stressing the necessity of examining first the extent to which non-compliance is voluntary or involuntary and second whether non-compliance amounts to a substantial challenge to the rule in question or whether it essentially involves a technical problem, differentiate four sources of non-compliance – cheating; ambiguity/impreciseness of a prescription; norm considered wrong; and lack of capacity to implement non-compliance – by examining first the extent to which non-compliance is voluntary or involuntary and second whether non-compliance amounts to a substantial challenge to the rule in question or whether it essentially involves a technical problem. Each of the four sources of non-compliance corresponds to a specific theoretical perspective on compliance – incentives, legalization, legitimacy and responsiveness – which has clear links to IR theories: incentives – rational institutionalism; legalization and legitimacy – liberalism and social constructivism; responsiveness – theories stressing discourses and communication. Sand (1996) mentions three primary reasons for non-compliance: treaty ambiguity – incertitude about treaty standards; lack of state capacity – the incapacity of states to meet treaty

1. *Ambiguity*: Treaty language and the complexity of the treaty can create ambiguity. If the norms, rules or procedures in the treaty are not clear, it requires interpretation, and when it is interpreted, different parties can adopt different positions on the same rules.
2. *Capacity limitations*: States may be out of compliance with an agreement because they lack the resources to establish domestic regulatory mechanisms. This is especially true of developing countries, with limited capacities to carry out their commitments. They lack the necessary scientific, technical, bureaucratic and financial resources to build effective domestic regulatory mechanisms.
3. *Temporal dimension*: Treaties are designed to persist over long periods of time, so there can be "considerable time lags" between commitments and their implementation (Chayes/Chayes 1995: 15). In order to adapt to changing situations, new instruments may be required, but states may not be able to respond to them instantly. So transition periods can be necessary to achieve a high level of compliance.

Secondly, like institutionalism, the management model underlines the importance of the role of the law in the performance of functions such as coordination, monitoring and enforcement, and stresses the importance of international institutions and treaties in performing these functions (Danish 2007: 222).

Moreover, the management model also has similar views to the constructivist approach. In fact, it is proper to argue that "constructivism provides a natural complement" to the management model (Brunnée 2003: 261). For example, Chayes and Chayes claim that states comply with an agreement's obligations not merely in return for its functional benefits, but also because participation in different international regimes and legal processes (as a result of "New Sovereignty" [Chayes/Chayes 1995: 123, 127; Chayes et al. 1995: 75]) can form an "iterative process of discourse" (Chayes/Chayes 1995: 25) or, "justificatory discourse" (Chayes/Chayes 1995: 26). The iterations from signature to ratification and also from the conclusion of a convention to the adoption of a protocol can be enough to provide adherence to commitments, and contributing to the transformation of states' identities and interests can lead to international cooperation (Setear 1996). So, even though international environmental agreements rarely incorporate enforcement mechanisms, compliance with them can be high (Barrett 2003).

The constructivist account of compliance, which is labelled a transformationalist model by Downs, Danish and Barsoom, can be found in much of the literature on international environmental regimes like Young's works which study not only the states' benefits from regimes (Downs et al. 2000), but also regimes' social and constitutive impacts (the impacts of legitimacy, capacity, socialization, legalization and standard operating procedures (SOPs) (Young 1999).

---

commitments; and changing circumstances – inflexibility of treaties in the face of changing circumstances. Mitchell (1996) perceives the sources of non-compliance to be preference, incapacity, or inadvertence.

Besides the features which make it a synthesizing theory, the other important aspect which renders it more important for this study is its proposal to establish a compliance management mechanism which has been quite influential on the design of the CMs under MEAs.

Their mechanism involves a transparent information system (reporting-monitoring-verification), dispute settlement, capacity building, adaptation and responses.

1. *Ensuring transparency*: Ensuring transparency firstly requires the development of data through the state's report on its compliance problems, which is not only crucial for resolving the existing non-compliance problems, but also the possible ones. By reporting, states open the way to resolve the potential problems before they occur. In fact, according to the type of problem that they have (e.g. financial, technical inadequacy etc.), they are granted remedies to resolve it before it results in non-compliance. In addition, transparency includes the assessment of information on actors' behaviours under the agreement, collected from various sources, including formal state verification systems and informal checks by other states, non-governmental organizations or other interest groups.

2. *Dispute settlement*: The improvement of dispute settlement procedures may both solve the problem of ambiguity in the meaning of norms and provide for compulsory conciliation, with non-binding recommendations which determine the required performance of the disputants in the particular cases. Thus, by both maintaining the principle of sovereignty, and not forcing parties to accept certain decisions, conciliation can be reached and a broad range of disputes can be addressed. In most cases, as international agreements involve a large number of "informal mediative processes" (Chayes/Chayes 1995: 24; Chayes et al. 1995: 84) or if agreement is managed by an organization, authoritative or semi-authoritative interpretation by a designated body – often the secretariat or a legal committee – to apply to dispute settlement is not required.

3. *Capacity-building*: This can involve technical and financial assistance and advice to help manage the capacity deficit in technical areas, bureaucratic capability and financial resources. Assistance and advice may be required to help states enact legislation in accordance with the provisions of an international agreement, to implement a new regulation, to improve scientific facilities, to enhance data systems, or to train national enforcement officials etc.

4. *Adaptation-revision*: Agreements need the capacity to be amended (whether through formal amendment or a protocol approach) in order to adapt to economic, technological, social and political changes. Parties who do not want the amendments to come into force can block them. In order to overcome this problem, agreements should make provision for adaptation and amendment without "long ratification delays" (Chayes et al. 1995: 87).

5. *The response system*: Producing different responses to different types of non-compliance can also be used as a way of eliciting changes in actors'

behaviours in order to induce compliance.[5] But management strategy is against "hard-edge enforcement tools," like sanctions, and adjudication-style dispute mechanisms (Danish 2007: 226). So the response to possible violations of norms should not be those type of penalties, but exclusion from the cooperation network. This is particularly because the most important influence on parties' compliance choices is to ensure their engagement as members of the network of interdependent regulatory regimes in this ongoing process of justification and persuasion (see also Slaughter et al. 1997). Also, enforcement tools prevent the dialogue required for "transformation, consensus-building and identity convergence" (Danish 2007: 226).

Chayes and Chayes argue that these all lead to a broader process of persuasion, into which the above-mentioned elements of management strategy have merged. In this process, attempts are made to persuade the violator parties to change their behaviours through dialogue between the parties, the agreement's organization, and the international community.

Moreover, for Chayes/Chayes (1995), strict compliance with an agreement is neither necessary nor possible on every occasion, as compliance is not an "on-off phenomenon" (Chayes/Chayes 1995: 17). Therefore, an acceptable level of compliance which includes "deviance within acceptable levels" can be allowed by the parties as long as this deviant behaviour does not prevent others benefiting from the agreement and does not result in the agreement's collapse (Chayes/Chayes 1995: 17).

Levels of compliance can be explained in four categories: weak compliance, moderate compliance, substantial compliance and no action taken to comply (which occurs very rarely in practice) (Jacobson/Weiss 1998a: 518–520). While weak compliance means that a country has taken some actions to implement and comply with the obligations accepted under the treaty, but those actions have fallen significantly short of what is required, moderate compliance means that the country has taken appropriate action to implement certain treaties and is in compliance with most of the obligations provided for by these instruments. Any shortfall is generally the result of ineffective or weak administration. Substantial compliance, finally, means basically fulfilling obligations though there may be minor infractions. However, it should be underlined that there is no objective standard for determining an acceptable level of non-compliance; it changes according to the agreement, the context, the behaviour of parties, time and the political process.

Overall, as Chayes and Chayes see ambiguity in agreements, incapacity and time lags between commitment and performance as the main sources of non-compliance, they argue that instead of focusing on enforcement, international institutions and international law should seek to manage compliance using normative and institutional tools to enhance the compliance capacity of states. They claim that "[a]s a practical matter coercive economic – let alone military – measures to sanction

---

[5]For the different types of response systems – punitive, remunerative, preclusive, generative, cognitive and normative systems – see Mitchell (1996).

violations cannot be utilized for the routine enforcement of treaties... The effort to devise and incorporate such sanctions in treaties is largely a waste of time" (Chayes/ Chayes 1995: 2), as "intentional non-compliance is infrequent" (Chayes et al. 1995: 80) and sanctions for coercive enforcement are too costly and too difficult to mobilize. Similarly, Brunnée/Toope (2000: 46) define them as "adversarial, backward-looking and coercive". So, for these authors, the enforcement model of compliance (or "sanctions-driven enforcement strategy" [Danish 2007: 226]) is misconceived, and should be replaced with their management model, relying on a problem-solving, cooperative approach instead of a coercive one.

Thus, discouraging conflict between parties and promoting cooperation instead, Chayes and Chayes have provided the momentum for a new development in international environmental law: compliance mechanisms.

3.3.2 Enforcement Model of Compliance

Challenging management theory (later, it also challenges transformationalist), the enforcement theory of compliance emphasizes the central role of enforcement in providing, maintaining and securing compliance (Downs/Jones 2002; Downs et al. 1996, 2000; Downs 1998).

Within this theory, it is argued that the necessity of enforcement varies according to the "nature of the underlying game" (Downs 1998: 322). In a coordination game, enforcement is not relevant, but, in contrast, it is relevant in connection with the repeated Prisoner's Dilemma or mixed-motive game where states benefit from collective cooperation, but also have an incentive to defect from it. This theory also suggests that the optimal enforcement strategy varies depending on the nature of the goods being regulated, the quality of available compliance data and utility certainty, so it does not assume that when enforcement provisions are present, they exactly embody tit for tat or any other principle of reciprocity, but, depending on the conditions, it is possible to embody different enforcement strategies (Downs 1998: 327–328).

Scholars adopting this theory, Downs et al. (2000), find it necessary to analyse management approach and transformationalists separately when explaining their responses to the critiques towards enforcement theory. This is because, for them, the theorists in the management tradition contend that "formal enforcement provisions are almost irrelevant"; those referred to as transformationalists, on the other hand, argue that "enforcement is worse than irrelevant" as it has a counterproductive impact on the "evolution of cooperation" by reducing the willingness of states to participate in the regimes and by developing an adversarial environment among member states of regimes (Downs 1998: 319).

To transformationalists, participation in a regime has the potential to change the preferences of states and thus lead them to socialize and prefer "ever-increasing levels of cooperation" (Downs 1998: 336). There are basically four design principles facilitating this transformationalist process: 1. increasing the number of members participating in the institution; 2. establishing only soft and unthreatening commitments as initial obligations; 3. requiring consensus for decision-making; 4. providing compliance control by dispute avoidance and negotiated compliance management rather than coercive enforcement mechanisms (Downs et al. 2000: 472–477; Downs 1998: 336). However, Downs et al. argue that there is little evidence showing that changes in states' preferences are usually brought by transformational forces (Downs et al. 2000; Downs 1998).

In response to the management approach, they also argue that many violations may be deliberate, as both ambiguity and incapacity, which are regarded as the main reasons for non-compliance by the management school of thought, are to some extent endogenous (Downs 1998; Downs et al. 1996), due to the fact that states can deliberately choose the ambiguity of the agreements that they make, and the incapacity that they employ in connection with a given agreement.

For the scholars advocating enforcement theory, if compliance with an international agreement is high, this implies that the agreement requires states to do little more than they would do in the absence of that agreement. Downs et al. (1996) specifically mention the Outer Space Treaty[6]. and the Seabed Arms Control Treaty[7] as examples to support their views on compliance. Regarding international environmental law, this theory cites the Montreal Protocol and the Mediterranean Plan/ Barcelona Convention[8] as evidence of the existence of "shallow cooperation" in international environmental law. These instruments have, in fact, involved no meaningful condition to change state behaviour on reducing chlorofluorocarbon emissions (under the Montreal Protocol) or dumping and pollution (Mediterranean Plan).

Victor (1999: 152) agrees with Downs et al. (1996) that some sort of enforcement measures can sometimes be required, particularly when the cooperation is deep and incentives to violate the agreement are high, although he stresses the importance of internal public pressure from environmental groups and robust legal systems (a transnational legal process) in encouraging compliance "to enforce international commitments from inside (ground-up) rather than outside (top-down)." In addition, like Downs et al., he regards the high compliance rate of MEAs as "shallow cooperation", and supports his claim by referring to the dominance of "the lowest common denominator" in international cooperation (Victor

---

[6]Treaty on Principles Governing the Activities of States in the Exploration and Use of Outer Space, Including the Moon and Other Celestial Bodies (London, Moscow, Washington, 27 January 1967). Available at: http://www.unoosa.org/oosa/en/ourwork/spacelaw/treaties/introouterspacetreaty.html.

[7]Treaty on the Prohibition of the Emplacement of Nuclear Weapons and Other Weapons of Mass Destruction on the Seabed and the Ocean Floor and in The Subsoil Thereof (London, Moscow, Washington, 11 February 1971). Available at: http://www.un-documents.net/seabed.htm.

[8]Convention for the Protection of the Mediterranean Sea Against Pollution (Barcelona, 16 February 1976). Available at: http://www.unepmap.org/. See also Skjærsh (1993) for an analysis of the effectiveness of the Mediterranean Action Plan from the stance of international relations theory.

1999: 153) and the 1985 Sulphur Protocol,[9] whose provisions on sulphur emissions were already adopted by several countries prior to its conclusion.

In short, this theory argues that high compliance with international agreements is the indicator of the shallowness of that agreement, as when an agreement calls for a significant change in behaviour – in other words, when a regime deepens and increases the demands – both the gains from cooperation and incentives to deviate from the agreement increase (Downs et al. 1996). As a result of increased gains and deviances, to deter non-compliance and to sustain cooperation, the free-rider problem has to be resolved through stronger sanctions (retaliatory, monetary, political or reputation-based punishments). So, in deeper regimes requiring deeper cooperation, which require states to do something differently – that is, much more than they would do in the absence of that agreement – stronger enforcement tools should be used to be effective (Downs et al. 1996). Deeper cooperation can be rendered without much enforcement only if "there [has been] less incentive to defect from a given agreement… [such as] changes in technology, relative prices, domestic transitions" (Downs et al. 1996: 397–398).

3.3.3 Management Model Versus Enforcement Model

Danish (1997) attracts attention to the fact that Chayes and Chayes's management approach, "rejecting reliance solely on a material enforcement strategy, …in favour of a social enforcement strategy" (Downs et al. 2000: 487), has some features which make it close to the enforcement approach. He argues that both approaches can use verification and deterrence, and both rely on the information produced from the regime to threaten the parties with losing their good reputation and standing in the international field, and to force them to comply. These can be defined as elements of enforcement rather than management (Danish 1997: 806–808).

Similarly, Alter (2003) stresses that, in both approaches, the necessity of enforcement tools to facilitate compliance is accepted. Many different types of international and decentralized sanctioning mechanisms, such as reciprocity, sanctions and the threat of withdrawing inducements, can meet the enforcement strategy's "political economy theory" (Downs 1998: 320). The difference is, however, that Chayes/Chayes (1995: 2) see using enforcement strategies such as coercive sanctions as "a waste of time." In addition, in both strands it is accepted that, in coordination problems, cooperative solutions can result in effective compliance without enforcement tools, as in these contexts, self-interest can be ensured by following the common standards to utilize the benefits of international coordination.

---

[9]Protocol to the 1979 Convention on Long-Range Transboundary Air Pollution on the Reduction of Sulphur Emissions or Their Transboundary Fluxes by at Least 30 Per Cent (Helsinki, 8 July 1985). Available at: http://www.unece.org/fileadmin/DAM/env/documents/2012/EB/1985. Sulphur.e.pdf.

As seen in those studies mentioned above, both approaches have similarities, but at the same time, basic differences, particularly regarding the reasons for non-compliance and the forms of sanctions (in the form of coercive economic, military or social enforcement) (Crossen 2003: 33). More importantly, there can be some missing elements regarding other forms of free-riding, like non-participation in both approaches' analyses. Barrett (2003, 2007) stresses this deficiency of both management and enforcement approaches. He argues that, while Chayes and Chayes consider compliance and participation to be separate problems, Downs et al. (1996) fail to distinguish the sanctions punishments necessary for participation from the sanctions punishments necessary for compliance. For Barret, participation and compliance can be different phenomena, but, if the purpose of an agreement is to change behaviour, it should be capable of deterring both non-compliance and non-participation. As a matter of fact, non-participation is more important; this is because, "if non-participation can be deterred then non-compliance can easily be deterred" (Barrett 2007: 251).

Therefore, when asked which approach is best at promoting compliance and thus improving environmental quality, it is not easy to give a definitive answer. In their comprehensive study "Analyzing International Environmental Regimes: From Case Study to Database", Breitmeier et al. (2006) try to test these different approaches empirically through an International Regimes Database (IRD).[10] As a result of their analyses, in accordance with the enforcement approach, they conclude that enforcement mechanisms increase compliance (Breitmeier et al. 2006: 78). Regarding the management approach, they find that the mechanisms produced by the approach, such as giving different obligations to different members, granting transition or grace periods and using capacity-building as a tool for improving compliance, do not facilitate compliance but, on the contrary, worsen it (Breitmeier et al. 2006: 104–110).

However, they argue that their results cannot be shown as a justification to dismiss the management approach, as their sample may be biased, or their variables may be inadequate to cover all aspects of the perspective, as they remain isolated from other potential influences on compliance (Breitmeier et al. 2006: 110). Furthermore, the queries that they made about the impact of compliance mechanisms on goal attainment and problem-solving show that the management approach to compliance is far more dominant in international environmental regimes than the

---

[10]The International Regimes Database (IRD) contains information on more than 50 states and the EU, and 23 regimes: Antarctic, Baltic Sea, Barents Sea Fisheries, Biodiversity, CITES, Climate Change, Danube River Protection, Desertification, Great Lakes Management, Hazardous Waste, Inter-American Tropical Tuna Convention, Conservation of Atlantic Tunas, International Regulation of Whaling, London Convention, ECE Long-Range Transboundary Air Pollution, North Sea, Oil Pollution, Protection of the Rhine Against Pollution, Ramsar (Wetlands), Protection of the Black Sea, South Pacific Fisheries Forum Agency, Stratospheric Ozone, and Tropical Timber Trade (Breitmeier et al. 2006).

enforcement approach[11] (regarding goal attainment in 94.1 per cent of the cases; regarding problem-solving in 89.4 percent of the cases) (Breitmeier et al. 2006: 182, 189–236). This implies that international environmental regimes do not rely heavily on enforcement measures to elicit compliance (they only make use of procedures for issuing notices of violation which do not usually lead to the imposition of sanctions), but, instead, they rely on capacity-building measures and the management approach, which lead to improvements in both problem-solving and goal attainment in most cases.

Young (1979) also thinks that the management approach to compliance should be preferable to enforcement, as it is the simple way to rely on enforcement rather than understanding the complex relationship between compliance and behaviour. For Young and his colleagues, the reliance of regimes on capacity building and the management approach rather than the enforcement approach supports their "social-practice perspective" on regimes, because this "reflect[s] a strategy designed to draw individual regime members into an increasingly dense network of relationships that give rise to a situation in which compliance becomes an automatic response rather than a matter requiring an assessment of costs and benefits on a case-by-case basis" (Breitmeier et al. 2006: 187).

In their analysis of international environmental regimes Young and his colleagues also identify several mechanisms which influence state behaviour (Breitmeier et al. 2006). Indeed, challenging the distinction between the logic of consequences (LoC) and the logic of appropriateness (LoA), they point to the significance of three sources of behaviour: the precepts of a knowledge system or a discourse; perceptions of legitimacy; and habits or standard operating procedures (Breitmeier et al. 2006: 234–35). They argue that all these mechanisms influence the effectiveness of regimes, actors are motivated by not only self interest, enforcement and inducements, but also pressure from society, legitimacy, knowledge system and habits, and that the degree of influence varies across different regimes.

Based on their findings, they conclude that neither the depth or shallowness argument of the enforcement model nor the management approach of Chayes and Chayes can render a full understanding of the patterns of compliance with international environmental regimes. "[A] composite perspective" (Breitmeier et al. 2006: 110) that integrates "incentive mechanisms, juridification, participation of transnational NGOs in the rule-making process, and a responsive approach to the development of compliance mechanisms over time" (Breitmeier et al. 2006: 112) arises as a necessary means of ensuring and strengthening compliance with international environmental regimes.

---

[11]Enforcement, in fact, is "a more multi-faceted concept than often assumed – it encompasses a wide spectrum of means for 'compelling compliance' with law" (Brunnée 2006: 258). However, it is generally and narrowly defined as "imposition of legal sanctions, or penalties" (Brunnée 2006: 258). It should be stressed that enforcement is here taken as "a matter of using threats and punishments to attain results, mechanisms featuring promises and rewards are not coded as cases of enforcement" (Breitmeier et al. 2006: 189).

In line with this approach, many international environmental regimes aim not only to promote their 'utilitarian' impacts on states, but also to deepen 'non-utilitarian,' impacts (legitimacy, capacity, socialization, legalization and standard operating procedures [SOPs]) to maximize the opportunities for shared normative understanding (Breitmeier et al. 2006: 149), normative interaction and processes for states to justify their conduct (Bodansky et al. 2007).

As a consequence of all those discussions on the issue, it can be argued that an attempt to test them, such as Breitmeier et al.'s study, cannot be admitted as fully adequate empirical evidence which expresses that one is better than the other, or is the most or least effective in promoting compliance, altering behaviour, and thus in providing environmental improvement. This is firstly because it is not easy "to establish causality in social sciences" (Klabbers 2007: 1,004) because of the complexity and pluralism inherited in social issues which, contrary to common assumptions, can sometimes involve multiple and irrational actors and so can feature processes that are not easy to understand. The scholars themselves already accept that their findings involve 'tensions' (Breitmeier et al. 2006: 236), and they express "the need to move beyond conventional assumptions about the behaviour of actors" (Breitmeier et al. 2006: 238). In addition, it is difficult to assess compliance in regulatory regimes, as there is no predetermined criteria valid for all regimes, and the present ones are "either quite vague or difficult to identify" (Mitchell 2007: 912). More importantly, to accept one of the approaches as the best one can result in "ignor[ing] a large variety of ways by which [M]EAs influence state behaviour" (Mitchell 2007: 911), as "in many cases, [M]EAs have components of both models and in others their components do not readily fit into these oversimplified categories" (Mitchell 2007: 911).

All these findings yield the following conclusion: it is not possible to reach a definitive conclusion that one approach is better than another at improving compliance. Depending on the problem and its conditions, each can explain different aspects of compliance processes. In some circumstances, only the management approach or the enforcement approach can be used, while in others they can both be used in a "proper balance" (Zaelke/Higdon 2006: 383). So, in some cases, instead of abandoning the one for the sake of the other, it appears more appropriate to put emphasis on interaction between these two approaches and try to synthesize them. However, in others, neither of them can be adequate to explain the situation in which pressure from society, legitimacy, knowledge system and habits or standard operating procedures all need to be taken into account in line with the views of Young's social-practice perspective.

In short, because of the differences between accords and countries, different strategies can work better in different circumstances. Therefore, different approaches can be required for different countries, depending on their features with respect to compliance. The most appropriate strategy to induce compliance by a particular country can also change over time if the country's features change. Yet it should be kept in mind that, in the current system, combining the two approaches (management approach-enforcement approach), dominantly leaning on the management approach in the short term, and thus making compliance internalized by the members in line with Young's social practice approach (Breitmeier et al. 2006; Young 1999), which can lead through the ways of management approach in the longer term, emerges as the best way, instead of merely leaning on harder mechanisms in line with the enforcement approach (Table 3.2).

Table 3.2 A comparative analysis on two models of compliance

Management model	Enforcement model
Similarities	
• Both are basically interest-based models (but the management model has been enriched through constructivist accounts) • Both approaches use verification and deterrence • Both rely on threats of losing reputation and standing in the international community • Both use regime-generated information to convince or shame offending states to force them to comply • Both accept the necessity of enforcement tools to facilitate compliance (but the management model is more facilitation-orientated) • Both accept that cooperative solutions to coordination problems can result in effective compliance without enforcement tools • Non-participation emerges as a missing element in both approaches	
Differences	
Facilitation-orientated: facilitation is they key to compliance	**"Sanction-oriented"** (Brunnée 2006: 254): using enforcement strategies such as coercive sanctions is the key to compliance
LoC + LoA: Both logics are applied	**LoC** is the dominant logic
Regarding CMs, it is expected to find a relationship between the success of MEA and the operation of CMs, not necessarily working through the use of penalties	**Regarding CMs**, it is expected to find a relationship between the success of MEA and the operation of CMs, which, as a matter of necessity, are capable of generating penalties
Reasons for non-compliance: In violations other than deliberate violation, non-compliance occurs mainly because of three reasons: the ambiguity and indeterminacy of the agreement language; incapacity; time lags between commitment and performance	**Reasons for non-compliance**: many violations may not be deliberate. Regarding both ambiguity and incapacity, which are perceived as the main reasons for non-compliance by the management school, they claim that states can deliberately choose the ambiguity and incapacity
Forms of sanctions: They argue against hard enforcement tools like sanctions and adjudication-style dispute mechanisms which prevent meaningful dialogues. For them, the sanction for violating the norms should not be those types of penalty, but to be excluded from cooperation networks	**Forms of sanctions**: They do not exactly embody tit for tat or any other principle of reciprocity, but, depending on the conditions, it is possible to embody different enforcement strategies which can also involve hard enforcement tools
Which one is best?	
Taking into account that the management model has been the most effective in the design of CM with specific constructivist insights that it has evolved over time, it seems appropriate to assess each case according to its own conditions, and to employ one or both models depending on those conditions	

Source The author

References

Abbott, K.W. (1999). International Relations Theory, International Law, and the Regime Governing Atrocities in Internal Conflicts. *The American Journal of International Law,* 93, 2, 361–379.

Alter, K.J. (2003). Do International Courts Enhance Compliance with International Law? *Asian and Pacific Studies,* 25, 51–78.

Barrett, S. (2003). *Environment and Statecraft: The Strategy of Environmental Treaty-making.* Oxford: Oxford University Press.

Barrett, S. (2007). An Economic Theory of International Environmental Law. Bodansky, B., Brunnée, J. and Hey, E. (Eds.), *The Oxford Handbook of International Environmental Law* (231–261). NewYork: Oxford University Press.

Breitmeier, H., Young, O.R. and Zürn, M. (2006). *Analyzing International Environmental Regimes: From Case Study to Database.* Cambridge: MIT Press.

Bodansky, D., Brunnée, J. and Hey, E. (2007). International Environmental Law, Mapping the Field. Bodansky, B., Brunnée, J. and Hey, E. (Eds.), *The Oxford Handbook of International Environmental Law* (1–28). New York: Oxford University Press.

Brunnée, J. (2003). The Kyoto Protocol: Testing Ground for Compliance Theories? *Zeitschrift für ausländisches öffentliches Recht und Völkerrecht (Heidelberg Journal of International Law)* 63, 2, 255–280. Available at: http://www.hjil.de/63_2003/63_2003_2_a_255_280.pdf.

Brunnée, J. (2006). Enforcement Mechanisms in International Law and International Environmental Law. Beyerlin, U., Stoll, P.T. and Wolfrum, R. (Eds.), *Ensuring Compliance with Multilateral Environmental Agreements, A Dialogue between Practitioners and Academia* (247–258). The Netherlands: Koninklijke Brill NV.

Brunnée, J. and Toope, S.J. (2000). International Law and Constructivism: Elements of an Interactional Theory of International Law. *Columbia Journal of Transnational Law,* 39. Available at: http://ssrn.com/abstract=1432539.

Byers, M. (1999). *Custom, Power, and the Power of Rules: International Relations and Customary International Law.* Cambridge: Cambridge University Press.

Carr, E.H. (2001). *The Twenty Year Crisis, 1919–1939: Introduction to the Study of International Relations;* reissued with a new introduction and additional material by Michael Cox. Houndmills, Basingstoke, New York: Palgrave.

Chasek, P.S., Downie, D.L. Brown, J.W. (2006). *Global Environmental Politics.* Boulder, CO: Westview Press.

Chayes, A. and Chayes, A. (1995). *The New Sovereignty: Compliance with International Regulatory Agreements.* Cambridge, MA: Harvard University Press.

Chayes, A., Chayes, A., and Mitchell, R.B. (1995). Active Compliance Management in Environmental Treaties. Lang, W. (Ed.), *Sustainable Development and International Law* (75–89). London: Springer.

Checkel, J.T. (1999). Norms, Institutions, and National Identity in Contemporary Europe. *International Studies Quarterly,* 43, 83–144.

Checkel, J.T. (2000). Compliance and Conditionality. *ARENA Working Paper,* 18. Available at: https://www.sv.uio.no/arena/english/research/publications/arena-working-papers/1994-2000/2000/wp00_18.htm.

Checkel, J.T. (2001). Why Comply? Social Learning and European Identity Change. *International Organization,* 55, 3, 553–588.

Convention for the Protection of the Mediterranean Sea Against Pollution. Available at: http://www.unepmap.org/.

Crossen, T.E. (2003). Multilateral Environmental Agreements and the Compliance Continuum. *The Berkeley Electronic Press Legal Series,* 36. Available at: https://www.ippc.int/sites/default/files/documents/1182330508307_Compliance_and_theory_MEAs.pdf.

Danish, K.W. (1997). Management v. Enforcement: The New Debate on Promoting Treaty Compliance. *Virginia Journal of International Law,* 37, 4, 789–819.

Danish, K.W. (2007). International Relations Theory. Bodansky, B., Brunnée, J. and Hey, E. (Eds.), *The Oxford Handbook of International Environmental Law* (205–230). NewYork: Oxford University Press.

Downs, G., Rocke, D. and Barsoom, P. (1996). Is the Good News about Compliance Good News about Cooperation? *International Organization,* 50, 3, 379–406.

Downs, G.W. (1998). Enforcement and the Evolution of Cooperation. *Michigan Journal of International Law,* 19, 319–344.

Downs, G.W., Danish, K.W. and Barsoom, P.N. (2000). Transformational Model of International Regime Design: Triumph of Hope or Experience. *Columbia Journal Transnational Law, 38,* 465–514.

Downs, G.W. and Jones, M.A. (2002). Reputation, Compliance, and International Law. *Journal of Legal Studies,* 31, 95–114.

Franck, T.M. (1988). Legitimacy in the International System. *The American Journal of International Law, 82,*705–759.

Franck, T.M. (1990). *The Power of Legitimacy Among Nations.* USA: Oxford University Press.

Franck, T.M. (1995). *Fairness in International Law and Institutions.* Oxford: Clarendon Press.

Guzman, A.T. (2001). International Law: A Compliance Based Theory. UC *Berkeley Public Law Research Paper,* 47. Available at: http://ssrn.com/abstract=260257.

Henkin, L. (1968). *How Nations Behave? Law and Foreign Policy,* New York: Published for the Council on Foreign Relations [by] F.A. Praeger.

International Network for Environmental Compliance and Enforcement (INECE) (2009). *Principles of Environmental Compliance and Enforcement Handbook.* Available at: http://www.themisnetwork.eu/uploads/documents/Tools/inece_principles_handbook_eng.pdf.

Jacobson, H.K. and Weiss, E.B. (1998a). A Framework for Analysis. Weiss, E.B. & Jacobson, H. K. (Eds.), *Engaging Countries: Strengthening Compliance with International Environmental Accords* (1–18). Cambridge, Mass.: MIT Press.

Jacobson, H.K. and Weiss, E.B. (1998b). Assessing the Record and Designing Strategies to Engage Countries. Weiss, E.B. and Jacobson, H.K. (Eds.), *Engaging Countries: Strengthening Compliance with International Environmental Accords* (511–554). Cambridge, Mass.: MIT Press.

Kennedy, D. (1999). The Disciplines of International Law and Policy. *Leiden Journal of International Law,* 12, 9–33.

Keohane R. O. (1997). International Relations and International Law: Two Optics. *Harvard International Law Journal,* 38, 2, 487–502.

Kızılsümer Özer, D. (2009). *Çok Tarafli Çevre Sözleşmeleri.* Ankara: Usak Yayınları.

King, Garry, Keohane, Robert O., Verba, Sidney (1994). *Designing Social Inquiry, Scientific Inference in Qualitative Research.* Princeton, New Jersey: Princeton University Press.

Kingsbury, B. (1998). The Concept of Compliance as a Function of Competing Conceptions of International Law. *The Michigan Journal of International Law,* 19, 2, 345–372.

Klabbers, J. (2007). Compliance Procedures. Bodansky, B., Brunnée, J. and Hey, E.(Eds.), *The Oxford Handbook of International Environmental Law* (995–1009). New York: Oxford University Press.

Koh, H.H. (1997). Why do Nations Obey International Law? *The Yale Law Journal,* 106, 8, 2,599–2,659.

Koh, H.H. (1999). How Is International Human Rights Law Enforced? *Indiana Law Journal,* 74, 1,397–1,417. Available at: https://www.repository.law.indiana.edu/cgi/viewcontent.cgi?article=2279 &context=ilj.

Koskenniemmi, M. (2007). The Fate of Public International Law: Between Technique and Politics. *Modern Law Review,* 70, 1, 1–30.

Ku, C. (2001). Global Governance and the Changing Face of International Law. Available at: https://scholarship.law.tamu.edu/cgi/viewcontent.cgi?article=1574&context=facscholar.

March, J.G., and Olsen, J.P. (1998). The Institutional Dynamics of International Political Orders. *International Organization,* 52, 4, 943–969.

Mickelson, K. (2007). Critical Approaches. Bodansky, B., Brunnée, J. and Hey, E. (Eds.), *The Oxford Handbook of International Environmental Law* (262–290). NewYork: Oxford University Press.

Mitchell, R.B. (1996). Compliance Theory: An Overview. Cameron, J. Werksman, J. and Roderick, P. (Eds.), *Improving Compliance with International Environmental Law* (3–28). London: Earthscan.

Mitchell, R.B. (2002). International Environment. Carlsnaes, W., Risse, T. and Simmons, B.A. (Eds.), *Handbook of International Relations* (500–516). London: Sage.

Mitchell, R.B. (2007). Compliance Theory, Effectiveness, and Behaviour Change in International Environmental Law. Bodansky, B., Brunnée, J. and Hey, E. (Eds.), *The Oxford Handbook of International Environmental Law* (893–921). NewYork: Oxford University Press.

Morgenthau, H.J. (1960). *Politics Among Nations: The Struggle for Power and Peace*. New York: Alfred A. Knopf.

Protocol to the 1979 Convention on Long-Range Transboundary Air Pollution on the Reduction of Sulphur Emissions or Their Transboundary Fluxes by at Least 30 Per Cent. Available at: http://www.unece.org/env/lrtap/status/lrtap_s.html.

Raustiala, K. (2000). Compliance and Effectiveness in International Regulatory Cooperation. *Case Western Reserve Journal of International Law*, 3, 2, 387–440.

Raustiala, K. and Slaughter, A. (2002). International Law, International Relations and Compliance. Carlsnaes, W., Risse, T. and Simmonz, B.A. (Eds.), *The Handbook of International Relations* (538–558). London: Sage.

Sachariew, K. (1991). Promoting Compliance with International Environmental Standards: Reflections on Monitoring and Reporting Mechanisms. *Yearbook of International Environmental Law*, 2, 1, 31–52. https://doi.org/10.1093/yiel/2.1.31.

Sand, P.H. (1996). Institution-Building to Assist Compliance with International Environmental Law: Perspectives. *Zeitschrift für Ausländisches Öffentliches Recht und Völkerrecht (Heidelberg Journal of International Law)*, 56, 3, 774–795.

Scott, S.V. (1994). International Law as Ideology: Theorizing the Relation Between International Law and International Politics. *European Journal International Law*, 5, 313–325.

Setear, J.K. (1996). An Iterative Perspective on Treaties: A Synthesis of International Relations Theory and International Law. Available at: http://faculty.virginia.edu/setear/cv/hilj.pdf.

Simpson, G. (2000). The Situation on the International Legal Theory Front: The Power of Rules and The Rule of Power. *European Journal International Law*, 11, 2, 439–464.

Skjærseth, J.B. (1993). The Effectiveness of the Mediterranean Action Plan. *International Environmental Affairs*, 5, 4, 313–334.

Slaughter, Anne-Marie (1993). International Law and International Relations Theory: A Dual Agenda, *American Journal of International Law*, 87, 205–239.

Slaughter, A., Tulumello, A.S. and Wood, S. (1998). International Law and International Relations Theory: A New Generations of Interdisciplinary Scholarship, *The American Journal of International Law*, 92, 3, 367–397.

Slaughter Burley, A. (1993). International Law and International Relations Theory: A Dual Agenda. *The American Journal of International Law*, 87, 205–239.

Treaty on Principles Governing the Activities of States in the Exploration and Use of Outer Space, Including the Moon and Other Celestial Bodies. Available at: http://www.un-documents.net/ost.htm.

Treaty on the Prohibition of the Emplacement of Nuclear Weapons and Other Weapons of Mass Destruction on the Seabed and the Ocean Floor and in The Subsoil Thereof. Available at: http://www.un-documents.net/seabed.htm.

Vezirgiannidou, S. (2009). The Climate Change Regime Post-Kyoto: Why Compliance is Important and How to Achieve it. *Global Environmental Politics*, 9, 4, 41–63.

Victor, D.G. (1999). Enforcing International Law: Implications for an Effective Global Warming Regime. *Duke Environmental Law & Policy Forum*, 10, 147–184.

Waltz, K. N. (1979). *Theory of International Politics*. Reading, Mass.: Addison-Wesley Pub. Co.

Young, O.R. (1979). *Compliance and Public Authority: A Theory with International Applications* (onlinebook). Available at: http://www.questia.com/read/24304033#.

Young, O.R. (1999). *The Effectiveness of International Environmental Regimes, Causal Connections and Behavioral Mechanisms*. Cambridge: MIT Press.

Zaelke, D. and Higdon, T. (2006). The Role of Compliance in the Rule of Law, Good Governance, and Sustainable Development. *Journal for European Environmental & Planning Law*, 3, 5, 376–384.

Chapter 4
Compliance Mechanisms: A General Overview

4.1 Introduction

In this part, two important points should be noted regarding the method of examination. Firstly, in line with the definition given in the second chapter, analysis of the CM will take into account the three integral components which complement and support each other: gathering information on the parties' performance, institutionalised multilateral NCPs, and multilateral response measures. Secondly, the purpose here is not to make an overall comprehensive examination of all MEAs and CMs established so far. This would not be feasible, as each 'self-contained'[1] (Churchill/Ulfstein 2000: 633; Dagne 2007: 23; Koskenniemi 1992: 34; 2006: 65) or "tailor-made" (Bodansky et al. 2007: 23; Széll 1995: 108) MEA has different characteristics, and the CMs in different MEAs can vary depending on the agreement itself and its particular characteristics (Montini 2009). Therefore, it is not possible to identify a common structure which applies to all MEAs and their CMs.

However, the CMs in different MEAs share some common elements in each component. Therefore, taking into account these common elements, the aim in this part is to present a general overview of the system of these new complex CMs and the way they function. However, before this overview, as background information, this chapter will survey why these CMs have emerged in the global agenda. While making this survey, this chapter will also outline the general institutional structure which provides the framework for the mechanisms and will discuss the limitations of the traditional methods of compliance and the basic features of the CMs which make them more acceptable to the MEAs' parties.

---

[1]See also Dagne (2007) for detailed information on the self-contained regime approach (together with the harmonized approach) as one of the approaches to settling disputes.

© Springer Nature Switzerland AG 2019
Z. Savaşan, *Paris Climate Agreement: A Deal for Better Compliance?*
The Anthropocene: Politik—Economics—Society—Science 11,
https://doi.org/10.1007/978-3-030-14313-8_4

4.2 Development of Compliance Mechanisms: Reasons for Their Development

The development of compliance mechanisms and reasons for their development can be demonstrated briefly through two types of impact: preparatory impacts (involving the change in the international system and the concept of sovereignty, features of IEL and MEAs) and triggering impacts (involving the limitations of traditional means and the features making CMs more attractive and preferable) (Table 4.1).

4.2.1 Preparatory Impacts

The instruments relied on to protect and enhance the environment have so far reflected the dominant position of sovereign states in both the development and implementation of environmental law and politics. However, in recent decades, the international system based on sovereign states and the concept of sovereignty has altered in many aspects. In fact, the evolution of international environmental law (IEL) over time has articulated that the concept of state sovereignty involving two notions, 'territorial integrity and sovereignty,' cannot be revoked absolutely in IEL. This is particularly because IEL does not reflect only the interests and concerns of states, but also the common concerns and interests of the whole world community. This results in the interdependence between states which necessities cooperation between them for addressing collective environmental concerns. It also allows the application of special treatment for developing and least-developed states through the principle of common but differentiated responsibilities (Bodansky et al. 2007; Hunter et al. 2002). That is, although it is generally accepted that the responsibility

Table 4.1 Reasons for CMs' development

Preparatory Impacts	Triggering Impacts	
Change in the system	Limitations of traditional means	**Compliance Mechanisms**
Features of International Environmental Law	Features making compliance mechanisms more attractive	
Nature of MEAs		

Source The author

for global environmental problems, such as global warming and ozone depletion, lies much more with industrialized countries than developing countries, the cooperation of all countries in the world (even if they have different responsibilities) is essential to lead to effective solutions to these problems. This cooperation is, to a large extent, provided by MEA-based regimes today (Bodansky et al. 2007). The main roles are granted to the states in these regimes as well. Indeed, even if IEL and IEP have, to a large extent, been shaped and improved through a process of inter-state negotiation, that is, "inter-State cooperation" (Hunter et al. 2002: 446), non-state actors and sub-state actors have also played significant roles in that process (Bodansky et al. 2007; Hunter et al. 2002).

In addition, because environmental law has distinct characteristics from other fields of international law (Bodansky et al. 2007; Hunter et al. 2002; Lang 1995), it has addressed not only local and national but also global problems. And if the issues at stake are of a common or even global nature, environmental problems can concern a great number of actors. Therefore, not only states, but a great many interested parties play important roles in its structure. Furthermore, in line with the fact that "many international environmental problems have significant impacts that are difficult to reverse" (Hunter et al. 2002: 457), a preventative approach – and so "preventative dispute avoidance" (Potzold 2009: 4) instead of "ex-post evaluation" (Enderlin 2003: 155) after the breach or damage has occurred – has become more important and become more favoured in principle in IEL. In fact, in contrast to the traditional methods of settling disputes, it "focuses on the management of environmental problems leading to disputes" (Romano 2007: 1,038).

Furthermore, the nature of the MEAs, which are often related to common environmental issues which produce global benefits or costs for the whole world community rather than individual interests or costs for any one state, has been another preparatory impact of the creation of the CMs within MEAs. Indeed, non-compliance with an MEA can often affect the interests of the whole world community, rather than those of any single state. So, the non-compliant act of the state cannot always be linked smoothly to the injury of another state. Within MEAs, it is indeed difficult, even impossible, to have recourse to "the traditional paradigm of bilateralism" (Handl 1997: 34) or to establish the "causal link" between the injured state or states and the non-compliant act of another state (Wang and Wiser 2002: 182). So MEAs do not involve reciprocal obligations as distinct from bilateral agreements.

4.2.2 Triggering Impacts

The insufficiency of traditional means of settling disputes and ensuring compliance, such as the Law of Treaties, state responsibility and DSPs and dissatisfaction with them, has increasingly formed the emergence and continuing growth of CMs with features which make them more attractive and preferable to the alternatives (Beyerlin et al. 2006; Churchill/Ulfstein 2000; Tanzi/Pitea 2009; Wang/Wiser

2002). Although it is beyond the scope of this study to examine all aspects of these traditional means in depth, a brief summary of the crucial aspects of their limitations will be provided here.

4.2.2.1 The Limitations of Traditional Means

Law of Treaties (LoT). The LoT in principle requires the unanimous suspension of the operation of a multilateral agreement if there has been a material breach by one of the parties. It also allows the party specially affected by the breach to suspend, in whole or in part, relations between itself and the defaulting state (Art. 60, para. 2, Vienna Convention on the Law of Treaties [VCLT]). However, it is generally not feasible to determine any one party specifically affected by a breach in MEAs. Additionally, whether non-compliance has been intentional or unintentional, suspension of the agreement is not a good way to bring back the non-compliant party to compliance, since MEAs generally address global environmental issues resting on all parties being part of the commitments adopted under the agreement. So, to take the non-compliant part out of the agreement is contrary to the system's founding objectives and also hinders its supervision in those matters regulated under the agreement (for further details see Savaşan 2017: 838–839).

Responsibility of States. It is also not possible to enhance compliance with MEA obligations in an effective manner by evoking the international responsibility of the state (Birnie et al. 2009; Crossen 2003; Ehrmann 2002; Fitzmaurice 2007; Koskenniemi 1992; Sand 1990). This is because, although all acts of a state which breach an international obligation under international law are the responsibility of that state (Art. 1/Art. 2, International Law Commission [ILC] Draft Articles on Responsibility of States for Internationally Wrongful Acts), a causal relationship between the state injured by violation and the state that has caused the damage cannot be identified in an easy manner (Crossen 2003, 2004; Ehrmann 2002; Koskenniemi 1992; Werksman 1999). Furthermore, the legal consequences of the responsibility, such as restitution, compensation, and satisfaction (Arts. 34–37), are not applicable in the case of irreversible/irreparable environmental damage (Koskenniemi 1992). There is also no specific institutional mechanism for compliance issues (Stephens 2009). In addition, there are very few examples in which states bring their environmental claims against other states. Of those, the most well-known is the Trail Smelter case (United States v. Canada) in which the Arbitral Tribunal held Canada responsible for the conduct of a Canadian corporation – the Trail Smelter – and also asked for future monitoring of the factory's activities to prevent possible future damage to the United States' environment (UN 2006). There are also well-known recent cases brought before the ICJ, such as those in between Argentina v. Uruguay and Costa Rica v. Nicaragua. The former is about Uruguay's authorization and construction of pulp mills along the Uruguay River, which forms the border area between Argentina and Uruguay. In 2010, the Court gave its decision on the case, concluding that environmental impact assessment is a must under international law (ICJ 2010). The latter is on certain activities carried out by Nicaragua along the San Juan River. In 2013, the ICJ linked the case with that of the construction of a road in Costa Rica along the San Juan River (Nicaragua v. Costa Rica). The Court decided on both cases in 2015, stressing the importance of conducting an environmental impact assessment to prevent transboundary environmental harm, in parallel to the former decision (ICJ 2015).

Finally, state responsibility covers only state activities which are responsible for environmental damage, not private activities; and civil liability regimes[2] are not appropriate for environmental issues involving transboundary concerns, because they need "a relationship between an offender and the one whose rights have been infringed" (Wolfrum 1999: 85).

With regard to the relationship between state responsibility and CMs, it can be questioned whether the CMs can be seen as special regimes of state responsibility, since Art. 55 of the ILC Draft articles (*les specialis*) primarily allows operation of the special rules of international law on state responsibility. While considering this question, it should be remembered that CMs do not seek to find the international responsibility of a state, but to reduce non-compliance, regardless of whether it is intentional or unintentional, and to facilitate compliance by encouraging and assisting parties to fulfil their obligations under the related MEA. Therefore, they should be regarded as "different, separate and independent" (Pineschi 2009: 489).

It can also be asked whether a state which is party to an MEA can invoke the responsibility of another state party under the general regime on state responsibility despite the establishment of a CM under the agreement. Apparently, there is no provision which excludes the application of the general regime on state responsibility in MEAs. However, to give a definitive answer to this question is quite difficult. The answer should be given after evaluating every case in relation to its specific conditions. Applying the rules on state responsibility after CMs have failed seems the "most plausible hypothesis, but there can also be cases which do not arise from this hypothesis" (Pineschi 2009: 485–486, 491, 497).

Dispute Settlement Procedures (DSPs). According to the Rio Declaration and the UN Charter,[3] it is possible to scrutinize DSPs in two groups[4]: 1. diplomatic means,

---

[2]It should be noted that there are also a few agreements which establish state liability, such as the Basel Convention Liability Protocol, and the Convention on International Liability for damages caused by space objects, and in the Antarctic. See Crossen (2003); Fitzmaurice (2007); Matz (2006); Sands (1996); Stephens (2009) for the details of state liability.

[3]Principle 26 of the 1992 Rio Declaration clearly sets out that states have to "resolve all their environmental disputes peacefully and by appropriate means in accordance with the Charter of the United Nations." Art. 33(1) of the UN Charter states that the settlement of disputes can be provided "by negotiation, enquiry, mediation, conciliation, arbitration, judicial settlement, resort to regional agencies or arrangements, or other peaceful means of their own choice."

[4]See Sands (1996: 71–80) for details of dispute settlement. See also Romano (2007: 1,045–1,054) for detailed information on the problems inherited within these means, such as 1. disharmonic dispute settlement clauses, 2. fragmentation and cluster litigation, 3. competing and parallel legal regimes, 4. multiplication of actors and levels.

such as negotiation, good offices, inquiry, mediation, conciliation,[5] and 2. 'judicial' (Ehrmann 2002: 381) or 'legal' (Sands 1996: 74) or 'adjudicative' (Romano 2000: 91) means, such as arbitration and judicial settlement.[6] However, those means, both diplomatic and judicial, are rarely applied in practice (Charney 1996). This is because they are confrontational and adversarial and designed for bilateral disputes (particularly judicial means, but, diplomatic means as well), so are inappropriate for environmental problems which are often multilateral in nature. If the violation stems from inability or incapacity, but not from deliberate non-compliance, addressing non-compliance through judicial means becomes even more problematic. Diplomatic means are not compulsory or binding in nature; and judicial means are not adequate for MEAs for the following reasons: 1. states do not want to challenge other states and take them before a court (Brownlie 2003; Faure/Lefevere 1999; Werksman 1996); 2. court proceedings are troublesome, expensive and slow (Charney 1996; Kolari 2002: 49); 3. there is no enforcement and monitoring mechanism for implementation of and compliance with the court's decisions; 4. the court's decisions do not have a preventative feature, so they only can restore the previous situation or provide compensation after the damage occurs, conflicting with the irreversible character of environmental damages.

The International Court of Justice. The ICJ, as a permanent court, is the principal judicial organ of the UN, and all members of the UN are automatically parties to its Statute (Arts. 92, 93(1), UN Charter). Its jurisdiction consists of "all cases which the parties refer to it and all matters specially provided for in the Charter of the United Nations or in treaties and conventions in force" (Art. 36(1), ICJ Statute). So it is competent on all aspects of IEL, and it is feasible to bring before it questions which arise on environmental issues. However, only states have direct access to the cases before the Court and it has no compulsory jurisdiction independent from the consent of the states involved in the dispute (Art. 34(1), ICJ Statute).[7] So, given the acceptance of its compulsory jurisdiction by only sixty-seven states[8] with reservations, it becomes clear that its success depends, to a large extent, on the attitude of individual states (Petersmann 1999). Granting access to the ICJ to non-state actors can be quiet influential for "depoliticizing international adjudication" (Petersmann 1999: 783), and likewise have a greater effect on societies. An example of this is the Urgenda climate change case brought to the Rechtbank Den Haag by an NGO.[9] While the Dutch court ruled that this NGO had a sufficient legal interest to bring the case, at international level, no NGO has such a standing to initiate proceedings before the ICJ. So it requires a reform of the ICJ Statute and of

---

[5]See Romano (2000: 46–65) for details of diplomatic means of DSPs.

[6]See Romano (2000: 91–129) for details of judicial means of dispute settlement.

[7]On consent to ICJ's jurisdiction, see Brownlie (2003: 682–690).

[8]For the Declarations Recognizing the Jurisdiction of the Court as Compulsory, see https://www.icj-cij.org/en/declarations.

[9]Rechtbank Den Haag (24 June 2015) C/09/456689. Available at: https://elaw.org/nl.urgenda.15.

the UN Charter, and such a reform "seems to be *unrealistic* at the moment" (Rest 2000: 43).

Its decisions, which are binding for the parties to the case, become "*functus officio*" when given (Arts. 59 and 60 of the Statute) (Bulterman/Kuijer 1996: 21). The Court does not have the power to rescind its decisions, in fact, there is no international mechanism which can exercise those functions in international law. Only in exceptional situations, it "require[s] previous compliance with the terms of the judgment before it admits proceedings in revision" (Art. 61(3), ICJ Statute).

It also issues non-binding advisory opinions on any legal question when requested by the General Assembly or authorized organs of the United Nations (Art. 96, UN Charter; Art. 65, ICJ Statute).[10] However, as they are not legally binding, they can only have "the practical value" for environmental issues if they are adopted and implemented in practice (Kolari 2002: 57).

The ICJ also has the power to indicate provisional measures which should be adopted immediately pending the final decision to preserve the rights of the parties to a dispute (Art. 41(1, 2), ICJ Statute). These measures can play quite an important role in cases concerning environmental protection due to the "irreparability of environmental damage" (Sands 1996: 77). However, it has usually been criticized for being unsuccessful in either dispute settlement or compliance of environmental questions (Hempel 1996; Schwabach 2005), even when both states are willing to undergo the process.

When all these issues are considered together with the fact that ICJ proceedings tend to be slow (Aust 2000; Kolari 2002; Wang/Wiser 2002), expensive (Downes/ Penhoet 1999) and "inherently confrontational, thereby posing political risks to bilateral relationships" (Wang/Wiser 2002: 182), the ICJ is often seen as having little or no effect in more specialized areas like protection of the environment.

In fact, although it established an ad hoc Chamber for Environmental Cases in 1993 (Art. 26(1) ICJ Statute), up to now no case has been brought to the Chamber, so the Court decided not to hold elections for a Bench for that Chamber in 2006.[11] The ICJ has considered relatively few cases with environmental aspects.[12] The first

---

[10]See Sands (2016) for the proposal on the prospects for an intervention by the ICJ or ITLOS in the form of an Advisory Opinion from those courts.

[11]Chambers and Committees, see at https://www.icj-cij.org/en/chambers-and-committees.

[12]By 2017, the ICJ had considered the following cases with environmental aspects: Contentious cases: Nuclear Tests (Australia v. France and New Zealand v. France) (1973); Judgment, Maritime Delimitation in the area between Greenland and Jan Mayen (Denmark v. Norway) (1988); Certain Phosphate lands in Nauru (Nauru v. Australia) (1989); Gabcikovo-Nagymaros Project (Hungary/ Slovakia) (1993); Fisheries Jurisdiction (Spain v. Canada) (1995); Request for an Examination of the Situation in accordance with par. 63 of the Court's Judgment of 20 December 1974 in the Nuclear Tests (New Zealand v. France) (1995); Pulp Mills on the River Uruguay (Argentina v. Uruguay) (2006); Aerial Herbicide Spraying (Ecuador v. Colombia) (2008); Certain Activities carried out by Nicaragua in the Border Area (Costa Rica v. Nicaragua) (2010); Whaling in the Antarctic (Australia v. Japan: New Zealand intervening) (2010); Construction of a Road in Costa Rica along the San Juan River (Nicaragua v. Costa Rica) (2011); Dispute over the Status and Use of the Waters of the Silala (Chile v. Bolivia) (2016). Available at: https://www.icj-cij.org/en/

time it handled an issue particularly relevant to environmental problems – the
Gabcikovo-Nagymaros case – was back in 1997.[13] But, contrary to expectations, in
this case the court failed to examine what the concepts and principles of environmental law are, and, referring to the principle of sustainable development, confined
itself to advising cooperation.

Other International Tribunals. It should additionally be known that a dispute which
has arisen in the area of international environmental law can be brought for solution
before international courts other than the ICJ. Different courts can have a role in
resolving disputes which have arisen in relation to international environmental
agreements. These include the Permanent Court of Arbitration (PCA-CPA), set up
under the Hague Convention for the Pacific Settlement of International Disputes of
1989; the International Tribunal on the Law of the Sea (ITLOS) established as a
permanent body under Annex VI of the United Nations Convention on the Law of
the Sea (UNCLOS)[14]; the European Court of Justice (ECJ); and courts established
by numerous regional human rights treaties – most notably the European Court of
Human Rights (ECtHR), and also the International Criminal Court (ICC) and the
World Trade Organisation (WTO) DSP.

The UNCLOS has "what is probably the most detailed, comprehensive, and
complex dispute settlement clause ever" (Romano 2007: 1,041). Indeed, it has a
'flexibility' and a "highly consensual nature of Part XV" which provides several
options for settling disputes between states through informal procedures, such as
conciliation and arbitration, or formal ones, like applying to the court (Schiffman
1998: 9).

It is argued that, compared to the ICJ, the International Tribunal on the Law of
the Sea (ITLOS) is faster at giving decisions and more eligible "to bear greater
specialist expertise" on the disputes concerning the interpretation or application of
the UNCLOS (Downes/Penhoet 1999: 16). The decisions of ITLOS, and also
decisions given under the second section – Art. 296(1), (2) – are also final.
Decisions made under Art. 33(1), ITLOS Statute[15] are also binding, but, like ICJ
decisions, only between the parties concerned and in respect of the particular dispute at issue (Art. 33(2), ITLOS Statute, Art. 59, ICJ Statute.). ITLOS can also

---

contentious-cases. Advisory Proceedings: Legality of the Use by a State of Nuclear Weapons in
Armed Conflict (1993); Legality of the Threat or Use of Nuclear Weapons (1995). Available at:
https://www.icj-cij.org/en/advisory-proceedings.

[13]ICJ, Gabcikovo-Nagymaros Case, Judgment of 25 September 1997, ICJ Report, 1997.
    https://www.icj-cij.org/files/case-related/92/092-19970925-JUD-01-00-EN.pdf. For the details
see Anlar Güneş (2006: 91–116).

[14]See the United Nations Convention on the Law of the Sea (UNCLOS), (Montego Bay, Jamaica,
10 December 1982). Available at: http://www.un.org/depts/los/convention_agreements/texts/
unclos/unclos_e.pdf.

[15]International Tribunal on the Law of the Sea (ITLOS) Statute (Annex VI: Statute of the
International Tribunal for the Law of the Sea), at http://www.un.org/Depts/los/convention_
agreements/texts/unclos/annex6.htm.

establish special chambers for particular categories of disputes, such as the Sea Bed Dispute Chamber for disputes regarding deep seabed mining (Arts. 14, 35–40, ITLOS Statute; Arts. 186–191, UNCLOS). The parties to the Convention can submit disputes concerning the protection of the marine environment (UNCLOS Part XI, XII) to the ITLOS (Art. 20, ITLOS Statute).

Due to these features, it is sometimes argued that judicial settlement can be an effective method, if it has compulsory and binding jurisdiction like the ITLOS (Guruswamy/Hendricks 1997).

In practice, ITLOS has decided a large number of cases regarding environmental issues, and over time its role on environmental questions has increased.[16] However, its contribution to the law remains restricted to the issues relating to prompt release cases, provisional measures pending the establishment of an ad hoc arbitral tribunal (Serdy 2005) and the nationality of ships (Rayfuse 2005).

Some MEAs can also rely on ad hoc arrangements like the Permanent Court of Arbitration (PCA-CPA) rather than standing bodies for dispute settlement. Its 'flexible' structure, offering four different dispute settlement methods – inquiry, mediation, conciliation and arbitration (see Art. 33, UN Charter) – can make it more favourable (Rest 2000: 51).

Its Optional Rules for arbitration of disputes relating to natural resources and/or the environment were issued in 2001 (and its Optional Rules for conciliation of disputes relating to natural resources and/or the environment were adopted in

---

[16]ITLOS had considered the following cases involving environmental questions by 2017: Cases 3 & 4: Southern Bluefin Tuna Cases (New Zealand v. Japan, Australia v. Japan) (provisional measures) (Order of 27 August 1999); Case No. 5: Camouco Case (Panama v. France) (prompt release) (Judgment of 7 February 2000); Case No. 6: Monte Confurco Case (Seychelles v. France) (prompt release) (Judgment of 18 December 2000); Case No. 7: Case concerning the Conservation and Sustainable Exploitation of Swordfish Stocks in the South-Eastern Pacific Ocean (Chile v. European Community) (provisional measures) (Order of 16 December 2009); Case No. 10: The MOX Plant Case (Ireland v. United Kingdom) (provisional measures) (Order of 3 December 2001); Case No. 11: Volga Case (Russian Federation v. Australia) (prompt release) (Judgment of 23 December 2002); Case No. 12: Land Reclamation by Singapore in and around the Straits of Johor (Malaysia v. Singapore) (provisional measures) (Order of 8 October 2003); Case No. 13: Juno Trader Case (St. Vincent and the Grenadines v. Bissau) (prompt release) (Judgment of 18 December 2004); Case No. 14: The 'Hoshinmaru' Case (Japan v. Russian Federation), (Judgment of 6 August 2007); Case No. 15: The 'Tominmaru' Case (Japan v. Russian Federation) (Judgment of 6 August 2007); Case No. 18: The M/V 'Louisa' Case (St. Vincent and the Grenadines v. Kingdom of Spain), (Judgment of 28 May 2013); Case No. 19: The M/V "Virginia G" Case (Panama/Guinea-Bissau) (Judgment of 14 April 2014); Case No. 20: The "Ara Libertad" Case (Argentina v. Ghana) (Order of 15 December 2002); Case No. 22: The "Arctic Sunrise" Case (Kingdom of the Netherlands v. Russian Federation) (Order of 22 November 2013); Case No. 23: Delimitation of the Maritime Boundary between Ghana and Côte d'Ivoire in the Atlantic Ocean (Ghana/Côte d'Ivoire), (provisional measures) (Order of 25 April 2015). Advisory Proceedings: Case No. 17: Responsibilities and obligations of States sponsoring persons and entities with respect to activities in the Area (Advisory Opinion of 1 February 2011); Case No. 21: Request for an Advisory Opinion Submitted by the Sub-Regional Fisheries Commission (SRFC) (Advisory Opinion of 2 April 2015). Available at: https://www.itlos.org/en/cases/list-of-cases/.

2002).[17] Some innovations have been also brought with its Draft Rules, such as making the rules available for the use of all parties who have agreed to apply them, providing for the optional use of a panel of environmental experts or a panel of arbitrators, and ordering any interim measures necessary to prevent serious harm to natural resources and the environment. They have arisen as supporting elements to the potential role of the PCA-CPA in contributing to an improved judicial settlement of environmental disputes. In considering the potential role of the PCA-CPA in the settlement of environmental disputes, some also argue that until an IEC with compulsory jurisdiction comes into existence, the PCA-CPA "could be the appropriate forum to settle environmental disputes" (Rest 2000: 50). However, these Draft Rules require further amendments, particularly on the problems of "exhaustion of local remedies", "waiver of immunity" and "legal access and *ius standi* of 'non-state actors' and individuals" (Rest 2000: 56).

The European Court of Justice (ECJ), also the Court of First Instance (CFI), has jurisdiction over environmental disputes[18] which can be brought before it by its member states, institutions (such as the Commission, Parliament) and even non-governmental organizations and individuals (see Arts. 251–281 on the Court and also Arts. 191–193 on the environment, Treaty on the Functioning of the European Union [TFUE]). Its decisions are legally binding, and in general it is observed that compliance with its decisions is highly ensured (Downes/Penhoet 1999). Nevertheless, according to the restricted regional field of application of European Law, the jurisdiction does not go as far as desirable for global environmental protection (Rest 2000; Romano 2007).

Several cases that had an environmental aspect have also appeared before the European Court of Human Rights (ECtHR). Any individual, NGO or group of individuals have the right to bring complaints about the violation of the Convention (Arts. 33, 34, ECHR). Its jurisdiction is compulsory and its decisions are binding for the parties to the Convention that are parties to the case (Art. 46, ECHR). However, the complainant must have exhausted all available local remedies prior to resorting to the Court (Art. 35(1), ECHR). Until this "time-consuming, thorny procedure" has been exhausted, the complainant cannot have access to the Court (Rest 2000: 48). More related to the environment issue, as the Convention does not include a provision particularly on the protection of the environment, when the judgements are examined, it may be observed that the Court has generally invoked articles 2 (right to life) and 8 (right to privacy) of the Convention, and has not involved different aspects which are peculiar to the right to environment and its protection.[19]

The International Criminal Court (ICC) also has the competence to decide on war-crime cases which are brought before the Court because of "widespread,

---

[17]See the details on PCA-CPA, at https://pca-cpa.org/en/about/.

[18]See the list of the leading cases and judgements of the ECJ on environment, at: http://ec.europa.eu/environment/legal/law/pdf/leading_cases_2005_en.pdf.

[19]See the factsheet on environment-related cases in the ECtHR's case law, June 2016. Available at: http://www.echr.coe.int/Documents/FS_Environment_ENG.pdf.

For the details on the ECtHR and the right to the environment, see also Turgut (2007).

long-term and severe damage to the natural environment" arising from a military attack (Art. 5; Art. 8, para. 2, (b), (iv), ICC Statute). However, it is argued that its competence on environmental matters remains 'theoretical' (Maljean-Dubois/ Richard 2004: 27). This is because it does not provide comprehensive protection of all elements of the environment or prosecution of environmental crimes.

There has recently been a promising development for the claims brought before the ICC regarding environmental destruction. According to a policy paper released by the Prosecutor on 15 September 2016, "particular consideration [will be given] to prosecuting Rome Statute crimes that are committed by means of, or that result in, *inter alia*, the destruction of the environment, the illegal exploitation of natural resources or the illegal dispossession of land" (Section V, para. 41).[20]

Thus it expands the focus of the Court to crimes regarding environmental destruction and exploitation and reactivates the debate over similar cases, such as land-grabbing in Cambodia.[21]

Yet, it is too early to know how cases with environmental aspects will be resolved before the Court after this policy paper, what its influence will be on the disputes/crimes arising in the area of international environmental law, and to what extent it will be successful in meeting the expectations.[22]

Despite these recent promising developments on environmental crimes, the recently announced withdrawals from the Court by three full members of the ICC – Burundi, Gambia and South Africa (see Article 127(1) of the ICC Statute for withdrawal procedure) – in October 2016, and then by Russia on 16 November 2016, have increased the fears that other states may also show their intention to pull out, putting the future existence of the Court can at risk.[23]

Given the fact that a number of countries, including the US, China, India, Pakistan and Indonesia, are still not signatories of the Statute and that some signatories, like Egypt, Iran and Israel, have failed to ratify it, only two of the five permanent members of the UN Security Council – the UK and France – are members of the Court; and those recent withdrawals from the Court prompt questions about how it can contribute to environmental protection while coping with its own problems.

Finally, the WTO DSP is also noteworthy, as it has an original character when compared with other DSPs. It is based on a system established by the WTO Understanding on the Settlement of Disputes (DSU), which renders compulsory

---

[20]Policy Paper on Case Selection and Prioritisation. https://www.icc-cpi.int/itemsDocuments/20160915_OTP-Policy_Case-Selection_Eng.pdf.

[21]For details of the Cambodian case debate, see Braithwaite (2016); Maza (2016).

[22]Similarly, a symbolic trial held in The Hague, Netherlands in October 2016, 'The Monsanto Tribunal,' also arises as one of the recent developments concerning the evolution of views on environmental crimes. In the present system, it is not possible to bring criminal charges against a company like Monsanto for their crimes against human health and the environment. Therefore, this symbolic tribunal – not governed by the United Nations or by the International Criminal Court – aims to contribute to the debate on a possible amendment of the Rome Statute to include the crime of ecocide in international criminal law. For details, see: http://www.monsanto-tribunal.org/.

[23]For the debate on withdrawals from the Court, see Akande (2016); Robinson (2016).

and binding the settlement of disputes arising under WTO agreements. Its aim is to lead to 'prompt' (Art. 3.3, DSU) solutions which are "mutually acceptable" to the parties and 'consistent' with the covered agreements (Art. 3.7, DSU). This aim is reflected in some of its features, such as: its strict time limits on the duration of proceedings (e.g. 12.8, DSU); specific deadlines for intermediate steps in the dispute settlement process (e.g. Art. 5.4, DSU, Art. 8.7, DSU, Art. 16.1, DSU, Art. 16.2, DSU); 'quasi-automatic[ity]' (González-Calatayud/Marceau 2002: 277; WTO 2001: 23) in adoption of panel reports, the Appelate Body, its well-qualified and experienced panelists (Art. 8.1, DSU) and members (Art. 17.3, DSU); and the possibility of imposition of the suspension of trade concessions or the provision of compensation as measures, provisions which aim to facilitate compliance, such as notification requirements, counter-notifications, transparency, and committees for the review of the operation of the related agreement. These features enable it to be viewed as one of the most effective DSPs when comparing the dispute settlement provisions of the WTO with those of the MEAs (WTO 2001).

However, this system has also some shortcomings and the most significant one for this study is its inadequacy in cases involving environmental issues.

Article XX of GATT (Art. XX, GATT, (b), (g), (chapeau)) provides for exceptions regarding environmental concerns. So, even before the WTO Agreement, there have been cases related to environmental issues, such as various Tuna cases (US vs. Canada; US vs. Mexico; US vs. EEC), the Salmon and Herring case (US vs. Canada), the Cigarettes case (US vs. Thailand), and the Automobiles Case (US vs. EEC). After the WTO agreement, the Gasoline case (US vs. Brazil, Venezuela), the Shrimp Turtle Case (US vs. Malaysia, Indian, Pakistan, Thailand [joint case]) and the Asbestos case (EC vs. Canada) can be counted as examples (WTO 2004). Of these, the Shrimp Turtle case should be particularly emphasized, as the WTO DSP has begun to take into account further environmental concerns in this case (Cameron 2005; UNEP 2005a; WTO 2008, 2011), while there was less tolerance towards environmental issues from 1990 to 1998 (e.g. Tuna Dolphin Case) (Charnovitz 2005).

However, the decisions of the WTO DSP on the issues related to the environmental concerns are generally criticized as insufficient attention is paid to them. To improve the WTO DSP in environment-related cases, there are several recommendations. Some of them are: the involvement of MEA secretariats, the use of environmental experts, the facility to refer to the ICJ where rights and obligations beyond the WTO sphere are applicable (however, this requires an amendment to Art. 23, DSU), increasing the use of Art. 5, DSU methods (mediation, conciliation and good offices) – as they require the agreement of all parties to the dispute, they are rarely used – and the establishment of an environment advisory board consisting of experts to which the parties to the dispute have to resort before formal DSPs (González-Calatayud/ Marceau 2002). However, all these recommendations still remain controversial.[24]

---

[24]The establishment of an International Environmental Court (IEC) with mandatory jurisdiction can also be raised as an alternative solution to the above-mentioned problems (Pauwelyn 2005; Rest 2000); however, there are arguments against it, so it requires further discussion (Hey 2000).

Table 4.2 Differences between DSPs and CMs

	DSPs	CMs
Main purpose	To settle the dispute between the related parties, finding the offender and penalizing it	To identify parties' compliance difficulties, to re-establish and facilitate better compliance
Main approach	Dispute settlement	Prevention/avoidance
Type of exercise	Bilateral exercise	Multilateral exercise
Terminology	– MEA's obligations – Breach/violation – Penalties or sanctions	– Parties' commitments to the MEA/requirements of the MEA – Compliance/non-compliance – Consequences/response measures
Basis for their application	The existence of a breach of any obligation of IEL, the causality between the breach and the damage	The existence of potential non-compliance, no requirement of breach or injury to a state party
Triggering of mechanism	The claims of one party are directed to another party; confrontational and adversarial	Not directed to any party, non-confrontational and non-adversarial
Institutional structure	Supported by judicial or quasi-judicial bodies	Non-judicial (except two samples) bodies
Decision-making	Judicial or quasi-judicial body makes a decision on the dispute in question	Political body; in some cases such as the Kyoto Protocol, quasi-judicial body
Outcome	Penalties, sanctions	Response measures (positive or negative)

Source Savaşan (2017: 842)

4.2.2.2 The Features Which Make CMs More Attractive

The distinct characteristics of the CMs (particularly the NCPs in these mechanisms) and many aspects of the DSPs (see Table 4.2)[25] also express those features of CMs which address the limitations of the traditional means mentioned above. These features can be summarized under nine dimensions: main purpose, main approach, type of paradigm, terminology, basis for application, initiation of the mechanism, institutional structure, decision-making, consequences (Savaşan 2017).

The main purpose of CMs is to identify the reasons for and difficulties of non-compliance, thus they are "forward- rather than backward-looking" to provide "the future integrity of the regime" (Handl 1997: 34).

CMs are based on a preventative approach which relies on cooperation between parties (Klabbers 2007; Tanzi/Pitea 2009). In practice, parties seldom fail to cooperate when proceedings are instigated against them, so, the process can be seen as "compulsory in nature" (Klabbers 2007: 999). The parties, usually developing states, are

---

[25]The features list is compiled utilizing the following articles: Beyerlin et al. (2006); Eritja et al. (2004); Handl (1997); Jacur (2009); Klabbers (2007); Montini (2009); Pineschi (2009); Tanzi/Pitea (2009); Treves (2009a, b); Ulfstein/Werksman (2005); Wolfrum (1999).

also provided with assistance in technology transfer or capacity-building to promote compliance or terminate the activities which result in non-compliance. This is also seen as "new conditionality", which has to be applied as the price paid for the participation of developing countries (Lang 1995: 288, 289). CMs are not based on the traditional paradigm of bilateralism, but on multilateral exercises; and their terminology is different from the traditional means (see Table 4.2). Potential non-compliance is sufficient to trigger these mechanisms; breach and direct damage are not required. Therefore, all parties (even the non-compliant one) can initiate the mechanism even if they themselves have not directly suffered as a result of infractions.

Even if they are legally structured through not only legal rules of COP/MOP decisions and Committee decisions, but also Treaty Law, a Committee established under NCPs of the CM may decide on the existence of non-compliance. So, although two of them are defined as quasi-judicial with their features similar to judicial procedures, such as the International Treaty on Plant Genetic Resources (ITPGRFA)[26] and CM of the KP with its enforcement branch (EB), their decision-making procedures are based on a political body entitled to apply consequences to facilitate compliance rather than enforcement.

4.3 Operation of Compliance Mechanisms: Fundamental Bodies

CMs generally operate through the Committees (ImplCom/ComplCom) created under the NCPs (which will be spelled out in Sect. 4.3.3) and the following institutions established under MEAs: a conference or meeting of the parties (COP/MOP), a secretariat and one or more subsidiary advisory bodies (technical or scientific bodies) and a financial mechanism.

4.3.1 The Conference of the Parties

The Conference of the Parties (COP) – the term 'Conference of the Parties' was first used in an MEA by CITES in 1973 (Churchill/Ulfstein 2000) – is the plenary and governing body set up by the related MEA. The Meeting of the Parties (MOP), on the other hand, is the plenary body of a Protocol attached to the MEA in question.

It does not have a permanent seat, but the parties regularly (annually or every other year) meet at different meeting places which alter according to their decisions. A bureau can also be established by the COP to provide help to it and lessen its workload.

---

[26]See International Treaty on Plant Genetic Resources (ITPGRFA) (2001). Rome, 3 November 2001. Available at: http://www.fao.org/3/a-i0510e.pdf.

The COP is the ultimate decision-making-body on all matters involving both substantive and procedural issues, and also both internal and external ones related to the functioning of the MEA in question. So it has a variety of tasks depending on its powers.[27] Its tasks can be summarized under two main spheres: the internal sphere (decision-making on procedural or substantive matters; controlling compliance of parties) and the external sphere (concluding agreements; capacity for making agreements).

4.3.1.1 Tasks and Authorities Within the Internal Sphere

Decision-making on Procedural Matters:

- Adopting decisions on the rules of procedure for itself and other subsidiary bodies and changing them where applicable.

Decision-making on Substantive Matters:

- Deciding on the working of the mechanism (regarding the time and place of its meetings, and the rights and obligations of contracting parties, establishing a bureau for itself and subsidiary bodies for the mechanism, electing representatives of parties for the ComplCom/ImplCom and determining observers from NGOs and officials for the secretariat...etc.).
- Deciding on financial matters (the legal basis of the COP on deciding these issues and determining the contributions of the parties to the budget can derive from the agreement itself. In fact, the environmental agreements can refer to the need to transfer financial resources in their preambles, or in their articles, and can entitle the COP to decide on these issues for implementation. If it is not explicitly in the agreement, it can also be assessed within the scope of "implied powers finding its roots in customary international law" (Jacur 2009: 425).
- Interpreting the texts of the relevant MEA and COP decisions, making new rules by specific provisions of MEAs (see e.g. Art. 17 of the Kyoto Protocol) or by relying on the doctrine of implied powers[28] developed for formal IGOs. Thus, the MEA can give authority to the COP to decide and act on additional issues, if necessary for the functioning of the agreement. Thus, the COP can decide on the

---

[27]For a discussion on the legal basis of its decision-making power, see Savaşan (2018).

[28]See: Reparation for Injuries Suffered in the Service of the UN, Advisory Opinion of 11 April 1949, ICJ Reports (1949).

Here, the ICJ for the first time questioned whether the functions and rights conferred to the United Nations by its constituent instrument (UN Charter) necessarily implied the attribution of international personality of the organization, and held that the UN could exercise powers which can be implied from the provisions of its Charter. Available at: https://www.icj-cij.org/en/case/4/advisory-opinions.

issues for which no explicit provision exists in the relevant MEA empowering the COP to decide (Churchill/Ulfstein 2000).

- Determining the rights and obligations of parties and making modifications to them in two main ways: 1. having power to make amendments or adjustments to modify existing obligations of the parties, e.g. the adjustment procedure of the Montreal Protocol (MP), Art. 2.9 and Art. 6; and 2. changing the existing provisions through the use of 'enabling clauses,'[29] e.g. the Montreal Protocol, Art. 8.
- Deciding on the consequences of non-compliance, hence on the measures (such as recommendation of guidelines for supporting compliance, action plans or suspension of their some rights) which are to be applied against the non-compliant parties.
- Adopting amendments or new protocols and annexes to the respective MEA, and thus creating and developing the regimes.

Decision-making on Controlling Compliance:

- The COP/MOP controls the parties' compliance, with the respective MEA monitoring their compliance. Based on the parties' self-reporting and the ComplCom/ImplCom's report and recommendation on this report, it decides on the possible outcomes of the non-compliance.

Tasks and Authorities at the External Sphere

Even if it is controversial whether the COP has powers to act on external matters such as concluding agreements with other entities of international law (states, organizations, etc), it is possible to argue that it can act on the basis of an examination on "the wording of MEAs, the doctrine of implied powers and the practice" (Churchill/Ulfstein 2000: 647).

4.3.2 Secretariat

It is possible to observe various kinds of secretariats established under different MEAs with different tasks and authorities. With respect to CMs and their functions in CMs, they can be examined in two distinct categories: secretariats that are set up as permanent or as interim bodies of the agreement (Churchill/Ulfstein 2000).

1. Secretariats that are set up as permanent (Churchill/Ulfstein 2000) bodies of the agreement can both provide services for COP, subsidiary bodies of the agreement and parties, and also design and implement programmes or projects at the regional and national levels. Secretariats of the Convention on the Control of Transboundary Movements of Hazardous Wastes and Their Disposal, known

---

[29]See Fitzmaurice (2009: 462–463) for details on enabling clauses.

simply as the Basel Convention, the United Nations Convention to Combat Desertification (UNCCD), regional seas conventions and action plans, the Convention on the Conservation of European Wildlife and Natural Habitats (the Bern Convention) and the Convention on the Conservation of Migratory Species of Wild Animals (also known as CMS or the Bonn Convention) are examples of this kind of secretariat.

2. Secretariats that are set up as interim (Churchill/Ulfstein 2000) bodies of the agreement can only provide services to the above-mentioned bodies, and the COP is authorized to give final decisions. The Vienna Convention, the Montreal Protocol, the Climate Change Convention, the Kyoto Protocol, the Biodiversity Convention, the Convention on Wetlands of International Importance, the Ramsar Convention, and the Stockholm Convention on Persistent Organic Pollutants (POPs) can be viewed as examples of this kind of secretariats.

It should here be underlined that the secretariats acting under an existing IGO generally set up the MEA secretariats as well, like the secretariat of the United Nations (based in Bonn, Germany) working as the secretariat of the Climate Change Convention, Kyoto Protocol, the secretariat of the UNEP working as the secretariats of the CITES and the Basel Convention (both located at Geneva), the Biodiversity Conventions (based in Montreal, Canada), the Montreal Protocol (based in Nairobi, Kenya)), the International Maritime Organization (IMO) (Convention on the Prevention of Marine Pollution by Dumping of Wastes and Other Matter, the London Convention, located in London). The Ramsar Convention's secretariat should be stressed here as a divergent case, since it conducts its tasks under a non-governmental organization (NGO), the International Union for the Conservation of Nature (IUCN) (Birnie et al. 2009; Chambers 2008).

If the secretariat is under both an existing organization and an MEA, it is under the authority of both the host organization and the bodies of the related MEA.

Regarding the host organization, as its staff is generally employed by the host organization, the host organization has powers over their appointment and termination.

With respect to the bodies of the MEA, the COP/MOP or the subsidiary bodies can instruct the secretariat in general. The secretariat is assigned to provide services to the bodies of the MEA concerning the necessary issues for implementing and promoting the provisions of the related agreement. It assists them in monitoring and evaluating the compliance with their MEA, in the preparations for the meetings or draft decisions on compliance, in receiving and analysing the reports on the implementation of commitments, and also in providing coordination with other international organizations, financial institutions such as the GEF, other MEAs, and so forth. It can also assist the parties in the preparation of national reports on their compliance with MEA obligations. For example, Art. 12(a–g), MP counts as its tasks the following: to arrange for and service MOPs (RoP 28(a–f)); to receive and make available data upon request by a party; to prepare and distribute reports on a regular basis to the parties; to notify the parties of any request for technical assistance; to encourage non-parties to attend the MOPs as observers and act in accordance with the provisions of the Protocol; to provide information and requests to such non-party observers; and to perform such other functions which can be assigned to it by the parties for the achievement of the purposes of the Protocol.

4.3.3 Subsidiary Organs

Subsidiary organs can be established by the MEA itself, or after the adoption of the agreement, by a COP decision (in rare cases, also before the adoption of the agreement, by a different organization, e.g. the Intergovernmental Panel on Climate Change established by UNEP and the World Meteorological Organization).

The size of the subsidiary bodies can be limited, and their members can require certain qualifications, such as to be a qualified expert on the issue (e.g. the Scientific Council of the Bonn Convention (Art. VIII(2)).

They are usually created with the aim of providing better functioning of the whole mechanism. So they can be defined as auxiliary bodies whose mandate is usually restricted to giving advice (e.g. the Subsidiary Body for Scientific and Technological Advice of the Climate Change Convention) or financial and technical assistance (e.g. the Multilateral Fund, the Global Environment Facility (GEF)). By way of example, the Montreal Protocol (Art. 6) provides for regular assessments which have been carried out through different panels of experts qualified in scientific, environmental, technical and economic fields (the Scientific Assessment Panel (SAP), the Environmental Effects Assessment Panel (EEAP), the Technology and Economic Assessment Panel (TEAP)). According to Art. 6 of the Montreal Protocol, beginning in 1990, and at least every four years thereafter, the parties should assess the control measures provided for in Art. 2A–2I on the basis of the available scientific, environmental, technical and economic information. At least one year before each assessment, the parties should convene panels of experts qualified in these fields which report their conclusions to the parties through the secretariat within one year of being convened. These panels have played an important role through providing assessments on the possible amendments to the Protocol (Potzold 2009; Weiss 1998) and on necessary technology for Article 5 countries[30] and also providing advice to the ImplCom about issues requiring expertise beyond its scope (Weiss 1998).

---

[30]Under the MP, there are Article 5 countries representing developing countries; and Non-Article 5 countries representing developed countries. The MP stipulated that the production and consumption of compounds that deplete ozone in the stratosphere, such as chlorofluorocarbons (CFCs) and other ODSs, should be phased out by 2000 in Non-Article 5 countries and 2010 in Article 5 countries (Art. 2A, Art. 5, MP). Methyl bromide, on the other hand, was scheduled to be phased out by 2015 and HCFCs by 2040 in Article 5 Countries (Art. 2F–2G–2H, Art. 5, MP). Through the Kigali (Rwanda) Amendment (14 October 2016) to the Protocol adopted at the MOP 28, all countries committed to legally binding targets which mandate gradual reductions in HFC consumption and production as well, starting in 2019 for Non-Article 5 countries and 2024 or 2028 for Article 5 countries (as they are divided into two groups) (MOP 28, Decision XXVIII/1, 2016).

Some of them can be entitled to exercise functions on implementation and compliance, such as the Subsidiary Body for Implementation (SBI) under the Climate Change Convention (Art. 10). Yet, in this case as well, their mandate does not go beyond giving advice and assistance.

Ad Hoc working groups can also be counted among the subsidiary bodies, such as the Ad Hoc Group on the Berlin Mandate of the UNFCCC, which guaranteed the adoption of the Kyoto Protocol.

Its subsidiary bodies comprise government representatives who are experts in the relevant field of climate change and have to report on all aspects of their work in a regular manner (Art. 9.1, Art. 10.1, UNFCCC). Moreover, they are both open to participation by all parties (Art. 9.1, Art. 10.1, UNFCCC). Yet the structure of the two bodies has also been criticized in terms of two aspects. First, the political nature of the structure affects their performance negatively and makes it extremely difficult to cope with questions about compliance issues. Second, the high number of participants affects the management of the bodies negatively, so restricting the number of representatives can be suggested to create more effective operating bodies (Werksman 1996). Yet, given the fact that providing an equitable balance in the election of the representatives who serve as their members among the parties can cause further problems in their operation, this way of allowing all parties to participate in the bodies can be considered a better solution.

4.3.4   Financial Mechanism

In CMs, financing is generally provided from the general budget prepared and adopted by the COP/MOP, which is the only body making decisions about financing issues. In addition to the contributions determined by the budget, parties' voluntary contributions also form the financing resources.

In the recent period, there are also new trends to improve financing resources. The Multilateral Fund (Art. 10, Montreal Protocol), the World Heritage Convention, the Ramsar Convention's Wetlands Conservation Fund established in 1990, the Global Environmental Facility (GEF) established in 1991 by the World Bank, the Special Climate Change Fund (SCCF), the Least Developed Countries Fund (LDCF), and the Adaptation Fund (AF), which are separate from the GEF Trust Fund (GEFTF) but are operated by the GEF, the Green Climate Fund (GCF) established as an operating entity of the financial mechanism of the UNFCCC (Art. 11, UNFCCC) through COP 16/Decision 1, the United Nations Environment Programme (UNEP) and the United Nations Development Programme (UNDP) as the funding mechanism for relevant projects under the issues related to climate change (Art. 11, UNFCCC, Art. 11, KP) are the most significant examples of this new trend (Handl 1994; Jacur 2009; Maljean-Dubois/ Richard 2004; Wolfrum/Matz 2003).

Furthermore, the flexibility mechanisms created for industrialized countries in the Kyoto Protocol aim to develop new ways of sourcing financing. Of these mechanisms, the Clean Development Mechanism (CDM) and Joint Implementation (JI) are 'project-based' mechanisms by which industrialized countries are encouraged to invest in projects in developing countries or countries with economies in transition that reduce greenhouse gas emissions in a manner capable of being proved (Jacur 2009: 433). The other one, namely, the International Emission Trading (IET), is a 'market-based' mechanism by which "industrialized countries may exchange and trade their entitlements to emit greenhouse gases" (Jacur 2009: 433).

However, there are still critical challenges to providing necessary funding and resources to the parties to mitigate their technical and financial capacity-building deficiencies.[31]

4.4   Main Components of CMs

4.4.1   Gathering Information on the Parties' Performance

There are different types of information-gathering, e.g. operational information gathered on environmental conditions, technologies etc; overall regime reviews focusing on the regime's overall performance rather than a particular party's performance; and performance review information gathered from the parties' reports on their own compliance situation (Sachariew 1991; UNEP 2005b: 24). All these approaches are interrelated; however, none can replace the performance review information, because none can provide an assessment of the national "legislative, administrative, technical and other measures" taken in response to the MEAs' obligations (Sachariew 1991: 43).

Performance review information, in general, is rendered by parties' self-reporting. When the reports are submitted, they are firstly sent "to a more technical body" – generally the secretariats. These bodies are not usually entrusted with gathering supplementary information through investigations to verify the information in state reports, or with taking measures against parties if they have significant shortcomings (Wolfrum 1999). They only have the authority to trigger NCP based on the reports (e.g., the Committee is entitled to gather information in the territory of that party, as under Art. 7(e), NCP, Montreal Protocol).

The 'consolidated' reports prepared by the secretariats are then discussed and assessed in the competent body of the MEA (MOP/COP or ComplCom/ImplCom) (Sachariew 1991: 47). In this phase, "a double-tiered procedure" (Beyerlin et al.

---

[31]See *infra* Chaps. 5–6 and also Savaşan (2015: 180–181, 186–187); Savaşan (2017: 849–850). For an overview of the potential weaknesses of compliance mechanisms in general, see Savaşan (2017).

2006: 364), in which the ImplCom/ComplCom firstly presents a report on the issue to the COP/MOP which has the right to decide on the final state of the assessment, is recommended. The competent body, relying on the relevant information sources, can assess the sufficiency of the party in adopting necessary measures to meet the standards of the related MEA rules. It can also request additional information and make recommendations for better compliance. In some cases, the role of the competent body can be more limited "to providing switchboard services between the parties, with no substantive examination of the report taking place" (Sachariew 1991: 45).

The functions of the performance review information can be outlined as follows (Beyerlin et al. 2006)[32]:

- Its results can make it easier to identify whether there is a violation of the related MEA's requirements or the threat of violation, whether targets of the related MEA have been met, and what the related party's past and present compliance problems are.
- It can lessen the free-rider[33] problem by making all parties aware of each other other's situation. In fact, through each party's reporting on its own performance, while the reporting party learns about its own situation, it also learns about the others' performance as well. This provides the information to the parties about whether their own compliance can be interpreted as or compromised by free-riding, or not.
- Its results can serve to determine future priorities of the party, enabling it to address any nascent problems and prevent them leading to non-compliance, providing a 'dialogue' between the competent body of the MEA and the reporting party (Beyerlin et al. 2006: 363).
- It produces the basic data for monitoring (observing the activities which form that data) and verification (controlling the accuracy of that data).

Monitoring and verification should be distinguished from reporting in that while reporting implies examining whether necessary measures are adopted by the related party to meet the MEA's requirements as detailed above, monitoring entails their "continuous observation" (Sachariew 1991: 34), and verification means evaluating the completeness, honesty and certainty of information gathered on compliance.

Through monitoring, the degree of compliance with international environmental requirements can be evaluated and existing environmental standards can be promoted based on the collected data which can be used as "scientific criteria" for that

---

[32]See Chaps. 5–6 *infra* and Savaşan (2015) for the challenges in gathering information and the ways of dealing with the findings.

[33]Free-riding implies deliberate attempts to escape the costs of meeting the requirements of the related MEA. It emerges in two ways: a state may prefer not to participate to that MEA to avoid incurring costs while benefiting from other parties' efforts (non-participation), or may prefer not to meet its requirements (non-compliance) (Hovi et al. 2007; Kolari 2002).

promotion (Sachariew 1991: 35). With the consent of the parties, on-site monitoring[34] can also be used for verification (Wang/Wiser 2002).

Finally, it should be underlined that, because of the non-adversarial/
non-confrontational character of the CMs, the aim of the reporting processes
(self-reporting, monitoring, verification) under MEAs is to find out the compliance
problems, to prevent them, or to help the non-compliant parties to cope with them
rather than to blame the parties for alleged non-compliance or determine the cases
of non-compliance and adopt specific measures towards the party in question
(Treves 2009a, b).

4.4.2   Non-Compliance Procedures[35]

4.4.2.1   Institutions Created Under the NCP: Committee

Regarding the composition of the Committee, different ways have usually been
adopted under different MEAs. They can be revealed in two main forms (Fodella
2009: 360):

- The first and generally adopted one, particularly in earlier MEAs, is the election
  of representatives of the parties to the MEA by the COP/MOP. In such a case,
  equitable geographical distribution can be stipulated as a requirement for
  membership (e.g. the Montreal Protocol, MOP 3, Decision III/20, the Kyoto
  Protocol, Section IV.I and V.1).
- The second one is formed by the election of independent experts serving in their
  personal capacity and elected by the COP/MOP. In this situation, the question
  on the expertise requirements of the members can be raised. In most cases, these
  requirements are not clearly demonstrated, for it is merely stated that members
  should be experts in the fields relating to the agreement. Although an evaluation
  conducted by independent experts is expected to ensure the "impartiality and
  objectivity" of the mechanism and thus to enhance its reliability (Fodella 2009:
  361), most of these bodies are rarely composed of independent experts. This is
  because the contracting parties of the related MEAs do not want to lose their
  "full control over the process" (Fodella 2009: 360). In addition, the only NCP in

---

[34]It should be noted, however, that on-site visits (on-site monitoring or on-site inspections) under
MEAs have not been very prevalent, as they are still heavily debated, due in particular to the
principle of state sovereignty (Faure/Lefevere 1999), or some other reasons depending on the
features of the related MEA, e.g. under the MARPOL, countries do not want inspections in their
ports since they make those ports less attractive to oil tankers than neighbouring ports (Mitchell
(1994). Yet, it is possible to see examples of them under some MEAs, such as the Ramsar
Convention on Wetlands, and the Montreal and Oslo Protocols. See also Bothe (2006) for the view
that as long as transparency increases and the role of NGOs in submitting information increases,
site visits become less important.

[35]For a discussion on the legal basis of the non-compliance procedures, see Savaşan (2018).

which the nomination of experts by NGOs as candidates for election to the ComplCom is possible is the Aarhus NCP (Stephens 2009; Treves 2009a), and in very rare cases, NGOs can also take part in the meetings of these bodies as observers.

- A third one in addition to these two main forms can also be identified. It represents the effort to find a way between these two paths. Here again the Committee is composed of representatives of the parties to the MEA, but the members of the body are usually expected to serve in their individual capacities (e.g. the Montreal Protocol) rather than as representatives of their states, or may be bound to "serve objectively and in the best interest" of the Convention (e.g. the Basel Convention, Terms of Reference, para. 5). Sometimes the avoidance of any possible "conflict of interest" during the performance of their duties within the Committee can be stipulated in the related agreement (e.g. the Kyoto Protocol, Section III; 2(d)).

In any case, in almost all MEAs, the members of the Committee are required "to have a specific legal or technical background in order to ensure that the requirements for a 'due process' are respected, in particular in terms of recognized competence and impartiality" (Montini 2009: 401).

These bodies can be composed of a limited number of members (as in the case of the Montreal Protocol, Art. 8: 10 parties elected by MOP) or open to the participation of all parties (as in the case of the Kyoto Protocol). Both methods can involve advantages and disadvantages. In a body which has a limited number of parties, reaching decisions can be easier than when a body consists of an unlimited number of parties. On the other hand, the greater the number of parties which have an opportunity to submit their views on the issue, the more the decisions taken by the body are seen as legitimate and transparent.

So, in the first option, in which a limited number of parties can be members of the Committee, it becomes essential to guarantee the legitimacy and the transparency of the decisions. To achieve this purpose, the Committee should only be entitled to make recommendations and the final decision should be given by the plenary organ, COP/MOP. Also, by adopting the principles of rotation and equitable geographical distribution for the members of the Committee, all parties' interests should be better protected (Ehrmann 2002). However, it is also necessary to take into account the nature of the matter that the body addresses. If that matter is basically bilateral in nature, a small body is sufficient, whereas if it has global impacts on the parties' interests, "an expanded, even open-ended committee" can function better (Handl 1997: 39).

With regard to the tasks and the powers of the Committee, it should first be stated that it has the right to regularly evaluate whether the parties comply with the MEAs in question. This evaluation should not be limited to compliance with procedural obligations, but should also involve compliance with substantive obligations (Fodella 2009). When a party's alleged non-compliance is referred to the body for consideration, in order to assess and report on it, if necessary it can seek additional information beyond that provided by the parties, either through the secretariat or directly from the party or parties concerned. It is also entitled to investigate the issue through site visits or "site fact-finding" (Marauhn 1996: 712), but only with the consent of the party concerned. If necessary, it can also cooperate and coordinate with other relevant international organizations and MEAs. Based on its investigations concerning the matter, it submits its findings and recommendations to the COP/MOP in a report. Thus the settlement of the issue is achieved by the Committee prior to its consideration by the COP/MOP, but the final decision is given by the COP/MOP.

4.4.2.2   Procedural Structure: Phases and Safeguards

In principle, the NCP is divided into four phases: submission (triggering) phase, preliminary phase, substantive phase (including consideration and recommendation phases) and final phase (including decision-making and final resolution phases). For instance, this is the case of the compliance mechanism of the Montreal Protocol. By contrast, different phases may be provided for by other MEAs (e.g. the Kyoto Protocol).[36]

Submission (triggering) Phase. In NCPs, it is possible to assert that the party or parties of the MEA do not comply with its obligations, and to bring this assertion regarding the parties' compliance problems to the attention of the related bodies of the MEA. This kind of assertion can be brought by the non-compliant party itself ('self-trigger') (Jacur 2009: 374), by the other parties, by the related bodies of the MEA (more often by the secretariat), and very rarely by third parties to the agreement.

The only example that allows the NGOs the right to submit communications concerning non-compliance is the Aarhus NCP, which has an opt-out procedure which allows parties to remain outside such complaints for up to four years (Stephens 2009). This is because it is perceived by state parties as a circumstance that can undermine the non-confrontational and cooperative nature of the mechanism, and also because, in practical terms, triggering by non-state actors can result in a huge increase in costs and the Committee's workload (Tanzi/Pitea 2009).

The opportunity given to the party with the compliance problems ideally reflects the non-judicial, cooperative character of the NCPs. Because the facilitative approach is dominant in these procedures, the parties do not hesitate to trigger the procedure against themselves. Although in some cases punitive measures can be adopted against the self-trigger (e.g. against the Russian Federation under the Montreal Protocol), the self-triggers are generally granted technical or financial assistance at the end of the procedure to improve their compliance situation.

Triggering by any other party to the MEA has so far been rarely used in the MEAs. This is because, the NCPs, as distinct from traditional means of ensuring compliance, do not have an adversarial character. Moreover, unless non-compliance

---

[36]See Chap. 6 *infra* on the Kyoto Protocol.

directly influences the triggering party, it is not easy to know whether the other parties comply with the MEA's obligations or not.

However, when a party brings an assertion against another party, the question arises as to whether it has to reveal that it has been directly affected by the non-compliance of the party, and so has a "specific interest" in triggering the procedure (Jacur 2009: 375). It is here essential to look to the agreement's character, in particular whether it is global or bilateral.

If it is global in character, that is, if it addresses global environmental problems and there are collective environmental interests of the parties at stake, it establishes "obligations *erga omnes partes*" (Jacur 2009: 375). In such a situation, the triggering party is not expected to reveal that it has been directly affected by the non-compliance of the other party, e.g. the Montreal and Kyoto Protocols.

If the agreement is global in character but involves obligations of a bilateral character, it is generally accepted that the triggering party must reveal its interest in bringing the case to the competent bodies of the MEA, e.g. under the Basel, Espoo, and Cartagena Conventions, only affected parties may trigger the procedure.

Yet the problem here is that the agreements involving bilateral obligations can also include some collective interests which need to be addressed (e.g. the Basel Convention, Art. 4.5, prohibits the trade of hazardous waste with non-parties). In such cases, allowing only the affected party to trigger the procedure can prevent assertions of non-compliance by the other parties and indirectly prevent the protection of the environment, which is the common aim of all MEAs.

With respect to the triggering by any other party, another issue that should be discussed is whether one party or more parties can trigger the procedure. If there is a clear expression in the relevant MEA about this issue, there is no problem. However, if not, the problem arises (e.g. triggering by South Africa on behalf of the Group of 77 and China, in the Facilitative Branch of the Kyoto Protocol) (Lefeber 2009). So, in order to inhibit these problems, specific procedural rules are required to be regulated in the MEAs, but the flexible nature of the NCPs can be damaged by strictly regulated rules governing these procedures. Therefore, the balance between strict regulation and the protection of flexibility should be safeguarded while making these regulations.

Triggering by MEAs' bodies like the COP/MOP, the ComplCom/ImplCom, or subsidiary bodies is not a frequently allowed method in MEAs. Triggering by the COP is only provided for in the International Treaty on Plant Genetic Resources for Food and Agriculture (ITPRGFA) NCP, section V.1.(c) and the London Dumping Convention NCP para. 4.1.1. Triggering by the Committee is, on the other hand, only seen in the Aarhus Convention NCP, para. 14, the Espoo Convention NCP, para. 6. and the Stockholm Convention Draft NCP, para. 17.

Finally, triggering by other subsidiary bodies' is similarly not provided for in any of the present MEAs, which only accept their submission of additional information to the secretariat while drafting its triggering report (the Montreal Protocol) or the report submission of expert review teams to the Committee (the Kyoto Protocol NCP, section VI.1).

Therefore, it can be argued that the only body of the MEA which has the power to trigger the procedure is the secretariat. In fact, due to the fact that an MEA secretariat has the power to collect and file the information received from the parties, it can provide the necessary information for analysing the compliance situation. Therefore, even if the respective MEA does not accept the triggering of the secretariat (e.g. the Cartagena Protocol, the Kyoto Protocol), it still plays an important indirect role in the triggering procedures.

However, it does not allow the IGOs and NGOs to trigger the NCP. Also, in contrast to the MCP referred to by the Convention (Art. 13), which contains triggering by the COP in addition to the party itself and by a party to another party (COP 4, Decision 10, 1998: para. 5), the NCP of the Protocol does not include initiation of the procedure by the COP/MOP.

Additionally, the secretariat is not entitled to trigger the procedure, thus is not empowered "with a stronger role regarding implementation supervision" (Oberthür/ Ott 1999: 214).

Instead of completely rejecting the right of the secretariat to trigger the procedure, it can be confined to the issues of compliance with certain obligations only, e.g. reporting obligations, or it can confine the sources of information to the parties' reports which the secretariat can rely on in the triggering procedure (Lefeber 2009). However, other sources of information aside from the parties' reports, such as data received from other bodies of the MEA, NGOs or IGOs, can also be used by the secretariat.

A controversial issue about the secretariat's gathering information should also be clarified here. It is whether the secretariat has to ask for additional information before triggering the NCP, when it has realized the possibility of a party's non-compliance with its obligations. This question must be assessed according to the related provisions of the MEA. If the MEA obliges the secretariat (by such wording as "it shall notify the party") to ask for information before triggering (e.g. the Barcelona Convention NCP, para. 23), this forms an obligation for the secretariat. However, in the opposite case, that is, if the MEA does not oblige notification, using instead phrases like "it may notify" or "request for information" (e.g. Convention on Long-range Transboundary Air Pollution (LRTA) NCP, para. 5), this confers a "discretionary power" which should be interpreted not as placing any obligations but as an option for the secretariat (Lefeber 2009: 383).

In MEAs which have a global character, triggering of NGOs or the public is not allowed. In these kinds of MEAs, NGOs only can join the COPs' meetings and submit any information they have on the non-compliance situation. If the COP finds the information 'worthy' of being taken into account, this can form an "indirect trigger" of the NCP (Lefeber 2009: 381).

Therefore, it can be said that triggering by NGOs, observers and very rarely the public (only in the Aarhus Convention NCP, para. 18) has been accepted only under MEAs with a regional scope or with a "particular vocation" – e.g. the Alpine Convention NCP, para. 2, the Aarhus Convention NCP, para. 18, the Kiev Protocol on Pollutant Release and Transfer Registers (PRTR Protocol) NCP, para. 18, and the Water and Health Protocol NCP, para. 16) (Lefeber 2009: 380).

Preliminary (notification) Phase. As with traditional DSPs, in which a prior consultation is made between the related parties, generally, in NCPs, the preliminary phase is also used between the related parties and the secretariat. After the submission phase, the secretariat sends a copy of that submission to the party alleged to be in violation of a particular provision of the relevant MEA. Then the concerned party has to send a reply. As soon as the secretariat receives the reply, it transmits the submission, the reply and other necessary information to the Committee for its consideration.

In the case of a secretariat' triggering, a time limit may also be imposed. To illustrate, in the Barcelona Convention NCP, para. 23, after notification by the secretariat to the non-compliant party about its situation and discussion on ways to resolve its problems, if the party cannot resolve the issues within the maximum time frame of three months, the party can trigger the NCP.

Substantive Phase. In this phase, the submission should be assessed by the Committee on the basis of the criteria (procedural and substantive) established under the related MEA, to decide on its admission. In order to clarify at least what the procedural and substantive criteria can be, a brief explanation is provided here:

With respect to the procedural criteria, examples are the admission of submissions only when they are made in writing or in a specific language. Some MEAs, like the Aarhus Convention and the Kyoto Protocol, can also determine detailed rules regarding the content of the submissions.

With regard to substantive criteria, MEAs may require submissions which indicate the specific obligations and corresponding MEA provisions, explain the causes of the non-compliance, and provide a list of information supporting the trigger's assertions. The prior exhaustion of local remedies and non-existence of parallel proceedings under other jurisdictions can also be expressed as substantive requirements of the admissibility of the triggering. "[T]he existence of a jurisdiction *ratione temporis* of the relevant compliance mechanism" can also be viewed as a substantive requirement (Lefeber 2009: 385). Yet, it should not be ignored that NCPs can address both actual and possible cases of non-compliance. So submissions should be evaluated case-by-case.

In the recommendation phase, the Committee drafts and adopts appropriate recommendations on the parties submitted for consideration, and then reports them to the COP/MOP.

Final Phase. After receiving the Committee's report, the COP/MOP decides upon the matter, and can apply positive or negative measures to bring the non-compliant party to full compliance.

Until the party in question has achieved full compliance, monitoring of the Committee continues, and it can repeat its consideration, recommendation and reporting phases to monitor the party's progress.

- *Procedural Safeguards*

Although NCPs developed under MEAs do not have "the characteristics of judicial proceedings," but rather the characteristics of administrative procedures, in order to ensure fairness they involve some procedural safeguards with "a different nature and degree than those normally available in judicial proceedings" (Montini 2009: 393).

With regard to these safeguards afforded to the parties involved in NCPs, first of all, it should be clarified that NCPs generally consist of two aspects – the facilitative aspect and the enforcement aspect – which are best reflected within the context of the Kyoto Protocol.

In the facilitative aspect, after examining the matter based on the information gathered and the report submitted by the party itself, the Committee decides to revoke the facilitative measures of the party whose compliance is at stake in order to resolve the matter of non-compliance.

If the facilitative measures do not work, and the non-compliance continues, in the second stage, the Committee can recommend to the COP the adoption of further measures involving sanctions.

In order to provide a 'due process' for the parties involved in these stages, some rights and safeguards are provided for the parties which can be highly influential in ensuring the parties' voluntary compliance, such as the rights provided to the parties under scrutiny and to the submitting parties; pre-determined deadlines for both submissions/decisions; the rights of transparency and publicity; the impartiality and independence of the Committee; fixed consequences; and the possibility of making an appeal.[37]

4.4.3 Non-Compliance Response Measures

In IEL, in contrast to other areas of IL, "sanctions are rarely used" and they can be identified as "largely irrelevant to and ineffective for environmental agreements" (Jacobson/Weiss 1998: 547). In fact, the UN Charter uses the word 'measure' instead of sanctions/penalties in Chapter II, Art. 39 onwards. Likewise, sanctions/penalties are not used in most MEAs, e.g. the Montreal Protocol (Art. 8) uses the word 'measures' when recommending an indicative list. It is possible to find different examples, like the Kyoto Protocol referring to the 'consequences' of non-compliance (Art. 18), instead of using the word 'measures'. Yet, there is no example of an MEA using sanctions/penalties. Therefore, in line with the UN Charter and the existing MEAs, this study employs the word 'measure' rather than sanctions/penalties in relation to the types of response to non-compliance.

After examining the situation of non-compliance in a CM, the Committee reports its findings and recommendations to the COP/MOP. Based on these findings, the COP/MOP firstly finds out whether a non-compliance situation has

---

[37]See *infra* Chaps. 5–6 on case studies for details.

emerged or not on the basis of a political assessment. The decision on the adoption of measures related to non-compliance is conducted ultimately by the COP/MOP; the Committee can only make recommendations to the COP/MOP on the adoption of such measures. In the Kyoto Protocol, the Compliance Committee (ComplCom), with its two branches (FB and EB), is entitled to decide on them (NCP, Section VI.7 and V.6), and in the Protocol on Water and Health to the Convention on the Protection and Use of Transboundary Watercourses and International Lakes, the Committee may issue cautions (Art. 15, KP, RoP XIII, Decision I/2, XI, 34(d)).

The CMs basically aim to warn parties about the consequences of non-compliance and endeavour to prevent their non-compliance by "making non-compliance ultimately costlier than compliance" (INECE 2009: 13). So response measures adopted in CMs tend to induce compliance with parties through either positive measures – "cooperative assistance-oriented measures" – or negative measures – "more severe measures" (Treves 2009a, b: 7). In general, if non-compliance stems from the lack of material, institutional or financial resources, supportive measures are seen as more appropriate ways to induce compliance. On the other hand, if there is deliberate and continued non-compliance, negative measures can also be imposed on the non-compliant party as a "last resort" to bring it back to compliance (Maljean-Dubois/Richard 2004: 25).

In this sense, response measures can be categorized under two different groups: positive measures and negative measures.

Positive measures imply soft measures consisting of recommendations made by the compliance bodies and technical/financial assistance. Differential implementation schedules and obligations (as set forth in Rio Principle 7) can also be assessed as positive measures, as they can remove political and economic barriers that might prevent some parties joining agreements and undertaking to comply with their requirements (Hunter et al. 2002).

Technical assistance generally involves the development of internal skills and expertise – in other words, capacity-building, and the development of technical, scientific and institutional conditions in which this capacity can be achieved effectively through the transfer of new technology and the exchange of information (Chapter 37, Agenda 21).

Financial assistance is ensured by funding mechanisms such as the Global Environment Facility (GEF) and the Multilateral Fund (MF) established under the Montreal Protocol (MP).

Thus, through both technical and financial assistance, the non-compliant party is encouraged to overcome the lack of human resources, and also the lack of material and financial resources.

Negative measures applied against a non-compliant party are mainly adopted to prevent future non-compliance. They can vary from shaming and/or imposing additional information to suspending the non-compliant party's rights.

References

Agenda 21 (1992). Available at: https://sustainabledevelopment.un.org/content/documents/
    Agenda21.pdf.
Akande, D. (2016). South African Withdrawal from the International Criminal Court – Does the
    ICC Statute Lead to Violations of Other International Obligations? 22 October 2016. Available
    at: www.ejiltalk.org/south-african-withdrawal-from-the-international-criminal-court/.
Anlar Güneş, Ş. (2006). Gabcikovo-Nagymaros Davasi. *Ankara Üniversitesi Hukuk Fakültesi
    Dergisi*, 55, 2: 91–116.
Basel Convention Terms of Reference (2003). Available at: http://www.basel.int/TheConvention/
    ImplementationComplianceCommittee/Mandate/tabid/2296/Default.aspx#para21.
Beyerlin, U., Stoll, P.T., Wolfrum, R. (2006). Conclusions from MEA Compliance. Beyerlin, U.,
    Stoll, P.T., Wolfrum, R. (Eds.), *Ensuring Compliance with Multilateral Environmental
    Agreements: Academic Analysis and Views from Practice* (359–369). Leiden: Koninklijke
    Brill NV.
Birnie, P.W., Boyle A.E. and Redgwell, C. (2009). *International Law and the Environment* (3rd
    edn.). New York, Oxford: Oxford University Press.
Bodansky, D., Brunnée, J. and Hey, E. (2007). International Environmental Law, Mapping the
    Field. Bodansky, B., Brunnée, J. and Hey, E. (Eds.), *The Oxford Handbook of International
    Environmental Law* (1–28). New York: Oxford University Press.
Bonn Convention (1979). Convention on the Conservation of Migratory Species of Wild Animals.
    Available at: http://www.cms.int/en/convention-text.
Bothe, M. (2006). Ensuring Compliance with MEAs – Systems of Inspection and External
    Monitoring. Beyerlin, U., Stoll, P.T. and Wolfrum, R. (Eds.), *Ensuring Compliance with
    Multilateral Environmental Agreements, A Dialogue between Practitioners and Academia*
    (247–258). Leiden: Koninklijke Brill NV.
Braithwaite, P. (2016). Environmental Crimes Could Warrant International Criminal Court
    Prosecutions, 1 October 2016. Available at: http://www.ipsnews.net/2016/10/environmental-
    crimes-could-warrant-international-criminal-court-prosecutions/.
Brownlie, I. (2003). *Principles of Public International Law* (6th edn.). Oxford, New York: Oxford
    University Press.
Cameron, J. (2005). Dispute settlement and conflicting trade and environmental regimes.
    Sampson, G. and Whalley, J.(Eds.), *The WTO, Trade and the Environment* (455–468).
    Cheltenham, UK, Northampton, MA: Edward Elgar Publications.
Chambers, B.W. (2008). *Interlinkages and the Effectiveness of MEAs*. Tokyo, New York: UN
    University Press.
Charney, J.I. (1996). The Implications of Expanding International Dispute Settlement Systems:
    The 1982 Convention on the Law of the Sea. *The American Journal of International Law*, 90,
    1, 69–75.
Charnovitz, S. (2005). The WTO and the Environment. Sampson, G. and Whalley, J. (Eds.), *The
    WTO, Trade and the Environment* (413–431). Cheltenham, UK, Northampton, MA: Edward
    Elgar Publications.
Churchill, R.R. and Ulfstein, G. (2000). Autonomous Institutional Arrangements in Multilateral
    Environmental Agreements: A Little Noticed Phenomenon in International Law. *The American
    Journal of International Law*, 94, 4, 623–659.
COP 7 (2001). Report of the COP on its Seventh Session. Part Two: Action taken by the COP at its
    Seventh Session. Volume III. Marrakesh, 29 October–10 November 2001. FCCC/CP/2001/13/
    Add.3. Available at: http://unfccc.int/resource/docs/cop7/13a03.pdf.
Crossen, T.E. (2003). Multilateral Environmental Agreements and the Compliance Continuum.
    The Berkeley Electronic Press Legal Series, 36. Available at: https://www.ippc.int/sites/
    default/files/documents/1182330508307_Compliance_and_theory_MEAs.pdf.

Crossen, T.E. (2004). The Kyoto Protocol Compliance Regime: Origins, Outcomes and the Amendment Dilemma. *Official Journal of the Resource Management Law Association of New Zealand Inc.*, I, XII, 1–6.

Dagne, T.W. (2007). Compulsory Dispute Settlement and Problems of Multiple Fora under International Environmental Law. Available at: http://papers.ssrn.com/sol3/papers.cfm?abstract_id=1460942.

Declarations Recognizing the Jurisdiction of the Court as Compulsory. Available at: https://www.icj-cij.org/en/declarations.

Ehrmann, M. (2002). Procedures of Compliance Control in International Environmental Treaties. *Colorado Journal of International Environmental Law and Policy*, 13, 2, 377–444.

Enderlin, T. (2003). Alpine Convention, A Different Compliance Mechanism, What is a Compliance Mechanism? *Environmental Policy and Law*, 33, 3–4, 155–162.

Eritja, M.C., Pons, X.F., Sancho, L.H. (2004). Compliance Mechanisms in the Framework Convention on Climate Change and the Kyoto Protocol. *Revue Generale de Droit*, 34, 51–105.

European Convention on Human Rights (ECHR) (1950). Available at: http://www.echr.coe.int/Pages/home.aspx?p=basictexts&c=.

Faure, M.G. and Lefevere, J. (1999). Compliance with International Environmental Agreements. Vig, N.J. and Axelrod, R.S. (Eds.), *The Global Environment: Institutions, Law and Policy* (138–156). Washington: CQ Press.

Fitzmaurice, M. (2007). International Responsibility and Liability. Bodansky, B., Brunnée, J. and Hey, E. (Eds.), *The Oxford Handbook of International Environmental Law* (1,010–1,035). New York: Oxford University Press.

Fitzmaurice, M. (2009). NCPs and the Law of Treaties. Treves, T., Tanzi, A., Pineschi, L., Pitea, C., Ragni, C. (Eds.), *Non-Compliance Procedures and Mechanisms and the Effectiveness of International Environmental Agreements* (453–482). The Hague: T.M.C. Asser Press.

Fodella, A. (2009). Structural and Institutional Aspects of NCMs. Treves, T., Tanzi, A., Pineschi, L., Pitea, C., Ragni, C. (Eds.), *Non-Compliance Procedures and Mechanisms and the Effectiveness of International Environmental Agreements* (355–372). The Hague: T.M.C. Asser Press.

González-Calatayud, A. and Marceau, G. (2002). The Relationship between the Dispute-Settlement Mechanisms of MEAs and those of the WTO. Available at: http://onlinelibrary.wiley.com/doi/10.1111/1467-9388.00326/pdf.

Guruswamy, L.D. and Hendricks, B.R. (1997). *International Environmental Law in a Nutshell*. USA: West Group.

Handl, G. (1994). Controlling Implementation of and Compliance with International Environmental Commitments: The Rocky Road from Rio. *Colo. J. Int'l Envtl. L. & Pol'y*, 5, 305.

Handl, G. (1997). Compliance Control Mechanisms and International Environmental Obligations. *Tulane Journal of International and Comparative Law*, 5, 29–51.

Hempel, L.C. (1996). *Environmental Governance: The Global Challenge*. Washington, DC: Island Press.

Hey, E. (2000). *Reflections on an International Environmental Court*. The Hague, The Netherlands: Kluwer Law International.

Hovi, J., Froyn, C.B. and Bang, G. (2007). Enforcing the Kyoto Protocol: Can Punitive Consequences Restore Compliance? *Review of International Studies*, 33, 435–449.

Hunter, D., Salzman, J. and Zaelke, D. (2002). *International Environmental Law and Policy*. New York: Foundation Press.

International Criminal Court (ICC) Statute (1998). Available at: https://www.icc-cpi.int/nr/rdonlyres/ea9aeff7-5752-4f84-be94-0a655eb30e16/0/rome_statute_english.pdf.

International Criminal Court (ICC) (2016). Policy Paper on Case Selection and Prioritisation. Available at: https://www.icc-cpi.int/itemsDocuments/20160915_OTP-Policy_Case-Selection_Eng.pdf.

International Court of Justice (ICJ) (1945). ICJ Statute. Available at: http://legal.un.org/avl/pdf/ha/sicj/icj_statute_e.pdf.

International Court of Justice (ICJ) (1949). Reparation for Injuries Suffered in the Service of the UN, Advisory Opinion of 11 April 1949, Available at: https://www.icj-cij.org/en/case/4/advisory-opinions.

International Court of Justice (ICJ) (1997). Gabcikovo-Nagymaros Case, Judgment of 25 September 1997. Available at: https://www.icj-cij.org/en/case/92/judgments.

International Court of Justice (ICJ) (2010). Pulp Mills on The River Uruguay (Argentina v. Uruguay), Judgement of 20 April 2010. Available at: https://www.icj-cij.org/en/case/135/judgments.

International Court of Justice (ICJ) (2015). Construction of a Road in Costa Rica along the San Juan River (Nicaragua v. Costa Rica). Judgement of 16 December 2015. Available at: https://www.icj-cij.org/en/case/152/judgments.

International Network for Environmental Compliance and Enforcement (INECE) (2009). *Principles of Environmental Compliance and Enforcement Handbook*. Available at: http://www.themisnetwork.eu/uploads/documents/Tools/inece_principles_handbook_eng.pdf.

International Law Commission (ILC) Draft Articles on Responsibility of States for Internationally Wrongful Acts (2001). Available at: http://legal.un.org/ilc/texts/instruments/english/commentaries/9_6_2001.pdf.

International Treaty on Plant Genetic Resources (ITPGRFA) (2001). Rome, 3 November 2001. Available at: http://www.fao.org/3/a-i0510e.pdf.

The International Tribunal For The Law Of The Sea (ITLOS) Statute (1982). Available at: https://www.itlos.org/fileadmin/itlos/documents/basic_texts/statute_en.pdf.

Jacobson, H.K. and Weiss, E.B. (1998). Assessing the Record and Designing Strategies to Engage Countries. Weiss, E.B. and Jacobson, H.K. (Eds.), *Engaging Countries: Strengthening Compliance with International Environmental Accords* (511–554). Cambridge, Mass.: MIT Press.

Jacur, F.R. (2009). Triggering Non-Compliance Procedures. Treves, T., Tanzi, A., Pineschi, L., Pitea, C., Ragni, C. (Eds.), *Non-Compliance Procedures and Mechanisms and the Effectiveness of International Environmental Agreements* (373–388). The Hague: T.M.C. Asser Press.

Klabbers, J. (2007). Compliance Procedures. Bodansky, B., Brunnée, J. and Hey, E. (Eds.), *The Oxford Handbook of International Environmental Law* (995–1009). New York: Oxford University Press.

Kolari, T. (2002). Promoting Compliance with International Environmental Agreements – An Interdisciplinary Approach with Special Focus on Sanctions. (Pro Gradu Thesis, Faculty of Social Sciences, Department of Law, University of Joensuu, 2002). Available at: http://www.peacepalacelibrary.nl/ebooks/files/C08-0029-Kolari-Promoting.pdf.

Koskenniemi, M. (1992). Breach of Treaty or Non-Compliance? Reflection on the Enforcement of the Montreal Protocol. *Yearbook of International Environmental Law*, 3, 1, 123–162.

Koskenniemi, M. (2006). Fragmentation of International Law: Difficulties Arising from the Diversification and Expansion of International Law. Report of the Study Group of the International Law Commission. Available at: http://legal.un.org/ilc/documentation/english/a_cn4_l682.pdf.

Kyoto Protocol to the United Nations Framework Convention on Climate Change (1997). Available at: http://unfccc.int/resource/docs/convkp/kpeng.pdf.

Lang, W. (1995). From Environmental Protection to Sustainable Development: Challenges for International Law. Lang, W. (Ed.), *Sustainable Development and International Law*. London: Springer.

Maljean-Dubois, S. and Richard, V. (2004). Mechanisms for Monitoring and Implementation of International Environmental Protection Agreements. Available at: http://halshs.archives-ouvertes.fr/docs/00/42/64/17/PDF/id_0409bis_maljeandubois_richard_eng.pdf.

Marauhn, T. (1996). Towards a Procedural Law of Compliance Control in International Environmental Relations, *Max-Planck-Institut für ausländisches öffentliches Recht und Völkerrecht*, 696–731. Available at: http://www.zaoerv.de/56_1996/56_1996_3_a_696_731.pdf.

Matz, N. (2006). Financial and Other Incentives for Complying with MEA Obligations. Beyerlin, U., Stoll, P.T., Wolfrum, R. (Eds.), *Ensuring Compliance with Multilateral Environmental Agreements: Academic Analysis and Views from Practice* (301–318). Leiden: Koninklijke Brill NV.

Maza, C. (2016). ICC move fuels debate on Cambodian case, 19 September 2016. Available at: http://www.phnompenhpost.com/national/icc-move-fuels-debate-cambodian-case.

Mitchell, R.B. (1994). Regime Design Matters: Intentional Oil Pollution and Treaty Compliance. *International Organization*, 48, 3, 425–458.

Montini, M. (2009). Procedural Guarantees in NCMs. Treves, T., Tanzi, A., Pineschi, L., Pitea, C., Ragni, C. (Eds.), *Non-Compliance Procedures and Mechanisms and the Effectiveness of International Environmental Agreements* (389–406). The Hague: T.M.C. Asser Press.

Montreal Protocol on Substances that Deplete the Ozone Layer (1987). Available at: https://treaties.un.org/doc/publication/unts/volume%201522/volume-1522-i-26369-english.pdf.

MOP 1 (2005). Report of the COP serving as the MOP to the Kyoto Protocol on its First Session. Part Two: Action taken by the COP serving as the MOP at its First Session. Montreal, 28 November–10 December 2005. FCCC/KP/CMP/2005/8/Add.3. Available at: http://unfccc.int/meetings/montreal_nov_2005/session/6260/php/view/reports.php.

MOP 3 (1991). Report of the COP serving as the MOP to the Montreal Protocol on its Third Session. Nairobi, 19–21 June 1991. UNEP/OzL.Pro.3/11. Available at: https://www.informea.org/en/event/third-meeting-parties-montreal-protocol-substances-deplete-ozone-layer.

MOP 4 (1992). Report of the COP serving as the MOP to the Montreal Protocol on its Fourth Session. Copenhagen, 23–25 November 1992. UNEP/OzL.Pro.4/15. Available at: https://www.informea.org/en/event/fourth-meeting-parties-montreal-protocol-substances-deplete-ozone-layer.

MOP 28 (2016). Report of the Twenty-Eighth Meeting of the Parties to the Montreal Protocol on Substances that Deplete the Ozone Layer. Kigali, 10–15 October 2016. UNEP/OzL.Pro.28/12. Available at: http://conf.montreal-protocol.org/meeting/mop/mop-28/final-report/English/MOP-28-12E.pdf.

Pauwelyn, J. (2005). Judicial Mechanisms: Is there a Need for a World Environment Court?. Chambers, W. B. and Green, J.F. (Ed.), *Reforming International Environmental Governance: From Institutional Limits to Innovative Reforms* (150–177). Tokyo: United Nations University Press.

Petersmann, E. (1999). Constitutionalism and International Adjudication: How to Constitutionalize the U.N. Dispute Settlement System? *New York University Journal of International Law and Politics*, 31, 4, 753–790.

Pineschi, L. (2009). Non-Compliance Procedures and the Law of State Responsibility. Treves, T., Tanzi, A., Pineschi, L., Pitea, C., Ragni, C. (Eds.), *Non-Compliance Procedures and Mechanisms and the Effectiveness of International Environmental Agreements* (483–498). The Hague: T.M.C. Asser Press.

Policy Paper on Case Selection and Prioritisation. Available at: https://www.icc-cpi.int/itemsDocuments/20160915_OTP-Policy_Case-Selection_Eng.pdf.

Potzold, C. (2009). Multilateral Environmental Agreements: Contributions to Global Environmental Governance. Are Non-Compliance Mechanisms a Viable Alternative to the More Traditional Form of International Dispute Settlement for the Resolution of Global Environmental Problems? *Society for Environmental Law and Economics*, UBC Conference Paper March 2009.

Procedures and Mechanisms relating to Compliance under the Kyoto Protocol (NCP) (2005). Decision 27/MOP 1. Available at: https://unfccc.int/files/kyoto_protocol/compliance/application/pdf/dec.27_cmp.1.pdf.

Procedures and Mechanisms relating to Compliance under the Montreal Protocol (NCP) (2006). *Handbook for the Montreal Protocol on Substances that Deplete the Ozone Layer* (419–421). Nairobi: UNEP.

Protocol on Water and Health to the 1992 Convention on the Protection and Use of Transboundary
    Watercourses and International Lakes (1999). Available at: http://www.unece.org/fileadmin/
    DAM/env/documents/2000/wat/mp.wat.2000.1.e.pdf.
Rayfuse, R. (2005). The Future of Compulsory Dispute Settlement under the Law of the Sea
    Convention. *VUWLR*, 36, 683–712. Available at: http://victoria.ac.nz/law/research/
    publications/vuwlr/prev-issues/pdf/vol-36-2005/issue-4/sea-rayfuse.pdf.
Rechtbank Den Haag (24 June 2015) C/09/456689. Available at: https://elaw.org/nl.urgenda.15.
Rest, A. (2000). The Role of an International Court for the Environment. *Working Paper for the
    Conference Giornata Ambiente 2000* (34–58). Available at: http://www.biotechnology.uni-
    koeln.de/inco2-dev/common/contribs/06_resta.pdf.
Rio Declaration on Environment and Development (1992). Available at: http://www.unesco.org/
    education/pdf/RIO_E.PDF.
Robinson, D. (2016). Feeling a Way Forward for International Justice-ICC, Africa and the World,
    22 November 2016. Available at: www.ejiltalk.org/feeling-a-way-forward-for-international-
    justice-icc-africa-and-the-world/.
Romano, C.P.R. (2000). *The Peaceful Settlement of International Environmental Disputes*.
    London: Kluwer Law International.
Romano, C.P.R. (2007). International Dispute Settlement. Bodansky, B., Brunnée, J. and Hey, E.
    (Eds.), *The Oxford Handbook of International Environmental Law* (1,036–1,056). New York:
    Oxford University Press.
Sachariew, K. (1991). Promoting Compliance with International Environmental Standards:
    Reflections on Monitoring and Reporting Mechanisms. *Yearbook of International
    Environmental Law*, 2, 1, 31–52. https://doi.org/10.1093/yiel/2.1.31.
Sand, P.H. (1990). Lessons Learned in Global Environmental Governance. *Boston College
    Environmental Affairs Law Review*, 18, 2, 213–277. Available at: http://lawdigitalcommons.bc.
    edu/ealr/vol18/iss2/2.
Sands, P. (1996). Compliance with International Environmental Obligations: Existing International
    Legal Arrangements. Cameron, J., Werksman, J. and Roderick, P. (Eds.), *Improving
    Compliance with International Environmental Law* (48–82). London: Earthscan.
Sands, P. (2006). Non-Compliance and Dispute Settlement. Beyerlin, U., Stoll, P.T., Wolfrum, R.
    (Eds.), *Ensuring Compliance with Multilateral Environmental Agreements: Academic Analysis
    and Views from Practice* (353–358). Leiden: Koninklijke Brill NV.
Sands, P. (2016). Climate Change and the Rule of Law: Adjudicating the Future in International
    Law. *Journal of Environmental Law*, 28, 19–35. https://doi.org/10.1093/jel/eqw005. https://
    academic.oup.com/jel/article-abstract/28/1/19/1748465/Climate-Change-and-the-Rule-of-Law-
    Adjudicating.
Savaşan, Z. (2018). Legitimacy Questions of Non-Compliance Procedures, Examples from the
    Kyoto and Montreal Protocol, *The Environment in International Courts and Tribunals:
    Questions of Legitimacy*, Christina Voigt (Ed.), Cambridge University Press (in process).
Savaşan, Z. (2017). Coping with Global Warming: Compliance Issue Compliance Mechanisms
    Under MEAs. Zhang, XinRong, Dincer, Ibrahim (Eds.), *Energy Solutions to Combat Global
    Warming*, Lecture Notes in Energy, 33. Switzerland: Springer International Publishing.
Savaşan, Z. (2015). Gathering Information under Compliance Mechanisms: Potential New Ways
    for Current Challenges. De Bree, M. and Ruessink, H. (Eds.), *Innovating Environmental
    Compliance Assurance* (171–194), The Netherlands: INECE.
Schiffman, H. (1998). The Dispute Settlement Mechanism of UNCLOS: A Potentially Important
    Apparatus for Marine Wildlife Management. *Journal of International Wildlife Law & Policy*,
    293–306.
Schwabach, A. (2005). *International Environmental Disputes: A Reference Handbook*. Santa
    Barbara, CA: ABC-CLIO.
Serdy, A. (2005). The Paradoxical Success of UNCLOS Part XV: A Half hearted Reply to
    Rosemary Rayfuse. *VUWLR*, 36, 713–722. Available at: http://victoria.ac.nz/law/research/
    publications/vuwlr/prev-issues/pdf/vol-36-2005/issue-4/sea-serdy.pdf.

Stephens, T. (2009). *International Courts and Environmental Protection*. Cambridge, New York: Cambridge University Press.

Széll, P. (1995). The Development of Multilateral Mechanisms for Monitoring Compliance. Lang, W.(Ed.), *Sustainable Development and International Law* (97–114). London: Springer.

Tanzi, A. and Pitea, C. (2009). Non-Compliance Mechanisms: Lessons Learned and the Way Forward. Treves, T., Tanzi, A., Pineschi, L., Pitea, C., Ragni, C. (Eds.), *Non-Compliance Procedures and Mechanisms and the Effectiveness of International Environmental Agreements* (569–580). The Hague: T.M.C. Asser Press.

Treaty on the Functioning of the European Union (TFUE) (2012). Consolidated version of the Treaty on the Functioning of the European Union, 26.10.2012/C 326/49. Available at: https://eur-lex.europa.eu/legal-content/EN/TXT/PDF/?uri=CELEX:12012E/TXT.

Treves, T. (2009a). Introduction. Treves, T., Tanzi, A., Pineschi, L., Pitea, C., Ragni, C. (Eds.), *Non-Compliance Procedures and Mechanisms and the Effectiveness of International Environmental Agreements* (1–10). The Hague: T.M.C. Asser Press.

Treves, T. (2009b). The Settlement of Disputes and Non-Compliance Procedures. Treves, T., Tanzi, A., Pineschi, L., Pitea, C., Ragni, C. (Eds.), *Non-Compliance Procedures and Mechanisms and the Effectiveness of International Environmental Agreements* (499–520). The Hague: T.M.C. Asser Press.

Turgut, N. (2007). The European Court of Human Rights and the Right to the Environment. *Ankara Law Review, 4*, 1–24.

Ulfstein, G. and Werksman, J. (2005). The Kyoto Compliance System: Towards Hard Enforcement. Stokke, O. S., Hovi J. and Ulfstein, G. (Eds.), *Implementing the Climate Regime, International Compliance* (39–64). USA: The Fridtjof Nansen Institute.

United Nations Convention on the Law of the Sea (UNCLOS) (1982). Available at: http://www.un.org/depts/los/convention_agreements/texts/unclos/unclos_e.pdf.

United Nations (UN) (1945). UN Charter. Available at: http://www.un.org/en/documents/charter/.

United Nations (UN) (2001). UN Juridical Yearbook. Available at: http://legal.un.org/docs/?path=../unjuridicalyearbook/pdfs/english/volumes/1993.pdf&lang=EF.

United Nations (UN) (2006). Arbitral Tribunal, Trail Smelter Case (United States, Canada), *UN Reports of International Arbitral Awards*, 1941. Available at: http://legal.un.org/riaa/cases/vol_III/1905-1982.pdf

United Nations Environmental Programme (UNEP) (2005a). *Environment and Trade, A Handbook*. Available at: http://www.iisd.org/pdf/2005/envirotrade_handbook_2005.pdf.

United Nations Environmental Programme (UNEP) (2005b). Comparative Analysis of *Compliance Mechanisms under Selected Multilateral Environmental Agreements*. UNEP: Nairobi. Available at: https://elaw.org/system/files/UNEP.comp_.mea_.compliance.pdf?_ga=2.50584041.1797524369.1551612177-1091855356.1551612177.

United Nations Framework Convention on Climate Change (UNFCCC) (1992). Available at: https://unfccc.int/resource/docs/convkp/conveng.pdf.

Urbinati, S. (2009). Procedures and Mechanisms relating to Compliance under the 1997 Kyoto Protocol to the 1992 UN Framework Convention on Climate Change. Treves, T., Tanzi, A., Pineschi, L., Pitea, C., Ragni, C. (Eds.), *Non-Compliance Procedures and Mechanisms and the Effectiveness of International Environmental Agreements* (63–84). The Hague: T.M.C. Asser Press.

Vienna Convention on the Law of Treaties (VCLT) (1969). Available at: https://treaties.un.org/doc/publication/unts/volume%201155/volume-1155-i-18232-english.pdf.

Wang, X. and Wiser, G. (2002). The Implementation and Compliance Regimes under the Climate Change Convention and Its Kyoto Protocol. *Review of European Community and International Environmental Law*, 11, 2, 181–198.

Weiss, E.B. (1998). The Five International Treaties: A Living Listory. Weiss, E.B. and Jacobson, H.K. (Eds.), *Engaging Countries: Strengthening Compliance with International Environmental Accords* (89–172), Cambridge, Mass.: MIT Press.

Werksman, J.D. (1996). Designing a Compliance System for the UNFCCC, Cameron, Cameron, J., Werksman, J. and Roderick, P. (Eds.), *Improving Compliance with International Environmental Law* (85–112). London: Earthscan.

Werksman, J.D. (1999). Procedural and Institutional Aspects of the Emerging Climate Change Regime: Improvised Procedures and Impoverished Rules? Available at: http://www.cserge.ucl.ac.uk/Werksman.pdf.

Wolfrum, R. (1999). *Recueil des cours: Collected Courses of the Hague Academy of International Law*, Vol. 272 (1998). The Hague, Boston, London: Martinus Nijhoff Publishers.

Wolfrum, R. and Matz, N. (2003). *Conflicts in International Environmental Law*. Berlin, Heidelberg: Springer.

World Trade Organization (WTO) (2001). Compliance and Dispute Settlement Provisions in the WTO and in Multilateral Environmental Agreements. World Trade Organization. Committee on Trade and Environment. WT/CTE/W/191, 6 June 2001 (1–2,811). Available at: http://www.unep.ch/etb/areas/pdf/wtoUNEPnoteDispSetPro.pdf.

World Trade Organization (WTO) (2004). Trade and Environment at the WTO. Available at: http://www.wto.org/english/res_e/booksp_e/trade_env_e.pdf.

World Trade Organization (WTO) (2008). Public Forum '08, Trading into the Future. Available at: http://www.wto.org/english/res_e/booksp_e/public_forum08_e.pdf.

World Trade Organization (WTO) (2011). *Understanding the WTO* (5th edn.). Available at: http://www.wto.org/english/thewto_e/whatis_e/tif_e/understanding_e.pdf.

World Trade Organization (WTO) Agreement, Annex 2, Understanding on Rules and Procedures Governing the Settlement of Disputes (DSU). Available at: http://www.wto.org/english/docs_e/legal_e/28-dsu.pdf.

Chapter 5
Case Study I: Ozone Layer Depletion

5.1 The 1987 Montreal Protocol Compliance Mechanism

The Vienna Convention for the Protection of the Ozone Layer (VC) was adopted in 1985 and came into force on 22 September 1988. As it is a framework convention, it only establishes a framework on parties' obligations. In fact, it includes no substantive detailed obligations (Arts. 2 and 3, VC), but does offer the possibility of adopting further protocols in the COPs of the Convention (Art. 8, Convention) when required to cope with issues regarding ozone depletion (see Annex A to this book for a list of COPs to the VC and MOPs to the MP, and Annex B for MOPs' decisions relating to compliance). So, it merely obliges parties to introduce measures to prevent depletion of the ozone layer which results from human activities, to protect human health and the environment (Art. 2.1, VC). But, it does not impose obligations to reduce the production or use of CFCs, "did not even mention CFCs" (Chasek et al. 2006: 109).

The Montreal Protocol on Substances that Deplete the Ozone Layer (MP),[1] adopted in 1987 and entered into force on 1 January 1989, has the same aim as the Convention: 'to prevent depletion of the ozone layer.' Yet, differently, it submits a more detailed perspective than the Convention.

First of all, it mandates phasing out the consumption and production of chemicals which destroy the ozone layer, known as ozone-depleting substances (ODSs). In line with its aim, it establishes precise targets for the production and use of ozone-depleting substances (ODSs) and specified time periods within which these targets should be met (Art. 2A–2I, MP). It also provides lists of controlled substances in Annexes. Annex A lists CFCs and halons. Annex D contains a list of products containing controlled substances specified in Annexes A, B (other halogenated CFCs, carbon tetrachloride and methyl chloroform), C (hydrochlorofluorocarbons and hydrobromofluorocarbons) and E (methyl bromide),

---

[1]For detailed information on the history of the creation of the Montreal Protocol, see Weiss (1998).

© Springer Nature Switzerland AG 2019
Z. Savaşan, *Paris Climate Agreement: A Deal for Better Compliance?*
The Anthropocene: Politik—Economics—Society—Science 11,
https://doi.org/10.1007/978-3-030-14313-8_5

in nine groups. Based on these groups, it also sets forth control measures to be applied by parties (Art. 2, MP), including the reduction of the levels of consumption and production of these substances.

The Protocol is also open to amendments based on regular assessments of the control measures at least once every four years on the basis of available scientific, environmental, technical and economic information (Art. 9, VC, Art. 6, MP). Following such assessments, it has been revised and thus improved several times through amendments made in numerous MOPs (Romanin, 2009). These revisions have provided numerous adjustments to the targets and timetable for phasing out the ODSs (like MOP 11), on the lists (such as MOP 2, Decision II/1; MOP 4, Decision IV/2–3; MOP 7). To illustrate, the Kigali (Rwanda) Amendment (14 October 2016) to the Protocol adopted at MOP 28 was a major step forward (MOP 28, Annex I, Kigali Amendment; MOP 29, Decision XXIX/3). Through this amendment, hydrofluorocarbons (HFCs) are addressed under the Protocol in addition to chlorofluorocarbons (CFCs) and hydrochlorofluorocarbons (HCFCs) through an agreement to phase out the emissions of HFCs – greenhouse gases with a very high global warming potential, though not as harmful to the ozone layer as CFCs and HCFCs. In other words, the aim of specific targets and timetables for reducing the production and use of these substances is to curb global temperature rise in line with the goals of the climate change agreements in conjunction with the Protocol's basic aim of protecting the ozone layer.

These revisions have also led to significant changes to the institutions of the Protocol, such as the development of non-compliance procedures (MOP 2, Decision II/5; MOP 3, Decision III/2; MOP 4, Decision IV/5; MOP 9, Decision IX/35; MOP 10, Decision X/10), including the establishment of an Implementation Committee (ImplCom) (MOP 3).

The NCP created under the MP was "the first" of this kind of procedure (Pineschi 2004: 244; Sands/MacKenzie 2000: 13; Wang/Wiser 2002: 183). It was set up by the MOP 4 (Decision IV/5) in 1992 pursuant to Art. 8 of the MP, which requires the parties to design and approve procedures and institutions for determining non-compliance with the Protocol's requirements and for dealing with parties' failures to comply with its terms.

Before its adoption in 1992, an Ad Hoc Working Group of Legal Experts was established to develop NCPs by the MOP 1, held in 1989 (Decision I/8). The draft procedure adopted by this working group was approved on an interim basis by the MOP 2 in 1990 (Decision II/5, Annex III). It was also decided to extend the mandate of the open-ended working group to elaborate further procedures on non-compliance and terms of reference for the ImplCom. In the MOP 3 (Decision III/2), the working group was requested to do some crucial tasks such as identify possible situations of non-compliance with the Protocol, develop an indicative list of advisory and conciliatory measures to encourage full compliance, and reflect the role of the ImplCom as an advisory and conciliatory body. At the same meeting a timetable was also adopted for finalization of the draft non-compliance procedures. The procedure was then finalized by the MOP 4, held in Copenhagen on 23–25 November 1992. The non-compliance procedure, as set out in Annex IV to the

Table 5.1 Development of the Montreal Protocol CM

MOP 1, Decision I/8 (Helsinki, 2–5 May 1989)	An open-ended Ad Hoc Working Group of Legal Experts was established to develop NCP
MOP 2, Decision II/5, Annex III (London, 27–29 June 1990)	The draft procedure adopted by the working group was approved on an interim basis
MOP 3 (Nairobi, 19–21 June 1991)	Decision III/2: further requests to the working group on substantive issues of non-compliance Decision III/17: requests on considering expedited procedures for amendment (Art. 9, VC) Decision III/20: increase in the number of the ImplCom from five to ten
MOP 4, Decision IV/5 (Copenhagen, 23–25 November 1992)	The procedure was finalized (Annex IV) and also the indicative list of measures (Annex V)
MOP 9, Decision IX/35 (Montreal, 15–17 September 1997)	Review on the procedure for the further elaboration and developing appropriate conclusions and recommendations
MOP 10, Decision X/10, Annex II (Cairo, 23–24 November 1998)	The NCP was reviewed and amended again
MOP 14 (paras. 83–88) (Rome, 25–29 November 2002)	The proposed amendments to the procedure raised by the US were withdrawn

Source The author

report of the MOP 4, and also the indicative list of measures that might be taken in respect of non-compliance, as set out in Annex V to the report of MOP 4, were adopted by the parties (Decision IV/5, Annex IV–V) based on the Ad Hoc WG Third meeting report, Annex I (NCP) and Annex II, Section II (indicative list of possible measures) (Table 5.1).

In 1997, as it was realized that the procedure needs to be reviewed regularly to ensure the effective operation of the Protocol, another Ad Hoc Working Group of Legal and Technical Experts composed of fourteen members (seven representatives from parties operating under para. 1 of Art. 5 and seven from parties not operating under Art. 5) was established to review, refine and strengthen the non-compliance procedure. In accordance with this aim, a timetable was adopted by the Working Group for reviewing the non-compliance procedure and it was decided that the findings of the Working Group would be considered and any appropriate decisions would be adopted at the MOP 10 (MOP 9, Decision IX/35 1997). By the MOP 10 in 1998, the NCP was reviewed and amended again (Decision X/10, Annex II). It is noteworthy that, in order to maintain the integrity of the Protocol, by the MOP 10, it was agreed that in the case of the existence of a persistent pattern of non-compliance by a party, the ImplCom should report and make appropriate recommendations to the MOP, taking into account the circumstances surrounding the party's persistent pattern of non-compliance. During this meeting, the parties also decided to consider again the operation of non-compliance procedure no later than the end of 2003. In 2002, an attempt to amend the procedure which aimed to make the ImplCom more effective could not be materialized due to the parties' lack of agreement on all elements of the proposal (MOP 14 2002: 13, paras. 83–88).

Overall, although the main structure of the procedure had been developed by the MOP 4 in 1992, the evolution of the NCP under the Protocol has not reached an end, but continues to develop by small amendments. So far, its development has experienced three crucial phases: the adoption of a draft NCP in 1990, the adoption of the current NCP in 1992, and its modification in 1998. But, in the future, it may need further modifications in line with changing needs, and the functioning of the mechanism will continue to improve through those modifications.

5.1.1   Main Components of the Montreal Protocol CM

5.1.1.1   Gathering Information on the Parties' Performance

The VC stipulates that parties should cooperate in gathering information from numerous sources for research and systematic observations.[2] It prescribes that the parties should initiate and cooperate in the conduct of research and scientific assessments on various issues related to the ozone layer, such as physical and chemical processes affecting the ozone layer and their effects on climate (Art. 3.1, VC). In addition, it allows the parties to promote or establish joint or complementary programmes for systematic observation of the ozone layer (Art. 3.2, VC), and to cooperate in ensuring the collection, validation and transmission of research and observational data through appropriate world data centres (Art. 3.3, VC).

Nevertheless, the collection of information from these sources does not yield particulars about the performance review of the Convention. For a performance review, it is necessary to refer to Art. 5 of the VC. According to this article, the parties are required to transmit information to the COP on the measures adopted by them in implementation of the Convention and of protocols to which they are party. In accordance with this article, at COP 1 (1989) it was decided that a summary of the measures adopted by the parties should be submitted every two years to the secretariat, which has also been charged with the preparation of a format for reporting and with the compilation and distribution of the reports to the parties (COP 1, Decision I/2 1989: 9, para. 39; Executive Director, 1989: 4, para. 19).

For the purposes of Art. 3 and Annex I (Research and Systematic Observations) to the Convention, these reporting provisions and obligations incorporated by the VC were all terminated at COP 3 (Decision III/4, 1990: 8), as it was found adequate to report data on ozone-depleting substances under the Montreal Protocol. In addition, at COP 2 (Decision II/2 1991: 8–9, para. 33), it was noted that the information exchange obligations under Annex II (Information Exchange) of the Convention would largely be fulfilled by reporting on data under Art. 7 and by exchanging information on activities in accordance with Art. 9 of the MP. It is

---

[2]See Savaşan (2012) for a comparative analysis of reporting under the Montreal and Kyoto Protocols.

therefore essential here to focus on the MP and its related provisions on gathering information, namely, Articles 7[3] and 9, MP[4] which require timely reporting of data and any other required information as a legal obligation for each party (MOP 6, Decision VI/2 1994: 15, para. 84).

First of all, all parties to the MP should provide their statistical data[5] of production, imports and exports of certain listed controlled ozone-depleting substances (ODSs) to the secretariat within three months of becoming a party (Art. 7(1)) or within three months of the entry into force of the relevant amendments with regard to the substances listed in the MP for that party (Art. 7(2)). The parties should also submit statistical data to the secretariat on their annual production of each of the controlled substances; for each substance, amounts used for feedstocks, destroyed by technologies approved by the parties, and also imports from and exports to parties and non-parties for the year during which related provisions concerning the substances entered into force for that party; and also statistical data on the annual imports and exports of each of the controlled substances in Group II of Annex A and Group I of Annex C that have been recycled (Art. 7(3). In accordance with the provisions of Article 9, they should cooperate in promoting research, development and exchange of information for improving the recovery, recycling, or destruction of controlled substances, possible alternatives to controlled substances, and relevant control strategies (Art. 9(1)) and awareness on the environmental effects of the emissions of ODSs (Art. 9(2)). They should also submit a biannual report to the secretariat, including a summary of their related activities (Art. 9(3)).

When these reports come to the secretariat, it makes them available to the parties and provides information to non-party observers (Art. 12c, 12f, MP). If it finds any signs of possible non-compliance by any party while preparing a summary report, it can ask for more information and data from the party concerned (NCP, para. 3) on condition that it will use the data provided by the party if an agreement cannot be reached on it (MOP 7, Decision VII/20: 37, para. 94). Additionally, the ImplCom can also request further information through the secretariat (NCP, para. 7c) and ask for site-visits (NCP, para. 7e). However, if the site-visits, which should be based on a set of rules (either drawn up separately for each visit or pre-determined by the

---

[3]The MOP 9 approved new formats for reporting data under Art. 7, MP. The old data formats used by the parties to report data were replaced, and since 1997 the revised formats have been used. In order to assist the parties in providing the data required by the revised formats, a handbook on data-reporting under the Montreal Protocol was prepared and distributed to all parties by the UNEP division of technology, industry and economics (UNEP-TIE).

[4]Besides Articles 7 and 9, MP, the parties have to send reports to the Secretariat under Art. 4b of the protocol. However, this article is about the implementation of licensing systems for the import and export of ozone-depleting substances, and it requires parties to report to the Secretariat regarding the establishment and operation of that system (Art. 4(3)); it is not directly related to gathering information in order to review national performance.

[5]To collect the required data, parties have relied heavily on customs statistics, organised in most countries according to the Harmonised Commodity Description and Coding System elaborated in the framework of the World Customs Organisation (WCO) on the basis of the Harmonised System Convention (Oberthür 2001).

Table 5.2 Gathering Information under MP

Initial Report (Art. 7 (1, 2))	Within three months of becoming a party or within three months of entry into force of the relevant amendments with regard to the listed substances to the Protocol for that party, the parties must submit their base year data on certain listed controlled ozone-depleting substances (ODSs) to the secretariat
Annual Report (Art. 7(3))	The parties must also submit their annual statistical data on the controlled substances to the secretariat
Biannual Report (Art. 9(3))	The parties must submit a biannual report to the secretariat, including a summary of the activities that they have undertaken pursuant to Art. 9, involving activities on research, development, public awareness and the exchange of information

Source The author

Committee for all visits) are not authorized by the related party, the ImplCom can just rely on the parties' reports.

Regarding developing countries, there is an important way of forcing them to give further information and to fullfil their data-reporting requirements. In fact, parties under Art. 5, MP can lose their status if they do not report their base-year data as required by the MP within one year of the approval of their country programme and their institutional strengthening by the Executive Committee (MOP 6, Decision VI/5 1994: 15–16, para. 84) (Table 5.2).

5.1.1.2 Non-Compliance Procedure (NCP)

The NCP of the MP was formulated with the aim of finding amicable solutions to matters involving possible non-compliance problems, pursuing a non-confrontational way, and so involves features appropriate to this aim. Through its features, it has had the potential to be followed and used as a reference by the other MEAs' CMs (Ehrmann 2002; Handl 1997; Pineschi 2004; Raustiala 2000; Sands/MacKenzie 2000; Wang/Wiser 2002).

In this part, the features which enable it to be accepted as a model for other CMs are analysed in detail, with the emphasis on ImplCom, procedural phases and the safeguards which function under it.

Institution created under the NCP: Implementation Committee. The creation of the ImplCom occurred after the convening of three MOPs: MOPs 2, 3 and 4. Firstly, MOP 2 (Decision II/5) decided to extend the mandate of the Ad Hoc Working Group of Legal Experts to elaborate further procedures regarding terms of reference for the Implementation Committee (ImplCom). MOP 3 (Decision III/2), on the other hand, went a step further and established the ImplCom as an advisory body, leaving the final decision to the MOP. Finally, under MOP 4 (Decision IV/5), the Implcom was established as a body which primarily assures the operation of the whole procedure.

The Committee can operate as both a "standing body" (Downes/Penhoet 1999: s24) meeting periodically (twice a year) (NCP, para. 6, MP) to address issues on compliance, and an "ad hoc mechanism" to address specific submissions regarding non-compliance and respond to them (Downes/Penhoet 1999: 24; Raustiala 2000: 418–419).

It consists of ten representatives of the parties elected by the MOP for two years. So, it is criticised that, it does not have experts from different areas, because an ImplCom composed only of lawyers, or only of scientists or technicians, or of diplomats and policy experts, can experience difficulties in dealing with cases involving matters beyond their own areas of expertise (Széll 1995).

It is based on equitable geographical representation of both developed and developing parties, and can elect its own President and Vice-President (the rapporteur of the ImplCom) to serve for one year (NCP, para. 5, MP). A party does not have the right to be re-elected for a third consecutive two-year term on the Committee; only after an absence of one year from the Committee does it become eligible to be re-elected for further terms of up to four years (NCP, para. 5, MP).

It should here be noted that its current size was agreed by the Ad Hoc WG report while acknowledging a possible need to increase the number of representatives in parallel with an increase in its workload in the future. So, although its current size was found appropriate for its "smooth functioning", amendments may be necessary in the future (Ad Hoc WG 1998: 4).

In principle, the ImplCom meets twice a year. If any interested party requests an additional meeting, the ImplCom can decide to arrange further meetings at different periods of time (NCP, para. 6, MP). In order to provide the continuity in attendance to the meetings, each party elected to the ImplCom should also notify the secretariat of the individual to represent it and should ensure that such representation remains throughout the duration of the term (NCP, para. 5, MP).

Its recommendations can be taken by a majority of the parties present and voting (RoP, 26.6b), i.e. the parties present at the voting session and casting an affirmative or negative vote; parties which abstain from voting are considered as not voting (RoP, 40.5). In the case of the adoption or rejection of a proposal, if the meeting decides in favour of reconsideration by a two-thirds majority of the parties present and voting, it may only be reconsidered at the same meeting (RoP, 38).

In the report of the Ad Hoc WG of 1998, a proposal making reference to a "specified majority vote" was raised for discussion. However, it was refused on the grounds that the RoP dealt satisfactorily with the decision-making rules for adoption of recommendations (Ad Hoc WG 1998: 7, para. 38).

The functions of the ImplCom are determined in para. 7, NCP as follows:

1. To receive, consider and report on any submission made by self-triggering, other parties, or the secretariat. It should be highlighted that although it has the authority to receive, consider and report on submissions, it does not have a function of initiating reports and recommendations for consideration by the MOP.

2. To receive, consider and report on any information or observations forwarded by the secretariat concerning compliance with the provisions of the Protocol. This function of the ImplCom also involves deliberating on the preparation of the parties' reports and discussing "the general quality and the reliability of the data" contained in these reports (Faure/Lefevere 1999: 153).

3. To request, where it considers necessary, further information on matters under its consideration through the secretariat. This implies that it is not entitled to request information directly, but is obliged to go through the secretariat. In addition, as it does not specify any bodies which can be used when gathering information, it allows information-gathering in an unlimited manner.

4. To identify the facts and possible causes relating to individual cases of non-compliance and to make appropriate recommendations to the MOP on the ways the party concerned can remedy the non-compliance.

   As can be inferred from these functions, the Committee has not been given the authority to determine non-compliance, but only to identify the facts and possible causes and to report to the MOP, including any recommendations it considers appropriate (NCP, para. 9, MP). Here, its purpose is to make recommendations relating to non-compliance not directly to the concerned party, but only to the MOP, which is empowered to determine the existence of non-compliance on the basis of the information gathered, and to take measures to assist a party's compliance and further the Protocol's objectives (Ad Hoc WG 1998: 5–6, paras. 27, 28). This is also confirmed by para. 14 of the procedure, which clarifies that the ImplCom does not have judicial or quasi-judicial powers.

5. To collect information relating to a party's compliance in the territory of that party only if that party has invited it to do so. Together with its function to request further information through the secretariat, this function can be identified as '[r]eactive' inspection, the most sophisticated form of investigation" (Maljean-Dubois/Richard 2004: 21).

6. To maintain an exchange of information with the Executive Committee of the Multilateral Fund to provide and arrange for assistance of financial and technical cooperation, and the transfer of technologies to Art. 5 parties of the Protocol.

Overall, on the basis of the above-mentioned functions, it can be argued that the ImplCom relies on facilitation rather than enforcement and employs an administrative rather than a judicial approach to non-compliance (Raustiala 2000).

Procedural Mechanism: Phases and Safeguards

Phases of the CM. Following the division of phases made in the primer for members of the ImplCom (Ozone Secretariat 2007), the procedure applied for the functioning of the compliance mechanism of the Montreal Protocol can be scrutinized in four phases: submission (triggering) phase, preliminary phase, substantive phase (consideration phase and recommendation-reporting phase), and final phase (decision-making phase and monitoring/final resolution phase). Based on this division, the procedure works in the following order:

Submission Phase: Under the Montreal Protocol, the NCP can be initiated by parties to the Protocol against another party, by the secretariat against any party to the Protocol or by self-triggering (that is, by the non-compliant party itself).[6]

Firstly, if one or more parties have 'reservations' (NCP, para. 1, MP) about another party's implementation of its obligations under the Protocol, they can apply to the secretariat in writing to trigger the NCP. While making their submission, they do not have to prove the causality between the non-compliance and its effect on them, but they do have to support their submission with "corroborating information" which has the potential to be "a useful tool to avoid abuse of the procedure" (Marauhn 1996: 702) (NCP, para. 1, MP). However, it should be clarified that, during the period between notification and the MOP's decision on the appropriate action, the NCP should not be invoked against a party which has notified that, having taken all practical steps, it is unable to implement any or all of the control measures due to the inadequate implementation of Articles 10 and 10A which are on financial and technical cooperation (Art. 5, para. 7, MP).

The submission by the secretariat, on the other hand, can be made if the secretariat becomes aware of possible non-compliance by any party with the Protocol's provisions, while it is preparing that party's report. In such a case, the secretariat can ask the concerned party for further necessary information about the matter. If the party does not reply in three months – or a longer period if specified – or if the issue cannot be resolved through administrative or diplomatic efforts, the secretariat has to include the matter in its report to the MOP and to inform the ImplCom (NCP, para. 3, MP).

Finally, the submission by the non-complying party in respect of itself can be raised when a party concludes that it is unable to comply with its obligations and provides some proof that it has done its best to comply. That party can then make a submission in writing with an explanation of the specific circumstances considered to be the cause of its non-compliance. The secretariat receiving this submission transmits it to the ImplCom for consideration (NCP, para. 4, MP).

Preliminary Phase: After the submission phase, the secretariat sends a copy of that submission to the party alleged to be in violation of a particular provision of the Protocol within two weeks of receiving a submission. Then, within three months of the secretariat sending the submission, the party concerned has to send a reply. Otherwise, the secretariat sends a reminder to make it reply. As soon as it receives the reply from the party, it transmits the submission, the reply and any other information which the secretariat has found to be beneficial for the ImplCom in its consideration of the party's situation to the ImplCom for the purpose of determining whether the concerned party is unable to comply with its obligations under the Protocol (NCP, para. 2, MP).

Substantive Phase: While considering the submission, the ImplCom uses the information provided by the parties under consideration, and can request any

---

[6]During the negotiations of the NCP, some other triggering methods were also debated, but were rejected (Romanin 2009).

additional information it finds necessary on matters under its consideration. It can also undertake information-gathering in the territory of the party concerned which invited it (NCP, para. 7, MP). In addition, it discusses the draft recommendations suggested by the secretariat on each matter under its consideration.

However, in some cases, through a process of "blanket approval" (Ozone Secretariat 2007: 14), the recommendation and reporting phase can be commenced omitting the consideration phase. In this process, the secretariat asks the ImplCom to inform it of the draft recommendations that it wants to review. Those that were not identified by the ImplCom are considered to be approved without additional deliberation, in other words, to have "blanket approval" (Ozone Secretariat 2007: 14). After asking the ImplCom whether it will make individual review or not, on the grounds that the secretariat receives further relevant information on the concerned party, the draft recommendation given "blanket approval" can still be put forward by the secretariat for individual review in the consideration phase (Ozone Secretariat 2007: 14).

As one of its main tasks is "to identify the facts and possible causes relating to individual cases of non-compliance" (NCP, para. 7d, MP), the ImplCom also drafts and adopts appropriate recommendations to the MOP on the parties submitted by the secretariat for consideration. While making recommendations and trying to find "an amicable solution of the matter" that can be accepted by all the parties involved (NCP, para. 8, MP), it performs the tasks of "factual and legal evaluation" (Marauhn 1996: 712) together. It should be stressed that this is not a strict legal evaluation, as the solution is found in accordance with the provisions of the Protocol (NCP, para. 8, MP), but not with general rules of international law (Ehrmann 2002). Therefore, even if it is not a member of the ImplCom or has not itself made such a submission, a concerned party is empowered to participate in the consideration of that submission (NCP, para. 10, MP), reflecting the existence of a dialogue with the ImplCom and the party concerned. However, no party involved in a matter under consideration by the ImplCom – not even a member of the ImplCom – can participate in the ImplCom during the phase of the elaboration and adoption of recommendations on that matter to be included in the report of the ImplCom (NCP, para. 11, MP). Moreover, the representatives of the Multilateral Fund and the implementing agencies do not automatically participate directly in the elaboration and adoption of the recommendations. Only if the ImplCom requests them to be present to answer questions relevant to the finalization of recommendations can they participate in this process (Ozone Secretariat 2007).

Recommendations adopted by the ImplCom can name specific parties or not name any specific party. Of those, the ones that name specific parties can involve requests for information from that party, proposals for the approval of that party's plan of action by the MOP, and acknowledgments of that party's progress in implementing its plan of action. The ones that do not name specific parties usually address compliance issues relevant to more than one party, such as the reporting of ODSs data, or notifying the secretariat of the establishment of an ODS import and export licensing system (Ozone Secretariat 2007).

On "less complex compliance matters" considered by the ImplCom regularly, the ImplCom uses "a set of standardized recommendations" which are known as "routine procedural matters of non-compliance" in order "to manage its increasing workload more efficiently and effectively and to ensure the equitable treatment of Parties in comparable circumstances" (Ozone Secretariat 2007: 16).

The ImplCom reports its recommendations to the MOP no later than six weeks before their meeting (NCP, para. 9, MP). This report is made available to any person on request, on condition that it does not contain any information received in confidence (NCP, para. 16, MP).

Final Phase: After receiving the ImplCom's report, the MOP decides upon the matter (RoP, 39–40) and can apply measures to bring the non-compliant party to full compliance with the Protocol (NCP, para. 9, MP). Indeed, even though the report and data provided to the MOP by the ImplCom constitutes a *de facto* determination of non-compliance and its recommendations have often been adopted without any change by the MOP – e.g. in the meeting of ImplCom 30 (2003: para. 57), the representative of the secretariat noted that "all past decisions of MOP on non-compliance had often been based on the ImplCom's recommendations and had never returned for revision"[7] – the final determination of non-compliance is considered to be a matter for the parties. So the MOP is the only body which has decision-making powers about non-compliance.

Each party in every MOP has one vote as a rule unless they are regional economic integration organizations which have the right to vote with a number of votes equal to the number of their member states which are parties to the Protocol (RoP, 39(1–2)). Again, as a rule, decisions of the MOPs on all matters of substance should be taken by a two-thirds majority vote of the parties present and voting (RoP, 40 (1)). Decisions on matters of procedure, on the other hand, should be taken by a simple majority vote of the parties present and voting (RoP, 40(2)).

Besides the functions granted in Art. 11.4(a–i) of the Protocol, the power "to consider and undertake any additional action that may be required for the achievement of the purposes of th[e] Protocol" is a granted under Art. 11.4(j).

The decisions taken by the MOP are sent to all parties in its report, at which point the decisions are adopted and are also published on the Ozone Secretariat's website. The secretariat also informs the parties concerned of the adopted decisions by letter as well as, when relevant, the secretariat of the Multilateral Fund and any implementing agencies assisting the party in returning to compliance (Ozone Secretariat 2007: 18).

After the decision-making phase, the monitoring/final resolution phase begins. The framework for addressing issues of non-compliance after the decision-making phase has been clearly identified in the report of the 20th Meeting of the ImplCom (ImplCom 20 1998: paras. 31–33). According to this framework, after the identification of a party in non-compliance, the ImplCom reviews the concerned party's

---

[7]See also Ozone Secretariat (2007: 18), as it clearly states that: "To date, virtually all draft decisions proposed by the Committee have been adopted by the Meeting of the Parties."

plan to achieve compliance. In the next step, it determines specific criteria, including specific policy measures or steps for reduction and phase-out for the concerned party to undertake in accordance with the proposed decision on that party. These benchmarks are used to monitor the party's efforts to achieve full compliance with the Protocol by both the ImplCom and the party in non-compliance. Throughout this process, the secretariat keeps the party fully informed of the reasons why it was identified as being in non-compliance. Finally, parties that have fully met agreed commitments within their plan are treated as being in compliance with the Protocol and are recommended for favourable consideration for financial assistance from GEF and the Multilateral Fund. Parties that have failed to meet their commitments can face additional measures, such as not being offered additional financial assistance or prohibition on exporting to that party until they meet their commitments or achieve full compliance with the Protocol (ImplCom 20 1998: paras. 31–33).

Overall, until the ImplCom records that the concerned party has complied with all requirements included in the decision pertaining to it, it continues to monitor the party's compliance status. Based on the secretariat's list of decisions about the action required by parties, the ImplCom can repeat its consideration, recommendation and reporting phases to monitor the party's progress. Even if the party returns to compliance, the monitoring continues if there are still some measures and "time-specific milestones" that should be completed in its plan of action (Ozone Secretariat 2007: 19).

Procedural Safeguards

The Rights of the Party Concerned. In the case of the MP's NCP, the party whose compliance is at issue has the right to receive a copy of the submission; it may reply and send any related information to the secretariat and to the other parties involved (NCP, paras. 2, 3, MP). In addition, when there is self-triggering by the party itself, it has the right to explain the specific circumstances that it considers to be the reasons for its non-compliance (NCP, para. 4, MP). Finally, it has the right to be represented during the consideration of the ImplCom of that submission (NCP, para. 10, MP), except during the elaboration and adoption of recommendations on that submission to be involved in the report of the ImplCom (NCP, para. 11, MP).

Predetermined Deadlines. There are some time restrictions, such as: sending a copy of submission to the concerned party within two weeks of receiving that submission by the secretariat (NCP, para. 2, MP); sending any reply and information to the secretariat and to the parties involved within three months of the date of the dispatch (or such longer period as the circumstances of any particular case may require – so not such a strict limitation) (NCP, paras 2, 3, MP); transmitting the submission, the reply and the information provided by the parties to the ImplCom as soon as the reply and information from the party are available and no later than six months after receiving the submission (NCP, para. 2, MP).

Impartiality and Independence. As mentioned above, the ImplCom consists of ten parties elected by the MOP for two years, based on equitable geographical distribution (NCP, para. 5, MP), so there are no experts serving in their personal capacity; in addition, there are no criteria in the NCP regarding the expertise required from parties' representatives. This situation raises questions about the independence of the ImplCom.

Regarding the impartiality of the work of the ImplCom, the procedure does not allow any concerned party, whether a member of the ImplCom or not, to take part in the elaboration and adoption of recommendations on that matter to be included in the report of the Committee (MOP 10 1998: paras. 10, 11; NCP, para. 11, MP).

Predetermined Consequences. The Protocol includes an indicative list of measures, yet, no list of possible situations of non-compliance. In the third meeting report, the Ad Hoc WG (1991) attempted to identify seven such situations which can be categorised as non-compliance. These situations are:

1. Non-compliance with provisions relating to control measures (Art. 2, MP).
2. Non-compliance with provisions relating to the control of trade with non-parties (Art. 4, MP).
3. Non-compliance with time schedules and non-reporting of specified data (Art. 7, MP).
4. Failure to report and co-operate in the activities under Art. 9, MP.
5. Failure to comply with the provisions concerning the operation of the financial mechanism and the payment of contributions to the financial mechanism (Art. 10, MP).
6. Failure to take necessary steps for the transfer of technology.
7. Non-compliance with the decisions of the MOP.

However, since the parties to the Protocol failed to agree on whether or not contributions to the financial mechanism should be regarded as non-compliance, the adoption of this list could not be achieved (Ad Hoc WG 1991: 5, para. 37). So the scope of the term 'non-compliance' has remained unclear with general reference to the MOP's decision (NCP, para. 9, MP) to bring the non-compliant party back to full compliance through measures to assist the parties' compliance with the Protocol, and to the Protocol's objectives (Ehrmann 2002; Marauhn 1996).

The Ad Hoc WG (1998: 7, para. 39) also received a proposal to revise this indicative list of measures, but the WG did not consider it due to the fact that revision of the indicative list had been excluded from its mandate by the MOP.

Proportionality. As there has been no consensus on determining the possible situations of non-compliance, it is stated in the Ad Hoc Report (Ad Hoc WG 1991: 6, para. 44) that in selecting appropriate response measures in practice, 'flexibility' should be taken as a basis in different situations of non-compliance. That is, while deciding on responses, the nature, degree and reason behind non-compliance, as well as the importance of the provision itself, should be assessed carefully. In addition, assistance measures should be applied prior to stronger measures for encouraging parties to comply with the Protocol.

Transparency. While the ImplCom's reports are available to anyone upon request, all information exchanged by or with the ImplCom relating to any of its recommendations to the MOP is also available to any party upon its request (NCP, para. 16, MP).

On the other hand, NGOs's participation in the ImplCom proceedings and in the processes of monitoring compliance with the Protocol have been limited. There is no formal way for NGOs to participate in the ImplCom proceedings, although with the secretariat's notification, any body or agency, whether national or international, governmental or non-governmental, can participate as 'observers' without having the right to vote (RoP, 6–7). These bodies should be qualified in fields relating to the protection of the ozone layer (RoP, 7) and there should be no objection from the parties present.

Appeal. Within the MP's NCP, appeal is not possible. If it has been made possible, it would be against the decisions of the MOP, not the decisions of the Committee, as distinct from the appeal process applied in the Kyoto Protocol's NCP.

5.1.1.3   Non-Compliance Response Measures

Once it has been determined that a party is non-compliant (Art. 8, MP), the MOP can take some measures based on the report of the ImplCom and the recommendations it considers appropriate in respect of that non-compliance (MOP 10, Decision X/10).

The indicative list of measures that might be taken by the MOP was set forth by Annex V (A-B-C) to Decision IV/5 (para. 3) of the MOP 4 (1992) pursuant to the Ad Hoc WG (1991) third meeting report, Annex II, Section II (indicative list of possible measures). This list includes both positive and negative measures:

A. Appropriate assistance, including assistance with the collection and reporting of data, technical assistance, technology transfer and financial assistance, information transfer and training.
B. Issuing cautions.
C. Suspension, in accordance with the applicable rules of international law concerning the suspension of the operation of a treaty [Arts. 60–64, VCLT], of specific rights and privileges under the Protocol, whether or not subject to time limits, including those concerned with industrial rationalization, production, consumption, trade, transfer of technology, financial mechanism and institutional arrangements.

Positive Response Measures

Appropriate Assistance to the Non-compliant Party: Even if there was no CM under the MP, and so no response mechanism, there has already been provision to support the parties' compliance efforts under the Convention and the Protocol. Originally, the purpose was not to ensure measures as a response to non-compliance by parties, but they have been used for this purpose after the creation of the CMs.

To illustrate, regarding technical assistance, the gathering and sharing of information is stressed "as an important means of implementing the Convention" (VC, Annex II, para. 1). In fact, Art. 4, VC obliges the parties to facilitate and encourage the exchange of scientific, technical, socio-economic, commercial and legal information (see Annex II, paras. 3–6 for the elaboration of these types of information) relevant to the Convention (Art. 4.1) and to cooperate in promoting the development and transfer of technology and knowledge (Art. 4.2). Article 9, MP includes similar provisions which require the cooperation of the parties in promoting research, development and the exchange of information on several issues which have the potential to improve the compliance of the parties. Article 10A, MP also provides that each party should take the necessary steps to ensure that the best available environmentally safe substitutes and related technologies are transferred to developing parties under fair and most favourable conditions.

With respect to financial assistance, a Trust Fund (later the 'Multilateral Fund') was established by COP 1 (1989) (Decisions I/5, I/9) in order to strengthen the capacities of developing countries. At MOP 1 (1989), the parties agreed to consider at MOP 2 (1990) the development of a programme of workshops, training courses, the exchange of experts and the provision of consultants on control options to promote the exchange and transfer of environmentally sound substitutes and alternative technologies for developing countries (Decision I/4).

Then, through the amendment of Article 10, MP at MOP 2 (see also MOP 4, Decision IV/18), it was decided to establish a mechanism for the purposes of financial and technical cooperation, including the transfer of technologies, to meet all the agreed incremental costs[8] for developing countries and to enable them to comply with control measures (Art. 10(1)).

It was decided that this financial mechanism would include a Multilateral Fund (MF) which would meet the agreed incremental costs (Art. 10.2a); possess finance clearing-house functions to identify the needs of developing parties and facilitate technical cooperation to meet these needs; distribute information; hold workshops and training sessions for developing parties; facilitate and monitor other multilateral, regional and bilateral cooperation for them (Art. 10.2b(i–iv)); and also finance the secretarial services of the Fund and related support costs (Art. 10.3c)).

In line with this article, a Multilateral Fund (MF) was established in 1990 (MOP 2, Annex II 1990) with the aim of strengthening the capacity of developing countries to comply with the obligations of the Protocol. In order to materialize its aim in practice, it provides financial assistance contributed by developed country parties for the incremental costs of developing country parties. That is, only countries that qualify as Art. 5 countries can benefit from its assistance after

---

[8]"[T]he term 'incremental costs' indicates that only the additional costs namely costs resulting from environmental efforts normally not undertaken will be covered by the solidarity of the community of states" (Wolfrum 1999: 136). On incremental costs, see Decision IV/18, Annex VIII (MOP 4 1992), which provides a list of incremental costs and allows the Executive Committee of the MF to interpret the costs and, if necessary, add to this list.

submitting a country programme (involving production and consumption of the regulated substances and an institutional structure) (Weiss 1998).

This Fund functions under the authority of an Executive Committee (Decisions II/8, IV/18) in which "industrialized and developing countries were participating as equal partners in financial decision-making" (Wolfrum 1999: 127). This Committee exercises its tasks with the cooperation and assistance of the World Bank, the United Nations Environment Programme, the United Nations Development Programme and other appropriate agencies (Art. 10.5).

When projects involving these programmes have been approved by the Executive Committee, countries which are eligible to receive assistance from the Fund have the opportunity "to develop programs for controlling CFC production and consumption and to move from use of ozone-depleting substances to use of ozone-friendly substitutes, or to close their production facilities" (Weiss 1998: 170).

The Global Environment Facility (GEF), on the other hand, was established in 1991 within the World Bank (it became separate institution in 1992 at the Rio Summit) to assist countries in their implementation efforts concerning not only ozone depletion, but also climate change, biodiversity and pollution of international waters. The list of parties recognized by GEF as eligible for assistance includes both developing countries (also eligible for MF assistance) and CEIT not classified as developing countries (not eligible for assistance by the MF) (Madhava 2005). However, the parties to the MP had already invited the GEF to support specifically the monitoring of ozone and related research in developing countries (COP 4, Decision IV/4 1996).

It should be stressed that the activities of the GEF should not compete with those of other MEA-specific funds, such as the Montreal Multilateral Fund, but they should supplement each other (Wolfrum/Matz 2003).

Negative Response Measures. Under the CM, the MOP has the power to issue warnings and also to suspend rights and privileges against the non-compliant party besides providing technical or financial assistance to that party.

During the 15th meeting of the open-ended working group (OEWG 15) of the parties to the MP, Canada also proposed an amendment to the Protocol to the effect that a non-compliant party with the Protocol should be treated as 'non-party' to the Protocol for the purposes of Art. 4 and so be subject to Art. 4 trade measures for the substance for which it is in non-compliance (OEWG 15 1997: 4). It elaborated its proposal by suggesting that this consequence, which would fall under item C of the list of indicative measures, would only be applied in the situations representing a "persistent pattern of non-compliance" (which should be defined based on principles such as frequency of non-compliance, length of non-compliance, and reasons for non-compliance [para. 19]) with key provisions (listed in para. 18) of the Protocol (OEWG 15 1997: 4, paras. 17–19). However, during discussions about this proposal, some representatives expressed concern that the application of sanctions (OEWG 15 1997: paras. 104–109) could threaten the Protocol's central goal of protecting stratospheric ozone. Therefore, they proposed, as "[a] better approach", "to focus, on a case-by-case basis, on the factors responsible for the instances of non-compliance, on the negative potential economic and social repercussions that sanctions could have, and on methods of assistance the Parties could provide to help address and alleviate the non-compliance" (OEWG 15 1997: 23, para. 107).

These discussions reinforced the main aim of the CM, which is not to impose any punitive measures against the non-compliant party, but to endeavour to bring it back to compliance (Ad hoc WG 1998: 2, para. 7). So this proposal was not accepted, and even the accepted negative measures have been rarely adopted against non-compliant parties by the MOP (see *infra*, the section on the CM of the MP in practice).

Issuing cautions: Under the CM, warnings can be issued by the MOP to non-compliant parties when the ImplCom finds that those parties have not made adequate efforts to comply with the obligations of the Protocol. The importance of this measure stems from the threat to itself which can result in the application of the measures consistent with item C of the indicative list of measures in the event that the parties have failed to return to compliance in a timely manner with regard to the phase-out of ODSs. ˙

Suspension of specific rights and privileges under the MP: The rights and privileges that can be suspended in the case of non-compliance are "those concerned with industrial rationalization, production, consumption, trade, transfer of technology, financial mechanism and institutional arrangements" (Annex II, Section II, para. C). This provision implies that non-compliant parties can lose their right to transfer their production rights for ODSs to each other, their consumption rights for HCFCs (Art. 2, paras. 5 and 5b, MP), their entitlement to receive assistance from MF or GEF (Arts. 10 and 10A, MP) and their rights under Art. 4, MP regarding trade.

With respect to the rights of developing parties to receive assistance, it should be emphasized that continuing to be eligible to receive assistance from the Fund has been made contingent on improving compliance. Nevertheless, "there is no measure for promoting compliance by developed countries equivalent to the placing of conditions on financial assistance that is employed with developing countries" (Downes/Penhoet 1999: 25). This can result in the view that the procedure has been applied "inequitably as between developed and developing countries" (Downes/Penhoet 1999: 25).

For developing parties, eligibility for financial assistance is applied in the following way: if a developing party fails to fulfil its obligations under the Protocol, to "report base-year data as required by the Protocol within one year of the approval of its country programme and its institutional strengthening by the Executive Committee" (MOP 6, Decision VI/5 (a-iii) 1994), and also to submit its complete data to the secretariat establishing that its annual calculated per capita level of consumption is below 0.3 kg" (MOP 2, Decision II/10 1990), its access to the Fund can be suspended, as it can lose its status of being developing and so not be eligible for Art. 5, para. 1 treatment. This kind of threat on the access of the developing party to the Fund can be applied effectively, forming conditionality between fund granted and compliant behaviour (Raustiala 2000). "[S]uch measures have been recommended only for failures to supply initial baseline data" (Downes/Penhoet 1999: 25), although Decision VI/5 of the MOP stated that cuts in funding would be made not only to parties which failed to report their baseline data within one year of their MLF country programme being approved, but also to parties which failed to report their progress in strengthening their institutional capacity (Victor 1996).

Regarding trade measures employed against non-compliant parties, a distinction between trade provisions employed for the operation of the related agreement and trade provisions used for enforcing it can be beneficial to understanding the characteristics of these measures adopted in the Protocol. The second ones have the characteristics of sanctions, and so they are 'punitive' (Jenkins 1996: 228) and "extend beyond the range of operation of the trade provisions" (Jenkins 1996: 228). But the first ones, incorporated into the MP, "which act as a disincentive" (Potzold 2009: 10) are only used as a means of operation, but not as trade sanctions as a means of enforcing it.

Article 4, MP involves provisions which require parties to the Protocol to ban the import of the controlled substances in Annexes A, B, C and E from any non-party to the Protocol (Art. 4(1)) from 1990, the import of products containing controlled substances in Annex A from 1992 (Art. 4(3)), the import of products produced with, but not containing, controlled substances in Annex A from 1994 (Art. 4(4)), and the export of any controlled substances in Annexes A, B, C and E to any non-party to the Protocol from 1993 (Art. 4(2)). Thus, it requires the parties to limit their trade in certain substances for a specified period with non-parties to the Protocol. However, it can also be used as "an important potential tool for use against non-complying parties" to the Protocol (Brack 2003: 220). In fact, if a party's rights under Art. 4 are suspended, other parties cannot trade in ODSs with that party. That party cannot export products containing CFCs (air conditioners, etc.) to those parties, e.g. Russia and the Ukraine , which have been subjected to trade measures (MOP 10, Decision X/26 1998).

Binding Effect of the Response Measures

In the first Ad hoc meeting report (Ad Hoc WG 1989: 4, para. 9i), it is suggested that decisions of the MOP should be recommendatory rather than mandatory. But, in the third report (Ad Hoc WG 1991: 6, para. 41), non-compliance with decisions of the MOP is also counted as one of the possible situations of non-compliance (Ad Hoc WG 1989: 10, situation vii, Annex II). This has caused discussions on the binding status of the MOP's decisions. While some representatives stated that the decisions of the parties do not have the same legal status as the provisions of the Protocol, others argued that not all decisions have a binding effect; some pointed out that, if they relate to matters of substance, they are binding. Since then, no consensus has been achieved on the binding status of the MOP's decisions.

5.1.2 Montreal Protocol's Compliance Mechanism in Practice

Up to the end of 2016, the MOP had held twenty-eight sessions, in some of which it dealt with decisions on non-compliance (or potential non-compliance) of the parties to the Protocol. The ImplCom, on the other hand, held fifty-six sessions in which it reported non-compliance cases to the MOP, reviewed the status of compliance with specific decisions of the parties and adopted its recommendations on non-compliance-related issues (see Annex C to the book).

Thus, up to now, both have examined issues related to non-compliance with procedural (failure to report fully or on a timely basis) or substantive obligations (failure to comply with control measures, targets and timetables) (see Annex D to the book for an overall list of decisions on compliance of all the parties to the MP).

Until 1995 when the formal procedure was first initiated through the acceptance of the joint statement made by the Russian Federation and others as a submission under para. 4 of the NCP of Art. 8 of the MP, the ImplCom had focused primarily on parties' non-reporting and incomplete or late reporting.[9]

From 1995 onwards, in addition to considering the issues with data reporting, the ImplCom has addressed individual cases of non-compliance and reviewed whether the parties fulfil their substantive obligations under the Protocol, such as those related to control measures, targets and timetables.

In the following part, the examination of these cases takes into account the ImplCom and the MOP reports. As there is no new information about some cases in recent reports, it was not possible to trace recent developments in those cases. So the explanation of each case was completed with the aid of the most recent report containing information about that case. If there is no recent information on the case, or on its consequence, that means there is no related information on that case in the subsequent reports.

The cases of non-compliance can be examined in three distinct groups which can be formed on the basis of three ways of the initiation of the NCP by parties: triggering by parties to the Protocol against another party (NCP, para. 1, MP), self-triggering (NCP, para. 4, MP), triggering by the secretariat against any party to the Protocol (NCP, para. 3) (see Table 5.3).

5.1.2.1 Triggering against Another Party (NCP, para. 1, MP)

There has been no submission by parties against another party to the Protocol, so this type of triggering has not arisen to date. The most important reason of this is that this type of triggering requires an "adversarial action" against another party (Faure/Lefevere 1999: 153), and parties in general remain reluctant to engage in this

---

[9]For detailed information on the early phases of the NCP, see Victor (1996).

Table 5.3 Montreal Protocol CM in practice

Triggering by parries to the Protocol against another party (NCP, para. 1)	Self-triggering (NCP, para. 4)	Triggering by the Secretariat against any party to the Protocol (NCP, para. 3)
No submission so far	1. The Russian Federation and others (Belarus, Bulgaria, Poland and Ukraine) 2. Latvia and Lithuania	All submissions have been brought to the ImplCom's agenda via the Secretariat (with the exception of self triggering by the Russia Federation and others, Latvia and Lithuania)

Source Prepared by the author on the basis of information/documentation given on the official web site of UNEP/Ozone Secretariat. See: http://ozone.unep.org/new_site/en/index.php

kind of action against other parties and to initiate proceedings against them, seeing it as a risk which can undermine their bilateral relationships with these parties.

5.1.2.2 Self-triggering (NCP, para. 4)

The Russian Federation and Others

The process started with a joint statement made by countries with economies in transition (CEIT), namely the Russian Federation, Belarus, Bulgaria, Poland and Ukraine, at the eleventh meeting of the Open-Ended Working Group (OEWG). It was about these parties's possible inability to comply with their obligations under the MP.[10] Through this joint statement, expressing their domestic difficulties, such as political, geopolitical and social change and the transition to a market economy, as a reason for their potential non-fulfilment of the requirements of the MP, the cited countries asked to be granted a five-year grace period, from 1 January 1996, for compliance with the obligations under the MP (ImplCom 11 1995: 4, para. 15; ImplCom 11 1995: 14–15; OEWG 12 1995; Annex II).

After discussing the matter, it was decided that this statement constituted a submission under para. 4 of the NCP of Art. 8, MP, according to which a party may trigger the procedure with respect to itself. So it was referred to the ImplCom by the secretariat for consideration of the matter. In the tenth ImplCom meeting, it was decided that the joint submission should not be considered as a joint case, but should be divided into five different cases. So the consideration first focused on the Russian Federation (the cases of the other four countries to be discussed at a later stage), because of the special conditions of the Russian Federation, such as being a major producer of controlled substances (producing and exporting them to the other

---

[10]Similarly, at the MOP 5, through a declaration, the heads of the delegations representing the governments of Belarus, Bulgaria, Romania, the Russian Federation and Ukraine had also asked the parties to decide at the MOP 6 on the question of the special status of countries with economies in transition which would require concessions and a certain flexibility in the fulfilment of their obligations under the MP (MOP 5 1993, Annex VII).

states of the Commonwealth of Independent States [CIS]) – whereas the other parties were primarily consumers – and being the only one among others that had not reported data in line with its reporting obligations under Art. 7 (ImplCom 10 1995: 7, para. 32).

The Russian Federation. Based on the recommendations of the ImplCom, which were the result of intensive consultation and of information and data submitted by the Russian Federation (ImplCom 11 1995; ImplCom 12 1995), the MOP decided that the Russian Federation was in compliance with its obligations under the MP in 1995, but, possibly, it would not be in 1996 (MOP 7, Decision VII/18 1995). So, despite the efforts of the Russian Federation to provide the necessary data requested by the ImplCom and the commitment for further action to phase out ODSs in production and consumption, it was agreed by the MOP that it should provide additional and more detailed information on the areas determined by the MOP. In addition, it should benefit from the international assistance of the GEF and the World Bank, provided that it complied with some obligations, such as submitting annual reports on progress in phasing out ODS in line with the schedule given by the ImplCom, which should be submitted in due time to enable the secretariat and the ImplCom to review them. Thus, international assistance was clearly contingent on reporting on progress in phasing out ODS and the settlement of related problems with the ImplCom. More importantly, the Russian Federation was allowed to export substances controlled under the MP to parties operating under Art. 2 of the Protocol that are members of the CISs, including Belarus and Ukraine. Thus, the MOP indirectly restricted the Russian Federation's exports to other countries in order to help it phase out exports in controlled substances, as required by the Protocol (MOP 7, Decision VII/18 1995).

The Russian Federation did not agree with paras. 8 and 9 of the draft decision, which were about restricting its trade and conditional international assistance, as they allowed the MOP to adopt "discriminatory measures and sanctions" against itself, and such "punitive measures" could not be accepted by it. The Federation also stressed that, according to the indicative list of measures that might be taken in respect of non-compliance, the first measure that the MOP could apply was appropriate assistance to the party in breach of its obligations, and it had sought such assistance from the GEF, the World Bank and other international financial organizations. Another measure was the issuance of a caution to the party con-cerned, but without applying that measure, the ImplCom recommended taking measures under item C (sanctions) of the indicative list. On the basis of these views, it proposed that paras. 8 and 9 of the draft decision should be deleted and replaced with the wording acceptable to the Russian Federation. It also underlined that it would vote against them, if they continued to stand by the decision.

Although the Russian Federation objected to the draft decision due to the reasons mentioned above, the decision was adopted by consensus, reflecting the position of the dissenting party in the meeting's report (MOP 7 1995: 53, para. 130). This procedure, called "consensus minus one" (Werksman 1996: 771), in which a draft decision can be adopted by consensus when only one party (identified as the non-compliant party in that decision) objects to it, has highly enhanced the ability of the MOP to address the non-compliance of a particular party, authorizing it to adopt the necessary decisions to override the objections of that party (Werksman 1999).

In brief, through its decision on the Russian Federation, the MOP came to an agreement on three elements for compliance: monitoring the Russian Federation regarding the phasing-out of ozone-depleting substances, restricting its trade, and imposing conditions on the provision of financial assistance. Thus "[t]he 'carrot' of financial assistance was offered in conjunction with 'stick' of financial condition-ality and trade restrictions to encourage Russia to fulfil its obligations" (Potzold 2009: 10).

On the basis of these three elements, the situation regarding the phase-out of ODSs was kept under review by the ImplCom and the MOP; the Russian Federation's trade was restricted and the financial assistance for ODS-phase-out continued to be contingent on developments with regard to non-compliance and the settlement of any problems with the ImplCom in subsequent years as well. However, it was usually considered favourably in order to provide funding for projects to implement the programme for the phase-out of the production and consumption of ODSs in the country.

Nevertheless, in 1996, the Russian Federation was found in a situation of non-compliance (ImplCom 16 1996; MOP 8, Decision VIII/25 1996) and also possible non-compliance with the Protocol in 1997 (MOP 9, Decision IX/31 1997), according to its own written submissions and the statements of its representative at the meetings of the ImplCom 13–19 (1996–1997), although it was acknowledged that the Russian Federation had made considerable progress in addressing non-compliance issues raised by the MOP 7.

In fact, the Russian Federation made great progress, particularly after Decision VIII/25. It reported detailed information on quantities of imports and exports of ODS and products containing such substances, data on the type of ODS (new, recovered, recycled, reclaimed, reused, used as feedstock), details of suppliers, recipient countries and conditions of delivery of the substances for 1996, details of imports and/or exports of ODS in 1996, provided by some parties mentioned in its submission to the ImplCom, on ways in which it was maximizing the use of its recycling facilities to meet internal needs and to diminish production of new CFCs. It also started implementation of its exports control of ODSs from July 1996 by not exporting any ODS (used, new, recycled or reclaimed) to any party, with the exception of Art. 5 parties and parties that are members of the CISs, including Belarus and Ukraine.

The appreciation of the MOP on the Russian Federation's efforts to come into compliance with the MP continued in its MOP 10, Decision X/26 (1998) as well. In that decision, it was underlined that it reduced consumption of CFCs from 20,990 ODP-tonnes in 1995 to a level of 12,345 ODP-tonnes, submitted a country pro-gramme in October 1995 (revised in November 1995) involving specific bench-marks and a phase-out schedule, and produced only 16,770 ODP-tonnes from Annex A, Group I, substances, which was below the benchmark of 28,000 ODP-tonnes determined in the country programme in 1996. In October 1998, through "Special Initiative for ODS Production Closure in the Russian Federation (Special Initiative)", it also made a commitment to reduce consumption of Annex A, Group I substances to no more than 6,280 ODP-tonnes in 1999, to reduce consumption of Annex A, Group II, substances to no more than 960 ODP-tonnes in 1999, to reduce consumption of Annex B, Group I, substances to no more than 18 ODP-tonnes in 1999, and to phase out the production of Annex A substances by 1 June 2000 and the consumption of Annex A and B substances by 1 June 2000 (Decision X/26). In the same decision, it was also noted that the Russian Federation should continue to be treated in the same manner as a "party in good standing" on the grounds that it was working towards and meeting the specific time-based commitments in its country programme and the Special Initiative and continuing to report annual data demonstrating a decrease in imports and consumption. If it could continue to meet these conditions, it would continue to receive international assistance to meet its commitments in accordance with item A of the indicative list of measures. Regarding this decision, it is also noteworthy that the Russian Federation was also cautioned through this decision in accordance with item B of the indicative list of measures. It was stressed in the decision that if the country failed to meet its commitments according to their time schedule, measures contained in item C of the indicative list of measures, i.e. suspension of specific rights and privileges under the Protocol, would be considered for the country.

When it came to 2000s, the Russian Federation failed to complete the closure of ODS production facilities by June 2000 as agreed, so it appealed to the parties for more understanding, stating that it would end all ODS production by 20 December 2000. The ImplCom agreed to give it until 20 December 2000 to halt all production of CFCs (ImplCom 25 2000: 2–3, paras. 9–11). Accordingly, by 20 December 2000, all ODS-producing enterprises in the country had completed closure activities and the Russian Federation achieved production sector phase-out (ImplCom 27 2001: 6, para. 32). Nonetheless, it was again found to be in non-compliance with the phase-out benchmarks for 1999 and 2000 for the production and consumption of the ozone-depleting substances in Annex A (MOP 13, Decision XIII/17 2001).

After the data reported by the Russian Federation for 2001 confirmed the complete phase-out of production and consumption of ODSs in Annexes A and B, as noted by Decision XIII/17 (ImplCom 29 2002: 12, para. 66; MOP 14, Decision XIV/35 2002), the MOP decided that it had returned to full compliance with its obligations (MOP 14 2002: 20, para. 133). Its success in complying with its commitments was used as an illustration of the success of the non-compliance system of the MP by the ImplCom (MOP 14 2002: 20, para. 133).

However, it was found to be in non-compliance with the provisions of Art. 4 of the MP prohibiting trade with non-parties to the MP, as it reported the export of 70.2 metric tonnes of Annex C, Group I HFCs in 2009 to a state classified as not operating under para. 1 of Art. 5 of the MP, and not party to the Copenhagen Amendment to the Protocol in that year (MOP 23, Decision XXIII/27 2011). It was decided to monitor its progress closely with regard to the implementation of its obligations under the Protocol.

Belarus. After the acceptance of the joint statement as a submission under para. 4 of the NCP, on the basis of the consultations of the ImplCom with the representatives of Belarus regarding possible non-fulfilment of that party's obligations under the MP (ImplCom 11 1995) and the information submitted by Belarus on related data and measures on the phase-out of ozone-depleting substances (ImplCom 12 1995), the MOP gave a decision on Belarus's status of compliance (MOP 7, Decision VII/17 1995).

Belarus was found in compliance with its obligations under the Protocol in 1995, but in possible non-compliance in 1996. So it was invited to provide information on its political commitment to the phase-out programme for ODSs, and if the ImplCom found necessary, additional information on further certain elements, such as the gradual achievement of the proposed phase-out plan. In addition, its entitlement to benefit from international assistance was bound to some conditions, such as submitting annual reports on its ODS phase-out progress in line with the schedule included in the country programme, and settling the questions related to the reporting requirements and the actions of Belarus with the ImplCom. It was not allowed to export any virgin, recycled or recovered substance controlled under the Protocol to any party operating under Art. 2 of the Protocol which was not a member of the CISs (MOP 7, Decision VII/17 1995).

A very short time after that decision, the ImplCom noted that, while the information available showed a situation of non-compliance by Belarus for 1996, it had taken significant steps in complying with Decision VII/17 and towards achieving full compliance with the control measures of the Protocol (ImplCom 13 1996). This status of non-compliance of Belarus continued till the 2000s, and was emphasized in different ImplCom reports (such as 15, 20, 21) and also Decision X/21. In Decision X/21 (MOP 10 1998) it was also noted that although Belarus submitted a list of specific projects with international financing that would reduce national consumption, it did not submit a phase-out plan with specific benchmarks to the Impl Com. Belarus' commitment to a phase-out in the consumption of Annex A and B substances by 1 January 2000 was also underlined. "Closely monitoring" the progress with regard to the phase-out of ODSs was another important point of the decision that should be mentioned. Finally, as in the case of the Russian Federation (MOP 10, Decision X/26 1998), the decision made it clear that to be treated "as a party in good standing" and to continue to benefit from international assistance, Belarus should continue to work towards and meet its specific time-based commitments and annually report related data demonstrating a decrease in imports and consumption. It cautioned Belarus that, if it failed to meet its commitments, the measures mentioned in item C of the indicative list of measures would be considered.

In 2002, the ImplCom found that Belarus was in non-compliance with the control measures under Art. 2, MP stressing that its reporting data on the consumption of substances in Annexes A and B to the Protocol in 2000 were above the control levels stated in Art. 2, MP. Therefore, "as a matter of urgency," it was requested to provide the requisite explanations for its non-compliance (ImplCom 29 2002; MOP 14, Decision XIV/28 2002). After just one year, Belarus reported the relevant data confirming that ODS consumption had been completely phased out, thus expressing that it had achieved full compliance with its phase-out plans (ImplCom 30 2003; para. 81).

Bulgaria. Bulgaria was also found in possible non-compliance in 1996 (despite its compliance status in 1995) in Decision VII/16 (MOP 7 1995), and it continued to be in non-compliance in 1997 (ImplCom 22; MOP 11, Decision XI/24 1999), despite its efforts to implement the Protocol fully.

However, in Decision XI/24 (MOP 11 1999), the MOP acknowledged that Bulgaria had acceded to the London and Copenhagen Amendments on 28 April 1999, had developed a national programme, and had established a phase-out plan that brought it into compliance with Protocol by 1 January 1998.

As in the cases of the Russian Federation and Belarus, the MOP emphasized the need for monitoring Bulgaria's progress towards meeting the specific commitments and treating the country in the same manner as a party in good standing provided that it was working towards meeting the specific time-based commitments and annually reported data demonstrating a decrease in imports and consumption.

In addition, as in these cases, Bulgaria was cautioned that if it failed to meet its commitments, it could encounter measures consistent with item C of the indicative list of measures, involving the possibility of actions available under Article 4, such as ceasing the supply of CFCs and halons and enabling the exporting parties to be in compliance. In 2001, it was again warned about its situation of potential non-compliance, and explanatory information about its consumption figures was requested by the ImplCom 26 (2001).

In 2002, in the context of the regular review of the ImplCom on the status of compliance with the decisions of the parties and the recommendations of the ImplCom on non-compliance issues, Bulgaria requested a revision of its baseline figure for methyl bromide consumption. It reported that it had been incorrectly reported as zero for 1991, causing it to be found to be in non-compliance (ImplCom 28 2002: para. 28). After gathering sufficient information from all relevant sources, the ImplCom considered Bulgaria's request its next meeting and decided to accept its request to change its baseline consumption data for Annex E substances in 1991 from zero to 51.78 ODP-tonnes (ImplCom 29 2002: 10–11, paras. 50–51, 53, 55, MOP 14, Decision XIV/... (requests for changes in baseline data) 2002: 24).

Poland. Due to its problems in reducing CFC consumption in the years 1994 and 1995, Poland wanted to join the limits for the two years and import the maximum possible in 1994 and 1995 to avoid tensions within its industry and the economy. Since it had not yet ratified the Copenhagen Amendment, it was not legally bound by Article 2, para. 3, of the MP. So it was decided by the ImplCom that once Poland had ratified the Amendment, it could explain its non-compliance under para. 4 of the NCP (ImplCom 9 1994).

After the acceptance of the joint statement as a submission under para. 4 of the NCP (ImplCom 10 1995), its compliance with the Protocol was also considered by the MOP 7 (1995) in Decision VII/15. In this decision, an assurance was given by Poland that it was in compliance with its obligations under the Protocol for the year 1995 and would likely be in compliance in 1996 as well. However, it was warned that if any doubts emerged about its compliance in the year 1996, it should submit the information to the secretariat as soon as possible so that the necessary action could be initiated.

After this decision, Poland ratified both the London and Copenhagen Amendments to the Protocol, and its consumption of CFCs in 1996 became below the level of the essential-use exemptions granted it by the MOP 6 (ImplCom 17 1997).

Ukraine. The MOP's first decision on Ukraine's status of compliance (MOP 7, Decision VII/19 1995) followed the same pattern, particularly its first decisions on the Russian Federation (MOP 7, Decision VII/18 1995) and Belarus (MOP 7, Decision VII/17 1995) and contained similar provisions, such as being obliged to provide information on the political commitment to the phase-out programme for ODSs and, if necessary, additional information on further certain elements, such as the gradual achievement of the proposed phase-out plan, being bound with some conditions, such as submitting annual reports on the ODS phase-out progress, etc. to receive international assistance. Unlike the Russian Federation, but as in the case of Belarus, it was not allowed to export any virgin, recycled or recovered substance controlled under the Protocol to any party operating under Art. 2 of the Protocol which was not a member of the CISs (MOP 7, Decision VII/19 1995).

After this decision, Ukraine continued to improve its compliance status by taking crucial steps in line with the decision (ImplCom 13 1996), so its development was welcomed by the ImplCom 15 (1996), as it was in the process of ratifying the London Amendment, had established a final phase-out for ODS in 1999, drafted provisions to license the import and export of ODS, submitted data on the consumption and production of ODS up to 1995 and also phased out the production of carbon tetrachloride, the only ODS produced in Ukraine, by 1994. However, in the ImplCom 20 (1998), it was listed as one of the parties whose data was in non-compliance in 1996 with the control measures in Art. 2 of the Protocol. Then, in Decision X/27 (MOP 10 1998), it was found in non-compliance with its control obligations under Articles 2A through 2E of the Protocol in 1996. In particular, the significant increase in consumption of ozone-depleting substances in Ukraine from 1995 to 1996 was stressed, and Ukraine's commitment to phase out the consumption of Annex A and B substances by 1 January 2002 was noted.

Due to the difficulties that could arise out in phasing out consumption in the domestic refrigeration sector phase-out by 1 January 2002, Ukraine was warned to plan carefully for its future refrigerant servicing needs and invite the Technology and Economic Assessment Panel to help it. As in the cases of the Russian Federation (Decision X/26), Belarus (Decision X/21) and Bulgaria (Decision XI/24), monitoring the progress towards meeting the specific commitments, treating the country in the same manner as a party in good standing in certain conditions and applying measures consistent with item C of the indicative list of measures if it failed to meet its commitments were again stressed. At its twenty-sixth meeting, while considering the status of compliance of Ukraine with Decision X/27, the ImplCom agreed that no action needed to be taken, as the series of benchmarks agreed between the ImplCom and Ukraine had not come into operation before 2000 (ImplCom 26, 2001: 8, para. 40). In its next meeting, the ImplCom found that Ukraine was in non-compliance with the control measures under Art. 2 of the Protocol in 2000, stressing that its reporting data on the consumption of substances in Annexes A and B to the Protocol in 2000 were above the control levels specified in Art. 2 of the Protocol. Therefore, "as a matter of urgency", it was asked to provide the requisite explanations for its non-compliance (ImplCom 29 2002).

Nevertheless, Ukraine was not included in Decision XIV/28 regarding non-compliance with consumption phase-out by parties not operating under Art. 5 in 2000, which included the countries Belarus and Latvia, due to the fact that under the terms of Decision X/27 it was still in compliance (MOP 14 2002: 21, para. 140). However, after ten years, in 2012, it was again found to be in non-compliance. In fact, in Decision XXIV/18 (MOP 24 2012), while appreciating its efforts to comply with the MP, the MOP stressed that Ukraine was still in non-compliance with the consumption control measures under the MP for HCFCs in 2010 and 2011; therefore, it had to follow up its commitments in its plan of action and also work with the relevant implementing agencies to implement them to phase out its consumption of HCFCs. In the same decision, it was also warned that its progress would be monitored, and informed that it would continue to be treated in the same manner as a party in good standing and continue to receive international assistance to the extent that it continued to comply with its commitments. Otherwise, it could encounter measures consistent with item C of the indicative list of measures.

In more recent reports, the ImplCom noted that Ukraine had submitted its data for 2014 and for 2015 in accordance with its obligations under Article 7, MP, and that the data indicated that it was in compliance with its commitments for those years, outlined in Decision XXIV/18 (ImplCom 55, Recommendation 55/1 2015; ImplCom 56, Recommendation 56/4 2016).

Latvia and Lithuania

In December 1995, the secretariat received a letter from Estonia, Latvia and Lithuania requesting a longer time scale for phasing out ODSs, showing the institutional and financial problems of their countries to justify their request. While Estonia was not taken into account by the secretariat - as the NCP was only applicable to parties, Latvia and Lithuania were requested to make a formal sub-mission under para. 4 of the NCP. In response to the secretariat's request of 27 December 1995 for information pursuant to para. 4 of the NCP, Lithuania submitted a plan of action to phase out ozone-depleting substances, while no response was received from Latvia (ImplCom 13 1996). After the requests of the ImplCom 13, Latvia also submitted a report on measures adopted to phase out ozone-depleting substances and its country programme for phasing-out ozone-depleting substances (ImplCom 14 1996). Although neither countries made formal submissions for triggering the NCP, the ImplCom considered their reports and action plans as a submission under para. 4 of the NCP, and thus the NCP was initiated.

The ImplCom reported that, based on the information given by the countries, they would be in a situation of non-compliance with the Protocol in 1996 and possible non-compliance in 1997. It also stressed that, even if the ratification of the London Amendment was a prerequisite for them to receive assistance from the GEF, international funding agencies should consider favourably the provision of financial assistance to them for the requisite projects to phase out ozone-depleting substances in their countries (ImplCom 14 1996). In line with the recommendations of the ImplCom, the MOP 8 (1996) decided that both countries should be kept under review with regard to ODS phase-out in their countries, that the provision of financial assistance, particularly by the GEF, should be considered favourably to provide funding for projects for phasing out ODSs to implement their programmes, and that they should be urged to ratify the London Amendment to the Protocol as soon as possible (MOP 8, Decision VIII/22–23 1996). In accordance with these decisions, the situation regarding ODS phase-out in these countries was kept under review (ImplCom 17 1997), and it was determined that they were in a situation of non-compliance in 1997 and would probably be in non-compliance in 1998 as well (ImplCom 18 1997; MOP 9, Decision IX/29–30 1997).

After these decisions, Lithuania acceded to the London and Copenhagen Amendments on 3 February 1998 and Latvia on 2 November 1998. Together with these positive developments, their other efforts to achieve compliance, like reducing their consumption of ODSs, were also indicated as signals of their commitment to become parties in full compliance with the Protocol through Decisions X/24–25 (MOP 10 1998).

In Decision X/24 on Latvia's situation, on the basis of Latvia's statements to the ImplCom, it was assumed that Latvia had made a commitment to observe the ban on the production and import of Annex A, Group II, substances imposed on 12 December 1997, to limit consumption of Annex A, Group I, substances to no more than 100 metric tonnes in 1999, and to ban the production and import of Annex A, Group I, and all Annex B substances by 1 January 2000.

In Decision X/25 on Lithuania's situation, it was also assumed that it had made a commitment to ban the import of CFC-113, carbon tetrachloride and methyl chloroform by 1 January 2000 and to reduce the consumption of Annex A and B substances by 86 per cent of its 1996 levels by 1 January 2000.

In both decisions, considering the plans produced by the two States, it was expected that they would be able to achieve a total phase-out of Annex A and B substances by 1 July 2001. In addition, in line with the decisions on the Russian Federation (Decision X/26), Belarus (Decision X/21), Bulgaria (Decision XI/24) and Ukraine (Decision X/27), monitoring the progress, treating them as parties in good standing as long as they worked towards and met specific time-based commitments, and applying measures consistent with item C of the indicative list of measures were again stressed.

The ImplCom continued its review of the status of the compliance of these parties with specific decisions and recommendations of the ImplCom on non-compliance in the coming years as well. In its twenty-ninth meeting, it clearly stated that Latvia was among the non-article 5 parties which were determined as being in non-compliance with the control measures under Articles 2A to 2H, stressing that Latvia had reported consumption of ODSs in 2000 in excess of the control limits, which could not be explained by allowed exemptions (ImplCom 29 2002: 11–12, para. 59). Therefore, it decided to forward the issue to the MOP 14, containing a draft decision, declaring that Latvia was in non-compliance with the control measures in 2000 (ImplCom 29 2002: para. 60). This draft decision was then approved in all respects by the MOP 14 (MOP 14, Decision XIV/28 2002). Monitoring of the country was continued in the subsequent year as well. Because its annual data for 2001 reported that its consumption of the controlled substance in Annex E (methyl bromide) was above the requisite 50% reduction, it was found to be in non-compliance with its obligations under Art. 2H of the Protocol for that year. Nevertheless, it was able to provide an explanation for its non-compliance and subsequently reported Annex E data for 2002, indicating its return to compliance (MOP 15, Decision XV/24 2003).

5.1.2.3   Triggering by the Secretariat

Countries with Economies in Transition (CEIT)

The CEIT which came to notice as non-compliant and so were reported to the ImplCom by the secretariat to date are Armenia (subsequently reclassified as developing country operating under Art. 5),[11] Azerbaijan, Czech Republic, Estonia, Kazakhstan, Tajikistan, Turkmenistan (subsequently reclassified as a developing country)[12] and Uzbekistan (see Annex E to the book on MOP decisions related to CEIT).

Armenia. Armenia was first found in non-compliance with the data reporting requirement under Art. 7 of the Protocol in 2001 through Decision XIII/18 (MOP 13 2001). In 2002 it was also found in non-compliance with the control measures under Art. 2 of the MP in 2000, as the data it reported on its consumption of substances in Annex A to the Protocol in 2000 were above the control levels

---

[11]Taking into account its difficult economic situation, its classification as a developing country by the World Bank and the UNDP and its low *per capita* consumption of ODSs, the MOP 14 decided to accept the application of Armenia to be listed as a developing country (MOP 14, Decision XIV/2 2002). So it is involved here in the section entitled 'Triggering by the Secretariat against any party to the Protocol, Non-compliance by the Countries with Economies In Transition (CEIT).

[12]Due to the fact that *per capita* consumption of Annex A and Annex B substances of the party were below the limits specified under Art. 5 of the MP and the party was classified as a low income country by the World Bank, MOP 16 decided to accept the application of Turkmenistan to be listed as a developing country (MOP 16, Decision XVI/39 2004; MOP 17 2005).

through Decision XIV/31 (MOP 14 2002). Through both these decisions and also Decision XV/27 (MOP 15 2003), Armenia was recommended to ratify the London Amendment in order to benefit from financial assistance for projects to phase out ODSs in that country. Then it applied for reclassification as a developing country operating under Art. 5 of the Protocol, and its application was accepted under Decision XIV/2 (MOP 14 2002). In addition, it ratified the London and also Copenhagen Amendments on 26 November 2003, and thereby attained the right to financial assistance to enable its compliance.

When it reported its annual consumption of the controlled substance in Annex E (methyl bromide) for 2004 to be 1.020 ODP-tonnes, it was again found in non-compliance with the control measures for methyl bromide under the Protocol in 2005 (MOP 17, Decision XVII/25 2005), as this exceeded its maximum allowable consumption level of zero ODP-tonnes for that controlled substance for that year. Therefore, the MOP asked Armenia to submit a plan of action with time-specific benchmarks – including the establishment of import quotas to support the phase-out schedule and requisite policy and regulatory instruments – to the ImplCom at its next meeting "as a matter of urgency". The MOP also decided (MOP 18, Decision XVIII/20 2006) to monitor its progress closely and to treat it in the same manner as a party in good standing, allowing it to continue to receive international assistance as long as it worked towards and met the specific Protocol control measures. In this decision of the MOP, and in accordance with item B of the indicative list of measures, Armenia was also cautioned that if it failed to return to compliance in a timely manner, it could face measures consistent with item C of the indicative list of measures, which could involve stopping the supply of the controlled substance in Annex E (methyl bromide) in accordance with Art. 4 (MOP 17, Decision XVII/25 2005).

After this decision by the MOP, Armenia submitted its plan of action for complying with the Protocol's methyl bromide control measures and committed itself to maintaining methyl bromide consumption at no more than zero ODP-tonnes from 2007, save for critical uses that might be authorized by the parties after 1 January 2015, and to introducing an import quota system for licensing the import and export of ODSs by 1 July 2007.

Its achievements and also its reporting on methyl bromide consumption for 2005 which showed its return to compliance in that year were welcomed by the MOP 18. However, as Armenia itself was not sure about its ability to sustain its phase-out of the consumption of methyl bromide, the MOP recommended it to work with the relevant implementing agencies to implement the remainder of the plan of action (MOP 18, Decision XVIII/20 2006). For 2006, Armenia submitted that, as preliminary data for 2006 showed, it would maintain total phase-out of methyl bromide consumption in that year (ImplCom 38 2007).

Regarding its commitment on introducing the licensing and quota system for the import and export of ODSs in accordance with Decision XVIII/20 (MOP 18 2006), it should be stated that it completed implementation of its commitment in 2007 in the last week of June (ImplCom 39, Recommendation 39/2 2007).

With respect to its commitment not to exceed zero consumption of methyl bromide from 2007, in accordance with Decision XVIII/20, Armenia reported that its methyl bromide consumption was zero ODP-tonnes in both 2007 and 2008, maintaining its compliance with the commitment recorded in Decision XVIII/20 (ImplCom 40, Recommendation 40/4 2008; ImplCom 42, Recommendation 42/4 2009).

Azerbaijan. Azerbaijan was first found in non-compliance with its control obligations under Articles 2A through 2E of the MP in 1996 through Decision X/20 (MOP 10 1998). Since it was believed by the country that this situation would continue at least throughout the year 2000, in order to achieve the virtual phase-out of CFCs and a complete phase-out of halons by 1 January 2001, it committed itself to manage the following measures:

- to phase out Annex A, group I substances (CFCs) by 1 January 2001 (save for essential uses authorized by the parties);
- to establish, by 1 January 1999, a system for licensing imports and exports of ODS;
- to establish a system for licensing operators in the refrigeration-servicing sector;
- to tax the imports of ozone-depleting substances to enable Azerbaijan to meet the year 2001 phase-out;
- to ban, by 1 January 2001, all imports of Annex A, group II substances (halons);
- to consider by 1999 a ban on the import of ODS-based equipment.

In line with similar decisions on non-compliant parties, in this decision it was again stressed that if Azerbaijan did not work towards meeting the above commitments within the pre-determined timescale, the parties would consider applying measures against it.

Based on the Decision X/20 (MOP 10 1998), Azerbaijan prohibited the import of halons in 1999; however, data submitted by Azerbaijan for 2001, 2002 and 2003 showed a level of CFC consumption which put Azerbaijan in non-compliance with its obligations under Art. 2A of the Protocol. It also failed to fulfil its undertaking, contained in Decision XV/28, to ban the consumption of CFCs from January 2003 (MOP 15, Decision XV/28 2003; MOP 16, Decision XVI/21 2004). Although it confirmed the introduction of a ban on the import of controlled substances in Annex A, group I (CFCs), unfortunately, as it did not achieve total phase-out of these controlled substances by 1 January 2001, 1 January 2003, or 1 January 2005, respectively, it was further cautioned that if it failed to achieve total phase-out of Annex A, group I controlled substances (CFCs) by 1 January 2006, implementation of item C of the indicative measures, which could include action available under Art. 4 to cease supply of Annex A, group I controlled substances (CFCs) to Azerbaijan, would be considered by the subsequent MOP (MOP 17, Decision XVII/26 2005). This caution was recalled through Recommendation 36/3 and Recommendation 37/2 of the ImplCom (ImplCom 36 2006; ImplCom 37 2006).

Azerbaijan then managed to return to compliance in 2006 with the consumption control measures for Annex A, group I controlled substances (CFCs) and also to achieve total phase-out of those substances by 1 January 2006. Nevertheless, it reported consumption of Annex B, group I (other CFCs) in that year for the first time, and this report showed that there was an apparent inconsistency with the Protocol's requirement to maintain total phase-out of those substances in 2006 (ImplCom 38, Recommendation 38/2 2007). In response, Azerbaijan explained that it resulted from a misclassification of imports and its corrected data expressed its compliance with the Protocol's requirement to maintain total phase-out of those substances in that year (ImplCom 39 2007).

With regard to receiving institutional strengthening assistance from the GEF, Azerbaijan was requested by ImplCom 38 (2007), through Recommendation 38/2, to submit to the secretariat as soon as possible, and no later than 1 August 2007, a status report on its efforts in conjunction with UNEP to expedite implementation of the additional institutional strengthening project. This project entailed national (such as the preparation of work plans and the acquisition of additional staff, expertise and equipment) and regional activities (such as activities conducted through the regional network of ozone officers for Eastern Europe and Central Asia) and the UNEP Green Customs Initiative on regional or transboundary issues (like illegal trade, stockpiling and the destruction of ODSs) for four countries with economies in transition – Azerbaijan, Kazakhstan, Tajikistan and Uzbekistan (approved by GEF on 9 April 2007) – to provide the institutional strengthening and customs officer training components of capacity-building assistance. It responded to this request of the ImplCom, but it could not manage to start implementing the project, so was advised to work in collaboration with UNEP to expedite the implementation (ImplCom 39, Recommendation 39/3 2007). In Recommendation 40/5 (ImplCom 40 2008), it was requested to complete the requisite operations for signing the pending agreement for full implementation of the project.

The party's submission of ODS data for 2012 showed that, Azerbaijan was in compliance with its hydrochlorofluorocarbon (HCFC) consumption obligations under the control measures of the Protocol. So it was decided that no further action was necessary to ensure compliance with the HCFC phase-out in 2012 and with the Protocol's control measures for HCFCs, but that it was necessary to monitor closely the party's progress with regard to the implementation of its obligations under the Protocol. In order to implement its plan of action for their consumption, it also warned Azerbaijan to work with the relevant implementing agencies (MOP 25, Decision XXV/10 2013; ImplCom 50, Recommendation 50/8 2013; ImplCom 51, Recommendation 51/2 2013).

Czech Republic. Because of the operation of special industrial cooling equipment for the chemical industry in the country, the Czech Republic was not able to be in compliance with the halon phase-out in 1994 or 1995 (MOP 8, Decision VIII/24 1996). It imported a total of 11.16 ODP-tonnes of methyl bromide, exceeding its freeze level of 6.0 ODP-tonnes. As only 7.9 ODP-tonnes of this amount was consumed in 1996 and no methyl bromide was imported in 1996, the average annual consumption for the two years remained below its freeze level (MOP 9, Decision IX/32 1997). Thus it escaped being faced with measures for non-compliance.

However, it was found in non-compliance with its control obligations under Articles 2A through 2E of the Montreal Protocol for 1996. This was named "technical non-compliance" due to the fact that the Czech Republic had imported ODSs in 1996 without obtaining essential-use authorization from the parties to the Protocol. So even though it claimed that some part of its CFC consumption was for essential use for metered-dose inhalers, as it had no authorization, it was still in state of technical non-compliance (MOP 10, Decision X/22 1998).

In the report of the ImplCom 23 the Czech Republic was praised because it had reported the seizure of illegal imports of ODSs in the country. It was also decided that all parties should be supported to participate in an active struggle against illegal trade in ODSs (ImplCom 23 1999). In 2001, the ImplCom finally decided that the Czech Republic was in compliance according to the consumption figures that it reported, so no action needed to be taken against it (ImplCom 26 2001).

Estonia. Despite its major efforts to achieve compliance with the MP and its consumption's reduction from 131 ODP-tonnes in 1995 to 36.5 tonnes in 1996 as a result of these efforts, Estonia was found in non-compliance with its control obligations under Articles 2A through 2E of the Protocol in 1996. Because this situation was likely to continue throughout at least the year 2000, it committed to a few reduction targets for complete phase-out (MOP 10, Decision X/23 1998). These interim targets were as follows:

- to reduce consumption to no more than 23 ODP-tonnes for Annex A and B substances by 1 January 1999;
- to establish a harmonized system for monitoring and controlling imports of ODSs in 1999;
- to completely phase out consumption of Annex B substances and to reduce consumption to no more than 14 ODP-tonnes of Annex A substances by 1 January 2000;
- to reduce consumption of CFC-12 to all but 1 tonne in 2001;
- to completely phase out Annex A substances by 1 January 2002.

As in other decisions of the MOP with regard to other non-compliant parties, in Decision X/23 (MOP 10 1998) Estonia was also cautioned that if it failed to meet the specific time-based commitments noted above, the parties would be entitled to consider measures which could include the possibility of ceasing the supply of the CFCs and halons that were the subject of non-compliance. In 2000, due to the fact that its consumption of Annex B/I substances increased in 1998 and that it did not meet its target for Annex A and B substances, the ImplCom decided to alert the country concerning its deviation from the reduction schedule and to ask for the secretariat to receive clarification on the reasons for the deviation (ImplCom 24 2000: 3, para. 10).

Finally, at its twenty-sixth meeting, the ImplCom decided to take no action on Estonia, as the consumption levels reported by the country fell within its specified commitments (ImplCom 26 2001: para. 36).

Kazakhstan. The first decision on Kazakhstan concerning its non-compliance with its control obligations under Articles 2A through 2E of the Protocol, was Decision XIII/19 (16–19 October 2001). This decision particularly indicated that the data reported by Kazakhstan for the years 1998–2000 clearly showed its non-compliance with the Protocol's requirements. It also recorded the party's commitment to return to compliance by, among other things (see MOP 13, Decision XIII/19 2001), phasing out its consumption of CFCs and methyl bromide by 1 January 2004 and establishing a ban on the import of equipment that used ODSs by 1 January 2003. According to para. 4 of the same decision, the MOP 13 agreed to monitor the progress of Kazakhstan with regard to the phase-out of ODSs, particularly towards meeting the specific commitments contained in the decision. As in other decisions of the MOP on non-compliance, this Decision contained the provision that if Kazakhstan failed to meet its commitments, it would not be treated in the same manner as a party in good standing and could be faced with measures consistent with item C of the indicative list of measures, which could include the possibility of actions specified in Art. 4. In 2004 its reported annual consumption of the controlled substances in Annex A, group I (CFCs) was not zero, despite the commitment made in Decision XIII/19 (MOP 17, Decision XVII/35 2005). In the subsequent decision on Kazakhstan, namely Decision XVII/35 (MOP 17 2005), the party was again reminded of para. 4 in Decision XIII/19 (MOP 18 2006). It was also underlined that, although Kazakhstan's reported CFC consumption for 2004 was less than its reported consumption in 2003, it was still inconsistent with the party's commitment, contained in Decision XIII/19, to reduce its consumption of the controlled substances in Annex A, group I (CFCs) to zero in 2004. Moreover, Kazakhstan was warned about its non-compliance on submitting the requested explanation for the deviation and reporting on its commitment to implement a ban on the import of equipment using ODSs (ImplCom 35, Recommendation 35/20 2005).

After this decision, with respect to its apparent deviation in 2004, Kazakhstan submitted an explanation referring to the administrative changes in the country which had resulted in delays in enforcement of the ban on the import of ODSs and products containing such substances. It also submitted its 2005 data, which were consistent with its commitment to return to compliance with the Protocol's control measures, reporting CFC and methyl bromide consumption of zero ODP tonnes, and completed implementation of its commitment to introduce a ban on the import of equipment using ODSs (ImplCom 36, Recommendation 36/23 2006).

Another important issue that should be mentioned about Kazakhstan is its non-compliance in trade with non-parties, in accordance with Art. 4 of the Protocol, as introduced by the 1999 Beijing Amendment. This issue was raised after the Republic of Korea exported to Kazakhstan HCFCs in 2008 and also in 2009, since under Art. 4 (para. 2), parties to the amendment were required to ban the export of HCFCs to any state not party to the Protocol. The definition of the term "state not party to the Protocol" should be clarified here in order to explain the situation of Kazakhstan. The term "state not party to the Protocol" is defined as a state that had not agreed to be bound by the control measures in effect for those substances.

In the case of HCFCs, control measures were imposed under the Copenhagen (controlling consumption) and Beijing (controlling production) Amendments (Art. 4, para. 9). It was defined by Decision XV/3 (MOP 15 2003) to include all other states that had not agreed to be bound by the Copenhagen and Beijing Amendments (Decision XV/3, para. 1b). This term did not apply to states operating under para. 1 of Art. 5 until 1 January 2013 (Decision XV/3, para. 1a, as amended by Decision XX/9), or to states which were determined to be in full compliance with Art. 2 (2A–2I) and Art. 4 by the MOP. However, it was ruled that the MOP should decide on the matter after a request from the state concerned (Art. 4, para. 8) for the exports of ODSs (like HCFCs) to be allowed to that state if it had submitted data in accordance with Art. 7 (Art. 4, para. 9).

When Kazakhstan's case is examined, it becomes clear that the Republic of Korea was bound by the requirements of the Beijing Amendment as a party to that amendment, while Kazakhstan was not a party to the Copenhagen or Beijing Amendments. However, it was a non-Art. 5 party, and its consumption of both HCFCs and methyl bromide for 2006–2008 was found to be above the specified control levels. This situation provided all the conditions to prevent Kazakhstan trading in ODSs, and particularly in HCFCs, with parties to the Protocol.

Therefore, the ImplCom recommended it to ratify the Copenhagen and Beijing Amendments to the Protocol (ImplCom 45, Recommendation 45/6 2010) and prepared a draft decision for consideration by the MOP 22 (2010).

As Kazakhstan's report on its annual consumption for the controlled substance in Annex E (methyl bromide) for 2011 was 6.0 ODP-tonnes, exceeding its maximum allowable consumption for that year, it was found to be in non-compliance with the consumption control measures under the Protocol for methyl bromide. Therefore, it was asked to submit an explanation for its excess consumption to the secretariat, together with a plan of action, as a matter of urgency and no later than 31 March 2014. It was also decided to monitor Kazakhstan's progress closely with regard to the phase-out of hydrochlorofluorocarbons and methyl bromide, and to warn it of the application of measures, including the possibility of actions available under Art. 4, such as ensuring a cease in the supply of the hydrochlorofluorocarbons and methyl bromide that were the subject of non-compliance so that exporting parties would not be contributing to a continuing situation of non-compliance (MOP 25, Decision XXV/12 2013; ImplCom 50, Recommendation 50/11 2013).

In the MOP 26 it was still in non-compliance with the consumption control measures under the Protocol for HCFCs with its annual consumption of the controlled substances in Annex C, group I (HCFCs), of 90.75 ODP-tonnes for 2011, 21.36 ODP-tonnes for 2012 and 83.32 ODP-tonnes for 2013, exceeding its maximum allowable consumption of 9.9 ODP-tonnes for those controlled substances. In addition, it was also in non-compliance with the consumption control measures under the Protocol for methyl bromide, as its annual consumption of the substance of 6.0 ODP-tonnes in 2011 and 19.0 ODP-tonnes in 2013 exceeded its maximum allowable consumption of zero ODP-tonnes.

Accordingly, it was decided that Kazakhstan should submit a plan of action to ensure its return to compliance, and work with the relevant implementing agencies to implement this action plan. Its progress on the implementation of its action plan and the phase-out of HCFCs and methyl bromide was to be monitored closely; and it was to be warned about the application of measures, including the possibility of actions available under Art. 4 (MOP 26, Decision XXVI/13 2014). Concerns about deviations from the Protocol's requirements regarding Kazakhstan's maximum allowable consumption had already been stressed through Recommendation 52/2 (ImplCom 52 2014) and Recommendation 53/1 (ImplCom 53 2014). In its 2016 report, the ImplCom stated that, in accordance with its plan of action set out in Decision XXVI/13, Kazakhstan had committed itself to reducing its HCFC consumption to no more than 9.9 ODP-tonnes and its methyl bromide consumption to zero ODP-tonnes in 2015, but had not submitted its data for 2015. In addition, it paid attention to the fact that, because no financial support was available to Kazakhstan, none of the implementing agencies were currently working with the country and no projects were being implemented. Therefore, it was decided that Kazakhstan should report its 2015 data on ODSs to the secretariat as soon as possible so that the Committee could assess its compliance status at the next ImplCom meeting (ImplCom 56, Recommendation 56/2 2016).

Tajikistan. Tajikistan was identified in Decision XIII/20 (MOP 13 2001) as being in non-compliance with its obligations under the Protocol (Articles 2A–2E). So to return to compliance, it committed to a series of measures (listed in para. 2 of the Decision) to achieve the near total phase-out of all Annex B substances by 1 January 2002, all Annex A substances by 1 January 2004, and the Annex E substance by 1 January 2005. As in the cases mentioned above, the decision (para. 4), also cautioned Tajikistan that the parties could consider measures, consistent with item C of the indicative list of measures, if it did not meet its specific time-based commitments. The ImplCom noted some inconsistencies in its progress report concerning the dates when some commitments were expected to be completed (ImplCom 28 2002: para. 71; ImplCom 29 2002: para. 70), and further noted that, regarding the progress made to implement these agreed benchmark commitments, no data had been received from the country (ImplCom 30 2003: para. 88). Its compliance with its commitments contained in Decision XIII/20 was declared after its submission of its data through the secretariat's report, and it was also congratulated on its complete phase-out of methyl bromide consumption even though it was not yet a party to the Copenhagen Amendment (ImplCom 31 2003: para. 77). However, despite its achievement on phasing out by 2003, as it failed to submit its 2003 data, its compliance status could not be evaluated (ImplCom 32 2004). After reporting data which met its benchmark for CFC consumption in 2003, its efforts and progress towards complying with the requirements of Decision XIII/20 were appreciated by the ImplCom (ImplCom 33 2004: para. 86). However, it was instructed by the Committee to submit its data for 2004 to the secretariat as soon as possible (ImplCom 34, Recommendation 34/41 2005).

With regard to its commitment to phase out the consumption of CFCs by 1 January 2004 contained in Decision XIII/20, Tajikistan succeeded in reporting zero CFC consumption in its 2004 data (ImplCom 35, Recommendation 35/38 2005). On its commitment to phase out its consumption of the Annex E controlled substance (methyl bromide) by 1 January 2005, it reported zero consumption of methyl bromide, but it had not submitted data for 2005. After it was recommended to submit it by the ImplCom 36 as soon as possible (ImplCom 36, Recommendation 36/43 2006) and submitted it based on this recommendation, it was found to have achieved total phase-out of methyl bromide in that year (ImplCom 37, Recommendation 37/35 2006). Thus, it completed the implementation of all its commitments contained in Decision XIII/20 to return to compliance with the Protocol (see also MOP 29, Decision XXIX/14; ImplCom 57, Recommendation 57/1 2016; ImplCom 58, Recommendation 58/1 2017).

Turkmenistan. Through Decision XI/25 (MOP 11 1999), Turkmenistan was declared by the MOP 21 (2009) to be in non-compliance with its control obligations under the Protocol in 1996. In this Decision, it also committed to a series of specific time-based commitments (para. 3) and was cautioned with the application of the measures consistent with item C of the indicative list of measures when it failed to meet its obligations (para. 4). After 10 years of this decision, when it came to the year 2009, this caution was repeated with regard to the phase-out of carbon tetrachloride in Decision XXI/25, paras. 3–4 (MOP 21 2009), as Turkmenistan reported its annual consumption for carbon tetrachloride (Annex B, group II) for 2008 to be 0.3 ODP-tonnes, which showed a deviation from its maximum allowable consumption of carbon tetrachloride of zero ODP-tonnes for 2008. After its submission of its plan of action, explaining that the imported amounts of 0.3 ODP-tonnes in 2008 and 0.7 ODP-tonnes in 2009 were necessary for the analysis of mineral oil in water, it requested that its situation should be re-examined in the context of Decision XXI/6 on the global laboratory-use exemption (which allowed Art. 5 parties to deviate from existing laboratory and analytical use bans in individual cases until 31 December 2010, and extended global laboratory-use exemption until 31 December 2014). On the basis of this decision, in Recommendation 44/5 the ImplCom 44 (2010) recommended Turkmenistan's compliance situation in respect of laboratory and analytical use exemption to be re-examined after 31 December 2010. There is no other information about the result of this examination in the forthcoming reports of the ImplCom.

Uzbekistan. Although Uzbekistan managed to decrease its consumption from an estimated 1,300 tonnes in 1992 to 275 tonnes in 1996, as it reported positive consumption of 272 ODP-tonnes of Annex A and Annex B substances for 1996, it was found to be in non-compliance with its obligations under the Protocol for that year through Decision X/28 (MOP 10 1998). The 'caution and monitoring clause' located in para. 4 of similar decisions of the MOP was repeated for Uzbekistan in this decision as well. The specific commitments included in its country programme, such as establishing policy instruments and regulatory requirements, import quotas, bans on imports of ODSs and equipment using and containing ODSs to support the phase-out schedule in 1999, and reducing consumption of CFCs, carbon tetrachloride, methyl chloroform completely by 2002, were to continue being monitored by the parties in the upcoming years as well. The ImplCom requested the secretariat to alert the country to its state of potential non-compliance because of its failure to report data for 1999. So it was alerted because of its failure to report data for 1999 (ImplCom 26 2001) and reminded to report data for the year 2000 (ImplCom 27 2001).

In the year 2007, it submitted all its outstanding data in accordance with its data-reporting obligations under the Protocol and Decision XVIII/34 (para. 2) requesting Uzbekistan to report its 2005 ODSs data. Therefore, it was declared to be in compliance with the Protocol's control measures in 2005 (ImplCom 38, Recommendation 38/49 2007). It was also found to be in compliance with its obligations under Art. 4B of the Protocol in 2008 (with, among others, Kiribati, Niue, Sao Tome and Principe, and the United Republic of Tanzania), due to its establishment and operation of a licensing system for the import and export of controlled ODSs (ImplCom 40 2008: paras. 248, 252).

Developing Countries

The MP invokes the principle of differentiated responsibility, granting a ten-year period for developing parties to implement the necessary control measures set forth in the Protocol (Art. 5.1, MP). In line with this principle, it also states that developed countries should provide financial cooperation (Art. 10, MP) and transfer technology (Art. 10A, MP) to support the developing countries' compliance and implementation (Art. 5.5, MP). Furthermore, it obliges the parties to cooperate by taking into account the needs of developing countries, particularly with regard to promoting research, development and the exchange of information on the best technologies for improving the containment, recovery, recycling, or destruction of controlled substances, possible alternatives to controlled substances, and the costs and benefits of relevant control strategies (Art. 9.1, MP).

Thus, through the principle of 'common but differentiated responsibilities', developing countries have been granted some privileges by the MP, such as being allowed to implement the control measures under a "grace period" (Art. 5); benefiting from the MF for their incremental costs (Art. 10) and from other parties for transference of the best available substitutes and technologies (Art. 10A); and being evaluated on the basis of the effective implementation of Articles 10 and 10A (Art. 5, paras. 5–6).

In line with the same principle, they were assessed on the substantive obligations under the Protocol in 1999, at the end of the grace period. Therefore, in Decision XI/23, the MOP (1999) requested the ImplCom to conduct a full review of their data before the MOP 12 (2000) through gathering information on failures to comply with obligations for 1998 and 1999. However, this could not be achieved due to the fact that countries did not reply to requests for data, and so the ImplCom advised the MOP to urge parties to send their data no later than 30 September 2001, as it would begin to review data relating to the first year of compliance for Article 5 Parties in 2001 (ImpCom 25 2000: para. 29, 31).

Thus, from the year 2001, Article 5 parties also began to be assessed according to their compliance status, as being in non-compliance or potential non-compliance (see Annex F to the book), greatly augmenting the number of cases that needed to be examined by the Committee.

To illustrate, in 2001, Bangladesh, Chad, Comoros, Dominican Republic, Honduras, Kenya, Mongolia, Morocco, Niger, Nigeria, Oman, Papua New Guinea, Paraguay, Samoa and Solomon Islands, were found to be above their individual baselines. As they did not respond to the request from the secretariat for data for the control period from 1 July 1999 to 30 June 2000, all were presumed to be in potential non-compliance with the control measures under the Protocol (MOP 13, Decision XIII/16 2001).

In 2002, potential non-compliance was identified for Guatemala, Malta, Pakistan and Papua New Guinea (MOP 14, Decision XIV/17 2002). Vietnam was also found to be in non-compliance with its obligations under Article 2B of the MP, with respect to its consumption of Annex A, group II substances, which was above its baseline (MOP 15, Decision XV/45 2003).

In 2003, potential non-compliance was also determined for Dominica, Haiti, Saint Kitts and Nevis, and Sierra Leone because of their consumption of CFCs (MOP 15, Decision XV/21 2003); Malaysia, Mexico, Nigeria and Pakistan because of their consumption of halons (MOP 15, Decision XV/22 2003); Morocco because of its consumption of hydrobromofluorocarbons (MOP 15, Decision XV/23 2003); and Barbados, Egypt, Paraguay, Philippines, Saint Kitts and Nevis, and Thailand because of their consumption of methyl bromide (MOP 15, Decision XV/25 2003).

In 2004, through the MOP Decisions XVI/19–20, Somalia (MOP 16, Decision XVI/19 2004), due to its consumption of Annex A, group II, ODSs (halons) for both 2002 and 2003, was found to be in potential non-compliance status, as were Bangladesh, Bosnia and Herzegovina, Ecuador (ImplCom 50, Recommendation 50/1 2013) and the Islamic Republic of Iran (which had requested a change to its baseline data for methyl chloroform) (MOP 16, Decision XVI/20 2004) because of their consumption of the controlled substance in Annex B, group III (methyl chloroform) for 2003.

From this point on, only the outstanding cases will be discussed (in an alphabetical order).

Albania. Through its plan of action, Albania committed itself to gradually reducing its CFC consumption from 69 ODP-tonnes in 2001 to 2.2 ODP-tonnes in 2008 (phasing out by 2009), to establishing a system for licensing imports and exports of ODS, including quotas, and to banning imports of ODS-using equipment by 2004 (MOP 15, Decision XV/26 2003). Through Recommendation 32/4 Albania was requested to report its outstanding data from the period 1999–2002, and it reported data for CFC consumption and all the outstanding data in 2003 (ImplCom 32, Recommendation 32/4 2004). Therefore, under Recommendation 33/1, it was noted that Albania had made progress towards complying with its CFC phase-out commitments regarding submission of its outstanding data. However, it was expected to continue its efforts to establish an ODS import and export licensing and quota system and a ban on the import of ODS-using equipment (ImplCom 33, Recommendation 33/1 2004). It could not meet the 2004 deadline, but expected to have a licensing and quota system operational by June 2005 and an import ban by January 2006 (ImplCom 34, Recommendation 34/2 2005). Its establishment of an ozone-depleting substance licensing and quota system and its implementation of a ban on the import of equipment using ODSs, in accordance with its CFC plan of action contained in Decision XV/26, was confirmed by Recommendation 35/2 (ImplCom 35, Recommendation 35/2 2005). It reported its data for the consumption of the controlled substances in Annex A, group I (CFCs) in 2005, ahead of its commitment to reduce its consumption of those ODSs to 36.2 ODP-tonnes in that year, and also ahead of its CFC phase-out obligations under the MP for 2005 (ImplCom 36, Recommendation 36/1 2006). It received a warning for the year 2006 because of not submitting its report on its 2006 data, though it had committed to reduce its consumption of the Annex A, group I controlled substances (CFCs) to no more than 15.2 ODP-tonnes in 2006, as recorded in Decision XV/26 (Recommendation 38/51, 2007). However, it then managed to implement its commitment to reduce CFC consumption to no more than 15.2 ODP-tonnes ahead of its obligations under the CFC control measures of the MP in that year (ImplCom 39, Recommendation 39/1 2007). The relevant data for 2007, 2008 and 2009 were also all reported by Albania, and this data placed it in compliance with its commitment contained in Decision XV/26 (ImplCom 40, Recommendation 40/3 2008; ImplCom 41, Recommendation 41/2 2008; ImplCom 42, Recommendation 42/3 2009; ImplCom 44, Recommendation 44/1 2010).

Bosnia and Herzegovina. Bosnia and Herzegovina specifically committed itself to reducing its CFC consumption from 243.6 ODP-tonnes in 2002 to 3 ODP-tonnes in 2007 (phasing out by 2008) and its methyl bromide consumption from 11.8 ODP-tonnes in 2002 to 5.61 ODP-tonnes in 2005 and 2006 (phasing out by 2007); to establishing a system for licensing imports and exports of ODS, including quotas by 2004; and to banning imports of ODS-using equipment by 2006 (MOP 15, Decision XV/30 2003).

Bosnia and Herzegovina reported annual consumption of the controlled substances in Annex C, group I (HCFCs) of 5.13 ODP-tonnes in 2013, which exceeded the party's maximum allowable consumption of 4.7 ODP-tonnes for those controlled substances for that year. So it was found to be in non-compliance with the consumption control measures under the Protocol for HCFCs. It submitted an action plan and an explanation for its non-compliance, confirming that it had introduced a comprehensive set of measures necessary to ensure future compliance; and its submission of ODS data for 2014 showed that it was in compliance with its HCFC consumption obligations under the control measures of the Protocol. Therefore, it was decided that no further action was necessary in view of the party's return to compliance with the HCFC phase-out in 2014 and its implementation of regulatory and administrative measures to ensure compliance with the Protocol's control measures for HCFCs for subsequent years (MOP 27, Decision XXVII/10 2015).

Botswana, Gambia, South Sudan. Botswana was decided as being in non-compliance with its obligations under Article 2H of the MP for 2002, as while it s baseline for the controlled substance in Annex E was 0.1 ODP-tonnes, its reported consumption in Annex E in 2002 was 0.6 ODP-tonnes. Its plan of action included to reduce methyl bromide consumption from 0.6 ODP-tonnes in 2002 to 0.2 ODP-tonnes in 2004 (phasing out by 2005), and to establish a system for licensing imports and exports of methyl bromide, including quotas (MOP 15, Decision XV/26 2003).

In 2012, through Decision XXIV/17 (MOP 24 2012), as Botswana had still not become a party to the Montreal Amendment to the Protocol or established a licensing system, it was encouraged to ratify the amendment and to establish a licensing system. In the same decision, South Sudan, which had just ratified the Montreal Amendment to the Protocol, was also requested to report to the secretariat on the status of its licensing system by 30 September 2013. Another party, Gambia, having a licensing system just controlling the imports of ODSs, was urged to restructure its system to include export controls as well. Therefore, in its Recommendation 50/12 (2013), the ImplCom, while appreciating Gambia for its licensing system operating in accordance with its obligations under Article 4B of the Protocol, encouraged Botswana and South Sudan to establish a licensing system. It repeated its request to both countries on the establishment of licensing systems and the submission of the related information on those systems, as called for in Decision XXV/15 (MOP 25 2013), through Recommendation 52/5 as well (ImplCom 52 2014). According to Recommendation 51/6 (2013), the ImplCom also invited both countries to submit information on the status of that system to the secretariat in 2014. It also agreed to instruct South Sudan to report its ODS data for 2012 to the secretariat as a matter of urgency, in line with Article 7, for consideration by the Committee at its next meeting (ImplCom 52, Recommendation 52/1 2014; ImplCom 54, Recommendation 54/8 2015; ImplCom 55, Recommendation 55/5 2015).

While Botswana's licensing system had become operational in 2014 (ImplCom 54, Recommendation 54/7 2015), South Sudan established a licensing system by ministerial order, though it could not enact legislation due to the continuing internal conflict in the country (ImplCom 56, Recommendation 56/8 2016).

Cameroon. Cameroon was stated as being non-compliant with its obligations under Article 2B and also under Article 2H of the MP in 2002, since its baseline for Annex A, group II substances was 2.38 ODP-tonnes, and for the controlled substance in Annex E 18.09 ODP-tonnes, whereas its reported consumption was 9 ODP-tonnes for Annex A, group II substances, and 25.38 ODP-tonnes for the controlled substance in Annex E. Therefore, besides its plan of action on the control measures for Annex A, group II substances, it was also requested to submit a plan of action with time-specific benchmarks with respect to its consumption of the controlled substance in Annex E (MOP 15, Decision XV/32 2003). As it had reported 2004 consumption data for halons well below the level permitted under its plan of action, and the data showed that it had already reduced its halon consumption by over 50 per cent, resulting in its removal from the list, it was stated that Cameroon continued to be ahead of its commitments to phase out halons (ImplCom 34, Recommendation 34/10 2005).

Democratic Republic of the Congo, Guinea-Bissau and Saint Lucia, Dominica, Somalia and Yemen. The Democratic Republic of the Congo was also determined as being in non-compliance with its obligations under Article 2B of the MP in 2002, as while its baseline for Annex A, group II substances was 218.67 ODP-tonnes, it reported consumption of 492 ODP-tonnes in 2002. It was requested to submit a plan of action incorporating the establishment of import quotas to freeze imports at baseline levels and support the phase-out schedule, application of a ban on imports of ODS-using equipment, and adoption of policy and regulatory instruments that could ensure progress in achieving the phase-out (MOP 15, Decision XV/33 2003).

Together with Guinea-Bissau and Saint Lucia, it justified the request for the revision of its consumption data for HCFCs for 2009, 2010 or both, which are part of the baseline for parties operating under Art. 5, para. 1. So the baseline HCFC consumption data of these countries for the respective years were revised in line with their request (MOP 25, Decision XXV/13 2013; ImplCom 50, Recommendation 50/3, Recommendation 50/4, Recommendation 50/5 2013; ImplCom 51, Recommendation 51/3 2013).

In Decision XXVII/9 (MOP 27 2015), the MOP stressed that the Democratic Republic of the Congo, together with Dominica, Somalia and Yemen, had not reported their 2014 data so they should report the required data as a matter of urgency. Among those, only Yemen did not report on time, mostly because of the political and security situation prevailing in the country. So Yemen was requested to report its data for 2014 as soon as possible for consideration by the Committee (ImplCom 56, Recommendation 56/1 2016).

Democratic People's Republic of Korea. The Democratic People's Republic of Korea was found to be in non-compliance with the consumption control measures under the Protocol for HCFCs due to its annual consumption of 90.6 ODP-tonnes, which exceeded its maximum allowable consumption of 78.0 ODP-tonnes for the controlled substances in Annex C, group I (HCFCs) for 2013, and also in non-compliance with the production control measures under the Protocol for HCFCs, due to its annual production of 31.8 ODP-tonnes in 2013, which exceeded its maximum allowable production of 27.6 ODP-tonnes for those controlled substances for that year.

When the MOP report was prepared, the republic had already submitted a plan of action to ensure its return to compliance with the Protocol's HCFC consumption control measures in 2015 and production control measures in 2016, and had made a commitment to reduce both its consumption and production of them. Appreciation for this was expressed at the MOP 26, and the Democratic People's Republic of Korea was also reminded that it should work with the relevant implementing agencies to implement its action plan, that its progress would be closely monitored, and that if necessary it would be faced with a cease in the supply of the HCFCs that were the subject of non-compliance, in accordance with item C of the indicative list of measures (MOP 26, Decision XXVI/15 2014; ImplCom 53, Recommendation 53/2 2014).

In 2015, the Committee noted that the Democratic People's Republic of Korea had submitted its data for 2014 in accordance with their obligations under Art. 7 of the MP and that the data indicated that it was in compliance with its commitments for that year (ImplCom 54, Recommendation 54/1 2015).

Ecuador. Ecuador, which was regarded as being in potential non-compliance with the control measures under the Protocol in 2003 (MOP 16, Decision XVI/20 2004), and also with the control measures for methyl bromide under the Protocol for methyl bromide in 2008 (MOP 20, Decision XX/16 2008), was recorded in Decision XX/16 as being in compliance with its commitment to limit its consumption of methyl bromide to no more than 52.8 ODP-tonnes in that year, according to its reporting data on its consumption of ODSs for 2011 (ImplCom 49, Recommendation 49/3 2012). Ecuador's data on consumption of ODSs for 2012 which showed that it was in compliance with its commitment (recorded in Decision XX/16) to limit its consumption of methyl bromide to no more than 52.8 ODP-tonnes in that year (ImplCom 51, Recommendation 51/1 2013). In 2015, in Recommendation 54/1, the Committee noted that Ecuador had submitted its data for 2014 in line with its obligations under Art. 7 of the MP and that the data indicated that it was in compliance with its commitments for that year (ImplCom 54, Recommendation 54/1 2015).

Ecuador's data for 2014, which was submitted on time, indicated that it was in compliance with its commitments for that year (ImplCom 54, Recommendation 54/1 2015).

Guatemala. Guatemala's reported consumption of Annex A, group I substances in 2002 and of the controlled substance in Annex E exceeded its baseline for these substances, so for 2002, Guatemala was found in non-compliance with its obligations under Articles 2A and 2H of the MP. However, its plan of action, under which it specifically committed itself to phase out CFC consumption by 1 January 2010 and methyl bromide consumption by 1 January 2015, was appreciated (MOP 15, Decision XV/34 2003).

In 2014, it was found to be in non-compliance with the consumption control measures under the Protocol for HCFCs because its consumption of 11.3 ODP-tonnes of the controlled substances in Annex C, group I (HCFCs) in 2013 exceeded its maximum allowable consumption of 8.3 ODP-tonnes. In line with its action plan, it had already made a commitment to reduce its HCFC consumption in 2014 below its allowable consumption by the excess amount consumed in 2013. Additionally, it was also urged to work with the relevant implementing agencies to implement its action plan to phase out consumption of HCFCs, and was reminded that its progress would be monitored closely and that it could encounter measures consistent with item C of the indicative list of measures (MOP 26, Decision XXVI/16 2014; ImplCom 53, Recommendation 53/5 2014).

In 2015, under Recommendation 54/1, the Committee also mentioned that Guatemala submitted its data for 2014 in line with its obligations under Art. 7 of the

MP and that the data indicated that it was in compliance with its commitments for that year (ImplCom 54, Recommendation 54/1 2015).

Guatemala reported that its annual consumption for the controlled substances in Annex C, group I (HCFCs) was 4.74 ODP-tonnes in 2014. This was inconsistent with its commitment, set out in Decision XXVI/16, to reduce consumption of HCFCs to no more than 4.35 ODP-tonnes in that year. So it was in non-compliance with the consumption control measures for that substance under the Protocol for that year. However, it corrected its HCFC consumption to 9.84 ODP-tonnes in 2013 and 4.74 ODP-tonnes in 2014, attributing the previous incorrect data to a technical error in computing the consumption of that substance in the country for those two years. Although it revised its 2013 data, it remained in non-compliance with its HCFC consumption obligations under the Protocol for 2013.

Guatemala's reported data for 2015 also indicated that it had already returned to compliance with the Protocol's HCFC control measures. Accordingly, it was decided that Guatemala should work with the relevant implementing agencies to complete the implementation process of its action plan and to continue to receive international assistance to meeting those commitments in accordance with item A of the indicative list of measures that may be taken by the MOP in respect of non-compliance (ImplCom 56, Recommendation 56/6 2016; MOP 28, Decision XXVIII/11 2016).

Libya. Libya was found to be in potential non-compliance, due to exceeding its 2009 maximum allowable consumption of zero ODP-tonnes of the controlled substances in Annex A, group II (halons), after reporting an annual consumption of 1.8 ODP-tonnes of them (MOP 23, Decision XXIII/23 2011). But, after only one year, it was in compliance with its commitments, recorded in Decisions XV/36 and XVII/37, to phase out its chlorofluorocarbon and methyl bromide consumption, according to its explanation of its halon consumption in 2009 and the submission of its ozone-depleting substance data for 2010 and 2011 (ImplCom 49, Recommendation 49/4 2012).

Along with Mozambique, Libya also submitted information in support of their request for the revision of the consumption data for HCFCs for 2010 and 2009, which are part of the baseline for parties operating under Art. 5, para. 1 (MOP 26, Decision XXVI/14 2014; ImplCom 50, Recommendation 50/6 2013; ImplCom 51, Recommendation 51/4 2013; ImplCom 52, Recommendation 52/3 2014).

In Recommendation 54/5, the Committee noted its concern over Libya's report, which was inconsistent with the Protocol's requirement to limit consumption of those substances to no more than 118.38 ODP-tonnes in those years. Additionally, it asked Libya to submit an updated plan of action with time-specific benchmarks for ensuring its prompt return to compliance (ImplCom 54, Recommendation 54/5 2015).

Because the annual consumption reported by Libya of the controlled substances in Annex C, group I (HCFCs), of 144.0 ODP-tonnes for 2013 and 122.4 ODP-tonnes for 2014 exceeded its maximum allowable consumption, it was declared to be in non-compliance with the consumption control measures under the

Protocol for HCFCs in 2015. Therefore, it was asked to work with the relevant implementing agencies to implement its plan of action to phase out the consumption of HCFCs. It was also warned that it could face measures consistent with item C of the indicative list of measures. Finally, as in all other decisions regarding non-compliance, it was stressed that if Libya demonstrably worked towards and met the specified Protocol control measures, it would continue to be treated in the same manner as a party in good standing and should continue to receive international assistance to meet those commitments in accordance with item A of the indicative list of measures (MOP 27, Decision XXVII/11 2015).

In more recent reports of the ImplCom, the Committee reiterated its concerns over Libya's reporting, showing that its consumption of the Annex C, group I controlled substances (HCFCs) of 144.0 ODP-tonnes in 2013 and 122.4 ODP-tonnes in 2014 were inconsistent with the Protocol's requirement of maximum annual consumption of no more than 118.38 ODP-tonnes of those substances in those years. Even if Libya submitted an explanation for its excess consumption in those years, and a plan of action for returning to compliance with the Protocol's control measures for HCFCs in 2018, it was urged to establish and implement a national import quota system for HCFCs as soon as possible (ImplCom 55, Recommendation 55/2 2015). Because it did not submit its data for 2015 on time, it was also instructed to report its ODSs data for 2015 to the secretariat as soon as possible, and no later than 15 September 2016 (ImplCom 56, Recommendation 56/3 2016).

People's Republic of China. China was declared to be in potential non-compliance due to the fact that its reported annual consumption of the controlled substances in Annex B, group I (other CFCs) for 2004 exceeded its maximum allowable consumption level for those controlled substances for that year (MOP 17, Decision XVII/30 2005). So it was expected to submit an explanation for the apparent deviation from its Protocol obligations (ImplCom 35, Recommendation 35/9 2005). In response, China disagreed with the designation as a Party in non-compliance, explaining that it had a different understanding of its baseline (ImplCom 36, Recommendation 36/10 2006). It was then agreed that the secretariat should report and review ODSs data submitted by the parties to one decimal place only, but it had been reported to and reviewed by the secretariat to three decimal places. According to this new calculation, China appeared to be in compliance with the Protocol's consumption control measures for other CFCs in 2004 (ImplCom 38, Recommendation 38/9 2007).

After that, there is no information on China in the reports regarding the consideration of non-compliance issues, but there is follow-up information on multiple issues such as production of ODSs, methyl bromide, HCFC, reporting on process-agent uses, phase-out projects and accounting reports for exemptions for essential uses of chlorofluorocarbons (CFCs) (ImplCom 49, para. 24; ImplCom 50, paras. 15, 23, 25; Implcom 51, para. 12; Implcom52, para. 14; ImplCom 53, paras. 16, 21, 23; ImplCom 54, paras. 16, 29; ImplCom 55, para. 14; ImplCom 56, para. 17, 33; MOP 29, Decision XXIX/5 2017).

Plurinational State of Bolivia. After Bolivia had been identified in Decision XIV/20 as being in non-compliance with the CFC consumption freeze for the control period 1 July 2000–30 June 2001, and had been asked to submit a plan of action to return to compliance (MOP 14, Decision XIV/20 2002), it promised to reduce its CFC consumption from 65.5 ODP-tonnes in 2002 to 11.35 ODP-tonnes in 2007 (phasing out by 2010), to monitor its system for licensing imports and exports of ODSs, including quotas, introduced in 2003 and its ban on imports of ODS-using equipment, introduced in 1997 for CFC-12 and extended to other ODSs in 2003 (MOP 15, Decision XV/29 2003). In the 33rd meeting of the ImplCom (ImplCom 33 2004: para. 44) it was confirmed that its reported data for CFC consumption in 2003 met its benchmark. The next year, it was ahead of its commitments to phase out CFC, as contained in decision XV/29 and prescribed under the Protocol (ImplCom 34, Recommendation 34/7 2005; ImplCom, Recommendation 35/5 2005). In the subsequent years, there has been some criticism due to delays in submitting the necessary data (ImplCom 36, Recommendation 36/6 2006; ImplCom 38, Recommendation 38/5 2007; ImplCom 46, Recommendation 46/2 2011; ImplCom 47, Recommendation 47/2 2011) and concerns over the inconsistency of the reported consumption of 0.1 ODP-tonnes of carbon tetrachloride in 2006 with the Protocol's requirement (Recommendation 39/5). However, its reported data for the consumption of the Annex A, group I controlled substances (CFCs) in advance of its commitment was appreciated, and it received recognition for remaining in advance of the CFC control measures of the MP and its report on updated disaggregated information on the establishment and operation of licensing systems in accordance with Article 4B of the MP (ImplCom 37, Recommendation 37/4 2006; ImplCom 40, Recommendation 40/8 2008; ImplCom 48, Recommendation 48/14 2012).

South Africa. South Africa reported zero consumption of the Annex B, group III controlled substance (methyl chloroform) for the years 2011–2013. However, it subsequently reported revised data on its consumption of methyl chloroform of 8.1 ODP-tonnes in 2011 and 3.6 ODP-tonnes in 2012, amounts that were inconsistent with the Protocol's requirement. However, its reported 2013 consumption data showed that the party was in compliance with its methyl chloroform consumption obligations under the control measures of the Protocol for that year. Therefore, it was asked by the Committee to submit an explanation for its excess consumption in previous years and, if appropriate, a plan of action with time-specific benchmarks for ensuring the party's prompt return to compliance (ImplCom 54, Recommendation 54/6 2015). Then, as South Africa declared that the apparent excess consumption was due to transcription errors in respect of customs codes when recording imports and that the correct figures for 2011 and 2012 should have been zero in each year, the Committee indicated that the party was in compliance with its methyl chloroform consumption obligations under the Protocol (ImplCom 55 2015: paras. 46–47).

Uruguay. In 2002, Uruguay was also amongst the developing parties found in non-compliance with its obligations under Article 2H of the Montreal Protocol, due to its consumption of the controlled substance in Annex E exceeding its baseline.

However, its plan of action, including commitments like phasing out methyl bromide consumption by 1 January 2005 and monitoring its system for licensing imports and exports of ODS, including quotas, was found promising for its future consumption (MOP 15, Decision XV/44 2003).

Uruguay's reported 2012 data confirmed that it had met its commitment contained in Decision XVII/39 (MOP 17 2005). All other parties with similar commitments, namely, Armenia (Decision XVII/28), Bosnia and Herzegovina (Decision XV/30), Fiji (Decision XVII/33), Guatemala (XVIII/26) and Honduras (Decision XVII/34), were considered under the same agenda item. The Committee agreed to commend them on their reported consumption of ODSs, which showed that they were in compliance with the commitments specified in the decisions applicable to them (ImplCom 50, Recommendation 50/2 2013).

Developed Countries

Israel. In Decision XV/24 (MOP 15 2003), Israel was found to be in potential non-compliance because of its excess consumption of the controlled substance in Annex E (methyl bromide) in 2002. Israel reported its annual data for 2002 to be above its requirement for a 50% reduction in consumption of the controlled substance in Annex E (methyl bromide), and it was considered to be in non-compliance with the control measures under the Protocol. Therefore, it was asked to submit an explanation for its excess consumption to the ImplCom as soon as possible, along with a plan of action which might involve the establishment of import quotas and policy and regulatory instruments. To the ImplCom report (ImplCom 32 2004), Israel subsequently submitted data for 2003 showing it to be in compliance and also clarified the 2002 situation, explaining that the excess consumption stemmed from one company's export of a quantity of methyl bromide to a non-party and the inclusion of this amount in the total consumption. The report also underlined that, after ratifying the Montreal Amendment – to which Israel was as not a party in 2002 – Israel was in no way allowed to export methyl bromide to non-parties to the Copenhagen Amendment.

In para. 5 of Decision XV/24 on Israel (MOP 15 2003) it is possible to see a 'monitoring and caution clause', which has become usual in the decisions of the MOP on non-compliance. In accordance with this paragraph, Israel, like other countries, was cautioned that if it failed to meet its time-specific benchmark commitments in its plan of action and thus failed to return to compliance in a timely manner, it could face measures consistent with item C of the indicative list of measures. However, unlike similar clauses relating to other parties, there is no statement in Decision XV/24 about Israel receiving international assistance if it continued to be treated in the same manner as a party in good standing, mostly because it had been defined as a developed country.

With regard to the issue of reporting on the stockpiling of ODSs, in Decision XVIII/17 (MOP 18 2006) the MOP requested the secretariat to maintain a consolidated record of cases in which parties had explained that excess consumption or production of ODSs in a given year had been the result of stockpiling of those substances under the circumstances described in the Decision. In Decision XXII/20

(MOP 20 2008), the MOP stated that no follow-up action from the ImplCom would be necessary in respect of any such consumption or production if the reporting party specified that it had in place the measures necessary to prohibit the use of the stockpiled ODSs for any purpose other than those described in Decision XVIII/17.

In accordance with Decision XXII/20, four parties – the European Union, France, Israel and the United States – had reported excess consumption or production of ODSs in 2012 under the circumstances described in Decision XVIII/17. Of those, the European Union, France and the United States had confirmed that they had in place measures to ensure that the substances were not used for purposes other than those specified in Decision XVIII/17. However, the fourth party, Israel, had not reported that it had in place the measures required by Decision XXII/20. The ImplCom then requested Israel to submit the outstanding information urgently (ImplCom 52, Recommendation 52/4 2014).

Regarding Israel's reporting of any data on its process agent uses, it should be first be stated that the reporting of process agent uses had been regulated by three MOP decisions:

- Decision X/14, defining process agent uses and permitted applications;
- Decision XVII/6, specifying that annual reporting of process agent uses was obligatory;
- Decision XXI/3, requesting the secretariat to bring non-reporting cases to the attention of the ImplCom for its consideration.

Because Israel failed to report any data on its process agent uses for 2010 and 2011, in 2013 the ImplCom agreed to ask Israel to submit, as a matter of urgency, the required outstanding information on process agent uses for 2010 and 2011 (ImplCom 50, Recommendation 50/10 2013). Israel also failed to report on its use of controlled substances as process agents in 2014. Therefore, in accordance with Decisions X/14 and XXI/3 mentioned above and also Decision XXIII/7 allowing Israel to use controlled substances as process agents, the Committee asked Israel to submit the outstanding information to the secretariat immediately (ImplCom 55, Recommendation 55/4 2015; ImplCom 56, Recommendation 56/5 2016).

Israel's failure to report the required information resulted in its non-compliance with its reporting obligations in 2016. Due to the fact that it had not provided the information on the measures that it had in place to avoid the diversion to unauthorized uses of the 17.3 ODP-tonnes of excess production of bromochloromethane stockpiled in 2014, and due also to its repeated failure to respond to the requests for information recorded in Recommendations 55/4, 56/5 and 56/7 of the ImplCom, it was asked to submit the related information on its use of controlled substances as process agents in 2014 and 2015, and on the measures it had put in place to avoid the diversion to unauthorized uses of the 17.3 ODP-tonnes of excess production of bromochloromethane stockpiled in 2014, to the secretariat as soon as possible, and no later than 31 March 2017 (MOP 28, Decision XXVIII/10 2016). In ImplCom 58 (para. 50), it was noted that Israel "express[es] commitment to ensure that it would in the future fully comply with its reporting requirements under the Protocol in a timely manner."

European Union. The European Union's case should also be noted here, as it was stated as being in non-compliance with Art. 4, MP, prohibiting trade with any state not party to the Protocol, due to its reporting the export of 16.616 metric tonnes of Annex C, group I HFCs in 2009 to a state classified as not operating under Art. 5, para. 1, of the MP that was also a state not party to the Copenhagen Amendment to the Protocol in that year (MOP 23, Decision XXIII/26 2011). Within the decision of the MOP on the case, it was stated that no further action was required on the basis of the party's implementation of regulatory and administrative measures to ensure its compliance. However, it was still decided to monitor its progress closely regarding the implementation of its obligations under the MP (MOP 23, Decision XXIII/26 2011).

France. The report of the MOP 25 found that France had reported the production in 2011 of 598.9 ODP-tonnes of the controlled substances in Annex C, group I (HCFCs), exceeding its maximum allowable production of 584.4 ODP-tonnes for those controlled substances for that year. Therefore, it was stated as being in non-compliance with the production control measures under the Protocol for HCFCs. But after that, through its action plan confirming its compliance with the Protocol's HCFC production control measures for 2012 and subsequent years, it was decided just to monitor its progress concerning the phase-out of HCFCs closely; to treat it in the same manner as a party in good standing as long as it worked towards meeting the specific Protocol control measures; and to warn it that it could face the actions available under Art. 4 if it failed to return to compliance in a timely manner (MOP 25, Decision XXV/11 2013; ImplCom 50, Recommendation 50/9 2013).

5.1.3   Lessons Learned: Some Tentative Conclusions

Based on the findings of the previous section on the functioning of the mechanism, it can be concluded that the first type of triggering – by another party to the Protocol – has not arisen to date, while the second type – self-triggering – was used by CEIT in the 1990s. But, the most used is the third type, triggering by the secretariat. Identifying the parties' situation as actual or potential non-compliance, the MOPs have usually recommended them to provide a detailed plan of action for phasing out ozone-depleting substances; to prepare a national programme for the phase-out of ODSs; to expedite the process of ratification of the necessary amendments to the Protocol (such as the London Amendment); to benefit from the GEF for the projects to be submitted; and to continue to provide the Committee with reports on progress made in phasing out ODSs in line with the schedule in their national programmes.

With respect to reporting issues, data on the parties' reporting explicitly indicate a general tendency towards compliance on reporting under Art. 7 by nearly all the

parties of the Protocol from its initial years onwards. In fact, according to the report
of 1999, of the 175 parties required to report, 152 parties (111 parties under article
5, 41 under non-article 5) in total reported data (MOP 13 2001), while in 1992, of
the 88 parties required to report, 59 parties (26 under article 5, 33 under non-article
5) and, in 1991, 68 parties (33 under Article 5, 35 under non-article 5) reported their
data (MOP 6 1994). According to the report of the secretariat on the reporting of
data by the parties presented for MOP 22 (2010), for the period 1986–2008, all
parties were in full compliance with their data-reporting obligations under Art. 7(3)
(ImplCom 45 2010; MOP 22, Decision XXII/14 2010). The total number of parties
to have reported their 2009 data under Art. 7(3) is 167 (123 parties operating under
Art. 5(1) and 44 not so operating) (MOP 22, Addendum 2010). This means that, of
the 196 parties required to report data, 167 of them had reported their data (at the
time of the preparation of the Implcom 45 report, 178 parties to the MOP 22,
Decision XXII/14, 196 parties of the 196 had reported). In addition, all parties
(including San Marino which had not submitted its base-year on 3 September 2010,
when the secretariat's report was presented) had submitted their base-year data as
required, fulfilling their data-reporting obligations under Art. 7(1, 2) (MOP 22,
Addendum 2010). It later became clear that 194 parties of the 196 that should have
reported data for 2011 had reported their data, the only exceptions being Mali, and
Sao Tome and Principe (MOP 23, Decision XXIV/13 2012). Only three parties
failed to report their data for 2012: Eritrea, South Sudan and Yemen (MOP 25,
Decision XXV/14 2013); for 2013, just the Central African Republic (MOP 26,
Decision XXVI/12 2014); for 2014, the Democratic Republic of Congo, Dominica,
Somalia and Yemen (MOP 27, Decision XXVII/9 2015); and for 2015, Iceland and
Yemen (MOP 28, Decision XXVIII/9 2016); all other parties did report their data
for those years in a timely fashion. All 197 parties reported their data for 2016; 180
of those parties had provided their reports by 30 September 2017, as required under
MP, Art. 7 (para. 3), and 130 of them had fulfilled their obligations by 30 June
2017, in accordance with Decision XV/15 (MOP 29, Decision XXIX/13 2017).

With respect to reporting, it should also be underlined that although there has
been an increase in the number of parties complying the requirements of the
reporting process, the situation is completely different for reporting under Art. 9. On
reporting under Art. 9, the Ozone Secretariat's website presents a list of the publications which parties have provided under Art. 9 to the Ozone Secretariat.[13] From
this list, it can be seen that the reporting under Art. 9 has not been carried on in a
regular manner, "probably due to the repetitive nature of the reports and because the
activities of the Multilateral Fund have fulfilled the needs of the Article 5 Parties for
information" (Madhava 2005: 306). This situation is also confirmed by the reports
of the COP/MOP. According to the reports of the COP/MOP (COP 7 2005; MOP
17 2005: 24, para. 181, MOP 22, Decision XVII/24 2010: 56), the number of
reports submitted by parties under Art. 9 had fallen over recent years. According to

---

[13]See Table on Article 9 reports submitted, at: http://ozone.unep.org/data_reporting/research_
development_public_awareness_and%20_information_exchange.shtml.

another report of the COP/MOP (COP 8 2008; MOP 20 2008: 17–18, para. 122; MOP 20, Decision XX/13 2008: 43), although it is a legal obligation to submit such a report every two years, the number of parties submitting them was relatively small; only 18 parties, most of which were Art. 5 parties, submitted such data in 2007 and 2008. This is particularly because of the informal understanding adopted by the ImplCom, indicating that this kind of reporting has become less crucial over time as its requirements have been "fulfilled by the regular reports from the Assessment Panels and WMO" (ImplCom 29 2002: para. 39).

The findings stemming from the analysis on the different non-compliance cases brought before MOP and specific decisions/recommendations given on the parties on non-compliance-related issues show that many parties, both developed and developing, have met their phase-out targets in line with their schedules. Thus, a high level of compliance rate on the commitments of the parties, contributing to a decline in the production and consumption of ODSs, has been achieved. Indeed, in the MOP 27, it is remarkably expressed that the parties have made significant progress in complying with their obligations to phase out ODSs, so it was necessary to prepare just two draft decisions dealing with cases of non-compliance (MOP 27 2015: para. 91). Regarding 2015 phase-out targets with a 10 per cent step-down target for Article 5 parties and a 90 per cent target for non-Article 5 parties, the same level of compliance has been attained, with only a few cases of non-compliance (MOP 28 2016: para. 97).

Overall, the findings stemming from the analysis of different non-compliance cases brought before the MOP and specific decisions/recommendations given on the parties on non-compliance-related issues indicate that many parties, both developed and developing, have met their phase-out targets in line with their schedules. Thus, a high level of compliance rate on the commitments of the parties, contributing to a decline in the production and consumption of ODSs, has been achieved to date, acknowledging the success of the MP in meeting its goals in phasing out ODSs. Indeed, the Protocol has achieved the phase-out of some 98 per cent (about 1.8 million ODP-tonnes or 2.5 million metric tonnes) of the production and consumption of 96 ozone-depleting substances globally, 2 per cent of which are mainly HCFCs (about 32,000 ODP-tonnes or 500,000 metric tonnes) (Ozone Secretariat 2015). According to the report of the MOP 27 (2015), due to the progress of the parties in complying with their obligations to phase out ODSs under the Protocol, just two decisions dealing with cases of non-compliance – by Bosnia and Herzegovina and Libya – were prepared by the Committee.

As a consequence, arguing that binding decisions/measures advance the success of the compliance mechanisms becomes controversial. The findings also indicate that measures applied against non-compliance are not the only factors that move a party forward towards the compliant behaviour. There are several other factors: the party's characteristics (social-economic conditions), the legitimacy of international norms, their implementation and adoption in the national systems etc. So the non-compliant party can resist being in non-compliance despite the existence of legally binding response measures, because of the other factors triggering the party towards non-compliance. Even if it has legally binding measures, there is no enforcement mechanism under the CM of the MP, likewise in other CMs, and also in traditional means of settling disputes under international law. That is, there are no provisions which can enforce the non-compliant party to comply with the consequences adopted within the mechanism. So the effectiveness of the responses in particular, and the success of the CM in general, depends to a large extent on the non-compliant party and its tendency towards complying rather than continuing to be non-compliant (Barrett 2003). The success of the MP shows that the parties of the Protocol tend to be in compliance rather than non-compliance with their commitments under the Protocol.

In brief, the CM of the MP incorporates a great number of lessons learned and practices experienced since its inception more than twenty years ago. It has, indeed, quite a successful track record in compliance matters. Therefore it has the potential to be an exemplar for other treaties' compliance mechanisms concerning ways to ensure and strengthen the compliance of the parties with their commitments. However, the success and high level of compliance under the Protocol need to be sustained to cope with the problems posed by the efforts to achieve the total global phase-out of ODSs. It is therefore necessary to continue to focus on ways to maintain and promote the current system, protect its strengths and overcome its weaknesses.

Annexes

A. The List of COPs to the Vienna Convention and MOPs to the Montreal Protocol.
B. The List of MOPs' Decisions Related to Compliance.
C. Summary on Implcom Reports on Compliance of the Parties to the Montreal Protocol.
D. MOP Decisions on Compliance of the Related Parties to the Montreal Protocol.
E. MOP Decisions on Compliance of the Countries With Economies in Transition (CEIT) to the Montreal Protocol.
F. MOP Decisions on Compliance of the Developing Country Parties to the Montreal Protocol.

References

Bafundo, N.E. (2006). Compliance with the Ozone Treaty. Weak States and the Principle of Common but Differentiated Responsibility. *American University International Law Review*, 21, 462–495.
Brack D. (2003). Monitoring the Montreal Protocol. *Verification Yearbook*, 209–226. http://www.vertic.org/media/Archived_Publications/Yearbooks/2003/VY03_Brack.pdf.

Chasek, P.S., Downie, D.L. and Brown, J.W. (2006). *Global Environmental Politics*. Boulder, Colo.: Westview Press.

Downes, D. and Penhoet, B. (1999). Effective Dispute Resolution, A Review of Options, For Dispute Resolution Mechanisms and Procedures. *Center for International Environmental Law (CIEL)*, Washington, D.C. Available at: http://www.ciel.org/Publications/effectivedisputere-solution.pdf.

Ehrmann, M. (2002). Procedures of Compliance Control in International Environmental Treaties. *Colorado Journal of International Environmental Law and Policy*, 13, 2, 377–444.

Faure, M.G. and Lefevere, J. (1999). Compliance with International Environmental Agreements. Vig, N.J. and Axelrod, R.S. (Eds.), *The Global Environment: Institutions, Law and Policy* (138–156). Washington: CQ Press.

Handl, G. (1997). Compliance Control Mechanisms and International Environmental Obligations. *Tulane Journal of International and Comparative Law*, 5, 29–51.

Jenkins, L. (1996). Trade Sanctions: Effective Enforcement Tools. Cameron, J., Werksman, J. and Roderick, P. (Eds.), *Improving Compliance with International Environmental Law* (221–229). London: Earthscan.

Maljean-Dubois, S. and Richard, V. (2004). Mechanisms for Monitoring and Implementation of International Environmental Protection Agreements. Available at: http://halshs.archives-ouvertes.fr/docs/00/42/64/17/PDF/id_0409bis_maljeandubois_richard_eng.pdf.

Marauhn, T. (1996). Towards a Procedural Law of Compliance Control in International Environmental Relations, *Max-Planck-Institut für ausländisches öffentliches Recht und Völkerrecht*, 696–731. Available at: http://www.zaoerv.de/56_1996/56_1996_3_a_696_731.pdf.

Oberthür S. (2001). Linkages between the Montreal and Kyoto Protocols, Enhancing Synergies between Protecting the Ozone Layer and the Global Climate. *International Environmental Agreements: Politics, Law and Economics*, 1, 3, 357–377.

Pineschi, L. (2004). Non-compliance Mechanisms and the Proposed Center for the Prevention and Management of Environmental Disputes. Available at: https://dadun.unav.edu/bitstream/10171/22204/1/ADI_XX_2004_05.pdf.

Potzold, C. (2009). Multilateral Environmental Agreements: Contributions to Global Environmental Governance. Are Non-Compliance Mechanisms a Viable Alternative to the More Traditional Form of International Dispute Settlement for the Resolution of Global Environmental Problems? Society for Environmental Law and Economics, UBC Conference Paper March 2009.

Raustiala, K. (2000). Compliance and Effectiveness in International Regulatory Cooperation. *Case Western Reserve Journal of International Law*, 3, 2, 387–440.

Romanin Jacur, F.R. (2009). The Non-Compliance Procedure of the 1987 Montreal Protocol to the 1985 Vienna Convention on Substances that Deplete the Ozone Layer. Treves, T., Tanzi, A., Pineschi, L., Pitea, C., Ragni, C. (Eds.), *Non-Compliance Procedures and Mechanisms and the Effectiveness of International Environmental Agreements* (11–33). The Hague: T.M.C. Asser Press.

Sands, P. and MacKenzie, R. (2000). Guidelines for Negotiating and Drafting Dispute Settlement Clauses for International Environmental Agreement. *The PCA/Peace Palace Papers. Foundation for International Environmental Law and Development (FIELD)*, University of London. Available at: https://pca-cpa.org/wp-content/uploads/sites/6/2016/01/Guidelines-for-Negotiating-and-Drafting-Dispute-Settlement-Clauses-for-International-Environmental-Agreements.pdf.

Savaşan, Z. (2012). Climate Change, Compliance and the Role of Reporting: A Comparative Analysis of Reporting under the Montreal and Kyoto Protocols. *World Congress on Water, Climate and Energy 2012, Conference Proceedings*, 13–18 May 2012. Dublin, Ireland. Available at: http://keynote.conference-services.net/resources/444/2653/pdf/IWAWCE2012_0737.pdf.

Széll, P. (1995). The Development of Multilateral Mechanisms for Monitoring Compliance. Lang, W.(Ed.), *Sustainable Development and International Law* (97–114). London: Springer.

Victor, D.G. (1996). The Early Operation and Effectiveness of the Montreal Protocol's Non-Compliance Procedure. Available at: https://pages.ucsd.edu/~dgvictor/publications/Victor_Article_1996_Early%20Operation%20and%20Effectiveness.pdf.

Wang, X. and Wiser, G. (2002). The Implementation and Compliance Regimes under the Climate Change Convention and Its Kyoto Protocol. *Review of European Community and International Environmental Law*, 11, 2, 181–198.

Weiss, E.B. (1998). The Five International Treaties: A Living Listory. Weiss, E.B. and Jacobson, H.K. (Eds.), *Engaging Countries: Strengthening Compliance with International Environmental Accords* (89–172), Cambridge, Mass.: MIT Press.

Werksman, J.D. (1996). Compliance and Transition: Russia's Non-Compliance Tests the Ozone Regime. *Zeitschrift für ausländisches öffentliches Recht und Völkerecht (ZaoRV)*, 56, 3, 750–773. Available at: http://www.zaoerv.de/56_1996/56_1996_3_a_750_773.pdf.

Werksman, J.D. (1999). Procedural and Institutional Aspects of the Emerging Climate Change Regime: Improvised Procedures and Impoverished Rules? Available at: http://www.cserge.ucl.ac.uk/Werksman.pdf.

Wolfrum, R. (1999). *Recueil des cours*: *Collected Courses of the Hague Academy of International Law*, Vol. 272 (1998). The Hague, Boston, London: Martinus Nijhoff Publishers.

Wolfrum, R. and Matz, N. (2003). *Conflicts in International Environmental Law*. Berlin, Heidelberg: Springer.

Official Documents Related to CM Under the Montreal Protocol

Montreal Protocol on Substances that Deplete the Ozone Layer (1987). Available at: https://treaties.un.org/doc/Treaties/1989/01/19890101%2003-25%20AM/Ch_XXVII_02_ap.pdf.

Ozone Secretariat (2007). Implementation Committee Primer, October 2007. Available at: http://42functions.net/Publications/ImpCom_Primer_for_parties.pdf.

Procedures and Mechanisms relating to Compliance under the Montreal Protocol (NCP) (2006). *Handbook for the Montreal Protocol on Substances that Deplete the Ozone Layer* (419–421). Nairobi: UNEP.

Rules of Procedure of the Compliance Committee of the Montreal Protocol (RoP) (2006). *Handbook for the Montreal Protocol on Substances that Deplete the Ozone Layer* (455–468). Nairobi: UNEP.

Vienna Convention for the Protection of the Ozone Layer (1985). Available at: https://treaties.un.org/doc/Publication/MTDSG/Volume%20II/Chapter%20XXVII/XXVII-2.en.pdf.

Reports of the COP

COP 1 (1989). Helsinki, 26–28 April 1989. UNEP/OzL.Conv.1/5. Available at: http://ozone.unep.org/meeting_documents/cop/1cop/1cop-5.e.pdf.

COP 2 (1991). Nairobi, 17–19 June 1991. UNEP/OzL.Conv.2/7. Available at: http://ozone.unep.org/meeting_documents/cop/2cop/2cop-7.e.pdf.

COP 3 (1993). Bangkok, 23 November 1993. UNEP/OzL.Conv.3/6. Available at: http://ozone.unep.org/meeting_documents/cop/3cop/3cop-6.e.pdf.

COP 4 (1996). San José, 25–27 November 1996. UNEP/OzL.Conv.4/6. Available at: https://unep.ch/ozone/Meeting_Documents/cop/4cop/4cop-6.e.pdf.

Reports of the MOP

MOP 1 (1989). Helsinki, 2–5 May 1989. UNEP/OzL.Pro.1/5. Available at: https://unep.ch/ozone/Meeting_Documents/mop/01mop/1mop-5e.shtm.

MOP 2 (1990). London, 27–29 June 1990. UNEP/OzL.Pro.2/3. Available at: https://unep.ch/ozone/Meeting_Documents/mop/02mop/MOP_2.shtml.

MOP 3 (1991). Nairobi, 19–21 June 1991. UNEP/OzL.Pro.3/11. Available at: https://www.informea. org/en/event/third-meeting-parties-montreal-protocol-substances-deplete-ozone-layer.

MOP 4 (1992). Copenhagen, 23–25 November 1992. UNEP/OzL.Pro.4/15. Available at: https:// www.informea.org/en/event/fourth-meeting-parties-montreal-protocol-substances-deplete-ozone-layer.

MOP 5 (1993). Bangkok, 17–19 November 1993. UNEP/OzL.Pro.5/12. Available at: https:// www.informea.org/es/event/fifth-meeting-parties-montreal-protocol-substances-deplete-ozone-layer.

MOP 6 (1994). Nairobi, 6–7 October 1994. UNEP/OzL.Pro.6/7. Available at: https://www.informea. org/en/event/sixth-meeting-parties-montreal-protocol-substances-deplete-ozone-layer.

MOP 7 (1995). Vienna, 5–7 December 1995. UNEP/OzL.Pro.7/12. Available at: https://www. informea.org/en/event/seventh-meeting-parties-montreal-protocol-substances-deplete-ozone-layer.

MOP 8 (1996). San Jose, 25–27 November 1996. UNEP/OzL.Pro.8/12. Available at: https:// www.informea.org/en/event/eighth-meeting-parties-montreal-protocol-substances-deplete-ozone-layer.

MOP 9 (1997). Montreal, 15–17 September 1997. UNEP/OzL.Pro.9/12. Available at: https://unep. ch/ozone/9mop-rpt.shtml.

MOP 10 (1998). Cairo, 23–24 November 1998. UNEP/OzL.Pro.10/9. Available at: https://www. informea.org/en/event/tenth-meeting-parties-montreal-protocol-substances-deplete-ozone-layer.

MOP 11 (1999). Beijing, 29 November–3 December 1999. UNEP/OzL.Pro.11/10. Available at: https://www.informea.org/en/event/eleventh-meeting-parties-montreal-protocol-substances-deplete-ozone-layer.

MOP 12 (2000). Ouagadougou, Burkina Faso, 11–14 December 2000. UNEP/OzL.Pro.12/9. Available at: https://www.informea.org/en/event/twelfth-meeting-parties-montreal-protocol-substances-deplete-ozone-layer.

MOP 13 (2001). Colombo, 16–19 October 2001. UNEP/OzL.Pro.13/10. Available at: https:// www.informea.org/en/event/thirteenth-meeting-parties-montreal-protocol-substances-deplete-ozone-layer.

MOP 14 (2002). Rome, 25–29 November 2002. UNEP/OzL.Pro.14/9. Available at: https://www. informea.org/en/event/fourteenth-meeting-parties-montreal-protocol-substances-deplete-ozone-layer.

MOP 15 (2003). Nairobi, 10–15 November 2003. UNEP/OzL.Pro.15/9. Available at: https:// www.informea.org/en/event/fifteenth-meeting-parties-montreal-protocol-substances-deplete-ozone-layer.

MOP 16 (2004). Prague, 22–26 November 2004. UNEP/OzL.Pro.16/17. Available at: https://www. informea.org/en/event/sixteenth-meeting-parties-montreal-protocol-substances-deplete-ozone-layer.

MOP 17 (2005). Dakar, 12–16 December 2005. UNEP/OzL.Conv.7/7–UNEP/OzL.Pro.17/11. Available at: https://www.informea.org/en/event/seventeenth-meeting-parties-montreal-protocol-substances-deplete-ozone-layer.

MOP 18 (2006). New Delhi, 30 October–3 November 2006. UNEP/OzL.Pro.18/10. Available at: https://www.informea.org/en/event/eighteenth-meeting-parties-montreal-protocol-substances-deplete-ozone-layer.

MOP 19 (2007). Montreal, 17–21 September 2007. UNEP/OzL.Pro.19/7. Available at: https:// www.informea.org/en/event/nineteenth-meeting-parties-montreal-protocol-substances-deplete-ozone-layer.

MOP 20 (2008). Doha, 16–20 November 2008. UNEP/OzL.Conv.8/7–UNEP/OzL.Pro.20/9. Available at: https://unep.ch/ozone/Meeting_Documents/mop/20mop/MOP-20-9E.pdf.

MOP 21 (2009). Port Ghalib, Egypt, 4–8 November 2009. UNEP/OzL.Pro.21/8. Available at: https://unep.ch/ozone/Meeting_Documents/mop/21mop/MOP-21-8E.pdf.

MOP 22 (2010). Bangkok, 8–12 November 2010. UNEP/OzL.Pro.22/9. Available at: https://unep. ch/ozone/Meeting_Documents/mop/22mop/MOP-22-9E.pdf.

MOP 23 (2011). Bali, 21–25 November 2011. UNEP/OzL.Conv.9/7–UNEP/OzL.Pro.23/11. Available at: https://www.informea.org/en/event/twenty-third-meeting-parties-montreal-protocol-substances-deplete-ozone-layer.

MOP 24 (2012). Geneva, 12–16 November 2012. UNEP/OzL.Pro.24/10. Available at: https://www.informea.org/en/event/twenty-fourth-meeting-parties-montreal-protocol-substances-deplete-ozone-layer.

MOP 25 (2013). Bangkok, 21–25 October 2013. UNEP/OzL.Pro.25/9. Available at: http://conf.montreal-protocol.org/meeting/mop/mop-25/report/English/MOP-25-9E.pdf.

MOP 26 (2014). Paris, 17–21 November 2014. UNEP/OzL.Conv.10/7–UNEP/OzL.Pro.26/10. Available at: http://conf.montreal-protocol.org/meeting/mop/cop10-mop26/report/English/MOP-26-10-COP-10-7E.pdf.

MOP 27 (2015). Dubai, 1–5 November 2015. UNEP/OzL.Pro.27/13. Available at: http://conf.montreal-protocol.org/meeting/mop/mop-27/report/English/MOP-27-13E.pdf.

MOP 28 (2016). Kigali, 10–15 October 2016. UNEP/OzL.Pro.28/12. Available at: http://conf.montreal-protocol.org/meeting/mop/mop-28/final-report/English/MOP-28-12E.pdf.

MOP 29 (2017). Montreal, 20–24 November 2017. UNEP/OzL.Conv.11/7–UNEP/OzL.Pro.29/8. Available at: http://conf.montreal-protocol.org/meeting/mop/cop11-mop29/report/English/COP-11-7-MOP-29-8E.pdf.

Reports of the ImplCom

ImplCom 9 (1994). Report of the 9th Meeting of the ImplCom. Nairobi, 3 October 1994. UNEP/OzL.Pro/ImpCom/9/2. Available at: https://unep.ch/ozone/Meeting_Documents/impcom/9impcom-2.e.pdf.

ImplCom 10 (1995). Report of the 10th Meeting of the ImplCom. Geneva, 25 August 1995. UNEP/OzL.Pro/ImpCom/10/4. Available at: https://unep.ch/ozone/Meeting_Documents/impcom/10impcom-4.e.pdf.

ImplCom 11 (1995). Report of the 11th Meeting of the ImplCom. Geneva, 31 August 1995. UNEP/OzL.Pro/ImpCom/11/1. Available at: https://unep.ch/ozone/Meeting_Documents/impcom/11impcom-1.e.pdf.

ImplCom12 (1995). Report of the 12th Meeting of the ImplCom. Vienna, 27 and 29 November and 1 December 1995. UNEP/OzL.Pro/ImpCom/12/3. Available at: https://unep.ch/ozone/Meeting_Documents/impcom/12impcom-3.e.pdf.

ImplCom 13 (1996). Report of the 13th Meeting of the ImplCom. Geneva, 18–19 March 1996. UNEP/OzL.Pro/ImpCom/13/3. Available at: https://unep.ch/ozone/Meeting_Documents/impcom/13impcom-3.e.pdf.

ImplCom 14 (1996). Report of the 14th Meeting of the ImplCom. Geneva, 23 August 1996. UNEP/OzL.Pro/ImpCom/14/4. Available at: https://unep.ch/ozone/Meeting_Documents/impcom/14impcom-4.e.pdf.

ImplCom 15 (1996). Report of the 15th Meeting of the ImplCom. San José, 18 November 1996. UNEP/OzL.Pro/ImpCom/15/3. Available at: https://unep.ch/ozone/Meeting_Documents/impcom/15impcom-3.e.pdf.

ImplCom 16 (1996). Report of the 16th Meeting of the ImplCom. San José, 20 November 1996. UNEP/OzL.Pro/ImpCom/16/1. Available at: https://unep.ch/ozone/Meeting_Documents/impcom/16impcom-1.e.pdf.

ImplCom 17 (1997). Report of the 17th Meeting of the ImplCom. Geneva, 15–16 April 1997. UNEP/OzL.Pro/ImpCom/17/3. Available at: https://unep.ch/ozone/Meeting_Documents/impcom/17impcom-3.e.pdf.

ImplCom 18 (1997). Report of the 18th Meeting of the ImplCom. Nairobi, 2 and 4 June 1997. UNEP/OzL.Pro/ImpCom/18/3. Available at: https://unep.ch/ozone/Meeting_Documents/impcom/18impcom-3.e.pdf.

ImplCom 19 (1997). Report of the 19th Meeting of the ImplCom. Montreal, 8–10 September 1997. UNEP/OzL.Pro/ImpCom/19/3. Available at: https://unep.ch/ozone/Meeting_Documents/impcom/19impcom-3.e.pdf.

ImplCom 20 (1998). Report of the 20th Meeting of the ImplCom. Geneva, 6–7 July 1998. UNEP/
OzL.Pro/ImpCom/20/4. Available at: https://unep.ch/ozone/Meeting_Documents/impcom/
20impcom-4.e.pdf.

ImplCom 21 (1998). Report of the 21st Meeting of the ImplCom Cairo, 16 November 1998.
UNEP/OzL.Pro/ImpCom/21/3. Available at: https://unep.ch/ozone/Meeting_Documents/
impcom/21impcom-3.e.pdf.

ImplCom 22 (1999). Report of the 22nd Meeting of the ImplCom. Geneva, 14 June 1999. UNEP/
OzL.Pro/ImpCom/22/4. Available at: https://unep.ch/ozone/Meeting_Documents/impcom/
22impcom-4.e.pdf.

ImplCom 23 (1999). Report of the 23rd Meeting of the ImplCom. Beijing, 27 November 1999.
UNEP/OzL.Pro/ImpCom/23/3. Available at: https://unep.ch/ozone/Meeting_Documents/
impcom/23impcom-3.e.pdf.

ImplCom 24 (2000). Report of the 24th Meeting of the ImplCom. Geneva, 10 July 2000. UNEP/
OzL.Pro.11/ImpCom/24/4. Available at: https://unep.ch/ozone/Meeting_Documents/impcom/
24impcom-4.e.pdf.

ImplCom 25 (2000). Report of the 25th Meeting of the ImplCom. Ouagadougou, Burkina Faso, 9
December 2000. UNEP/OzL.Pro/ImpCom/25/2. Available at: https://unep.ch/ozone/Meeting_
Documents/impcom/impcom_reports_index.asp.

ImplCom 26 (2001). Report of the 26th Meeting of the ImplCom. Montreal, 23 July 2001. UNEP/
OzL.Pro/ImpCom/26/5. Available at: https://unep.ch/ozone/Meeting_Documents/impcom/
26impcom-5.e.pdf.

ImplCom 27 (2001). Report of the 27th Meeting of the ImplCom. Colombo, 13 October 2001.
UNEP/OzL.Pro/ImpCom/27/4. Available at: https://unep.ch/ozone/Meeting_Documents/
impcom/impcom_reports_index.asp.

ImplCom 28 (2002). Report of the 28th Meeting of the ImplCom. Montreal, 20 July 2002. UNEP/
OzL.Pro/ImpCom/28/4. Available at: https://unep.ch/ozone/Meeting_Documents/impcom/
28impcom-4.e.pdf.

ImplCom 29 (2002). Report of the 29th Meeting of the ImplCom. Rome, 23–24 November 2002.
UNEP/OzL.Pro/ImpCom/29/3. Available at: https://unep.ch/ozone/Meeting_Documents/
impcom/29impcom-3.e.pdf.

ImplCom 30 (2003). Report of the 30th Meeting of the ImplCom. Montreal, 4–7 July 2003.
UNEP/OzL.Pro/ImpCom/30/4. Available at: https://unep.ch/ozone/Meeting_Documents/
impcom/30impcom-4.e.pdf.

ImplCom 31 (2003). Report of the 31st Meeting of the ImplCom. Nairobi, 5–7 November 2003.
UNEP/OzL.Pro/ImpCom/31/3. Available at: https://unep.ch/ozone/Meeting_Documents/
impcom/31impcom-3.e.pdf.

ImplCom 32 (2004). Report of the 32nd Meeting of the ImplCom. Geneva, 17–18 July 2004.
UNEP/OzL.Pro/ImpCom/32/6. Available at: https://unep.ch/ozone/Meeting_Documents/
impcom/32impcom-6.e.pdf.

ImplCom 33 (2004). Report of the 33rd Meeting of the ImplCom. Prague, 17–19 November 2004.
UNEP/OzL.Pro/ImpCom/33/4. Available at: https://unep.ch/ozone/Meeting_Documents/
impcom/33impcom-4.e.pdf.

ImplCom 34 (2005). Report of the 34th Meeting of the ImplCom. Montreal, 2 July 2005. UNEP/
OzL.Pro/ImpCom/34/6. Available at: https://unep.ch/ozone/Meeting_Documents/impcom/
34impcom-6.e.pdf.

ImplCom 35 (2005). Report of the 35th Meeting of the ImplCom. Dakar, 7–9 December 2005.
UNEP/OzL.Pro/ImpCom/35/10. Available at: https://unep.ch/ozone/Meeting_Documents/
impcom/ImpCom-35-10E.pdf.

ImplCom 36 (2006). Report of the 36th Meeting of the ImplCom. Montreal, 30 June–1 July 2006.
UNEP/OzL.Pro/ImpCom/36/7. Available at: https://unep.ch/ozone/Meeting_Documents/
impcom/ImpCom-36-7E.pdf.

ImplCom 37 (2006). Report of the 37th Meeting of the ImplCom. New Delhi, 25–27 and 30
October 2006. UNEP/OzL.Pro/ImpCom/37/7. Available at: https://unep.ch/ozone/Meeting_
Documents/impcom/IMPCOM-37-7E.pdf.

ImplCom 38 (2007). Report of the 38th Meeting of the ImplCom. Nairobi, 8–9 June 2007. UNEP/
    OzL.Pro/ImpCom/38/5.   Available   at: https://unep.ch/ozone/Meeting_Documents/impcom/
    ImpCom-38-5E.pdf.
ImplCom 39 (2007). Report of the 39th Meeting of the ImplCom. Montreal, 12–14 September
    2007. UNEP/OzL.Pro/ImpCom/39/7. Available at: https://unep.ch/ozone/Meeting_Documents/
    impcom/ImpCom-39-7E.pdf.
ImplCom 40 (2008). Report of the 40th Meeting of the ImplCom. Bangkok, 2–4 July 2008.
    UNEP/OzL.Pro/ImpCom/40/6.   Available   at: http://42functions.net/Meeting_Documents/
    impcom/IMPCOM-40-6E.pdf.
ImplCom 41 (2008). Report of the 41st Meeting of the ImplCom.Doha, 12–14 November 2008.
    UNEP/OzL/ImplCom/41/8. Available at: https://unep.ch/ozone/Meeting_Documents/impcom/
    IMPCOM-41-8E.pdf.
ImplCom 42 (2009). Report of the 42nd Meeting of the ImplCom. Geneva, 20–21 July 2009.
    UNEP/OzL.Pro/ImpCom/42/5.   Available   at: https://unep.ch/ozone/Meeting_Documents/
    impcom/IMPCOM-42-5E.pdf.
ImplCom 44 (2010). Report of the 44th Meeting of the ImplCom. Geneva, 21–22 June 2010.
    UNEP/OzL.Pro/ImpCom/44/5.   Available   at: https://unep.ch/ozone/Meeting_Documents/
    impcom/IMPCOM-44-5E.pdf.
ImplCom 45 (2010). Report of the 45th Meeting of the ImplCom. Bangkok, 4–5 November 2010.
    UNEP/OzL.Pro/ImpCom/45/5.   Available   at: https://unep.ch/ozone/Meeting_Documents/
    impcom/IMPCOM-45-5E.pdf.
ImplCom 46 (2011). Report of the 46th Meeting of the ImplCom. Montreal, 7–8 August 2011.
    UNEP/OzL.Pro/ImpCom/46/5.   Available   at: http://conf.montreal-protocol.org/meeting/
    impcom/46impcom/draft-reports/English/IMPCOM-46-5E.pdf.
ImplCom 47 (2011). Report of the 47th Meeting of the ImplCom. Bali, Indonesia, 18–19
    November 2011. UNEP/OzL.Pro/ImpCom/47/6. Available at: http://conf.montreal-protocol.
    org/meeting/impcom/47impcom/draft-reports/English/ImpCom47-6-E.pdf.
ImplCom 48 (2012). Report of the 48th Meeting of the ImplCom. Bangkok, 29–30 July 2012.UNEP/
    OzL.Pro/ImpCom/48/5. Available at: http://conf.montreal-protocol.org/meeting/impcom/48imp
    com/draft-reports/English/IMPCOM-48-5E.doc.
ImplCom 49 (2012). Report of the 49th Meeting of the ImplCom Geneva, 8–9 November 2012
    UNEP/OzL.Pro/ImpCom/49/5/Rev.1. Available at: http://42functions.net/Meeting_Docum
    ents/impcom/IMPCOM-49-5R.doc.
ImplCom 50 (2013). Report of the 50th Meeting of the ImplCom Bangkok, 21–22 June 2013
    UNEP/OzL.Pro/ImpCom/50/4. Available at: http://42functions.net/Meeting_Documents/imp
    com/IMPCOM-50-4R.doc.
ImplCom 51 (2013). Report of the 51st Meeting of the ImplCom. Bangkok, 18–19 October 2013.
    UNEP/OzL.Pro/ImpCom/51/4.   Available   at: http://conf.montreal-protocol.org/meeting/
    impcom/51impcom/draft-reports/English/IMPCOM-51-4E.pdf.
ImplCom 52 (2014). Report of the 52nd Meeting of the ImplCom. Paris, 9–10 July 2014. UNEP/
    OzL.Pro/ImpCom/52/4.   Available   at: http://conf.montreal-protocol.org/meeting/impcom/
    impcom52/draft-reports/English/IMPCOM-52-4E.pdf.
ImplCom 53 (2014). Report of the 53rd Meeting of the ImplCom. Paris, 14–15 November 2014.
    UNEP/OzL.Pro/ImpCom/53/4.   Available   at: http://conf.montreal-protocol.org/meeting/
    impcom/impcom53/recommendations/English/IMPCOM-53-4E.pdf.
ImplCom 54 (2015). Report of the 54th Meeting of the ImplCom. Paris, 27–28 July 2015. UNEP/
    OzL.Pro/ImpCom/54/4.   Available   at: http://conf.montreal-protocol.org/meeting/impcom/
    impcom54/final_report/English/IMPCOM-54-4E.pdf.
ImplCom 55 (2015). Report of the 55th Meeting of the ImplCom. Dubai, United Arab Emirates,
    28 October 2015. UNEP/OzL.Pro/ImpCom/55/4. Available at: http://ozonecell.in/wp-content/
    themes/twentyseventeen-child/Documentation/assets/pdf/IMPCOM-55-4E.pdf.
ImplCom 56 (2016). Report of the 56th Meeting of the ImplCom. Vienna, 24 July 2016. UNEP/
    OzL.Pro/ImpCom/56/4.   Available   at: http://conf.montreal-protocol.org/meeting/impcom/
    impcom56/report/English/IMPCOM-56-4E.pdf.

ImplCom 57 (2016). Report of the 57th Meeting of the ImplCom. Kigali, Rwanda, 9 October 2016. UNEP/OzL.Pro/ImpCom/57/4. Available at: http://conf.montreal-protocol.org/meeting/impcom/impcom57/report/English/IMPCOM-57-4E.pdf.

ImplCom 58 (2017). Report of the 58th Meeting of the ImplCom. Bangkok, 9 July 2017. UNEP/OzL.Pro/ImpCom/58/4. Available at: http://conf.montreal-protocol.org/meeting/impcom/impcom58/report/English/IMPCOM-58-4E.pdf.

Reports of the Ad Hoc Working Group (WG) on Non-Compliance

Ad Hoc WG (1989). Report of the 1st Meeting of the Ad Hoc WG. Geneva, 14 July 1989. UNEP/OzL.Pro.LG.1/3.

Ad Hoc WG (1991). Report of the 3rd Meeting of the Ad Hoc WG. Geneva, 9 November 1991. UNEP/WG.3/3/3.

Ad Hoc WG (1998). Report on the Work of the Ad Hoc WG of Legal and Technical Experts on Non-compliance with the Montreal Protocol. Geneva, 3–4 July 1998 and Cairo, 17–18 November 1998. UNEP/OzL.Pro/WG.4/1/.

Reports of the Open-Ended Working Group (OEWG)

OEWG 12 (1995). 12th Meeting of the Open-ended Working Group of the Parties to the Montreal Protocol. Geneva, 28 August-1 September 1995. UNEP/OzL.Pro/WG.1/12/4. Available at: https://unep.ch/ozone/Meeting_Documents/oewg/12oewg/12oewg-4.e.shtml.

OEWG 15 (1997). Item 3 of the Provisional Agenda, Consideration and Consolidation of the Amendments and Adjustments Proposed by Parties. Note by the Secretariat, Addendum, Proposal by Canada. Nairobi, 3–6 June 1997. UNEP/OzL.Pro/WG.1/15/2/Add.5. Available at: https://uncp.ch/ozone/Meeting_Documents/oewg/15oewg/15oewg-2-add5.e.pdf.

Chapter 6
Case Study II: Climate Change

6.1 The Compliance Mechanism Created Under the 1997 Kyoto Protocol

The UN Framework Convention on Climate Change (UNFCCC) was signed at the UN Conference on Environment and Development (UNCED) held in Rio, Brazil in 1992 and entered into force in 1994. The Kyoto Protocol, which sets out more detailed policies and measures that may be implemented by each party to achieve their commitments, was adopted at the third Conference of the Parties (COP 3 1997) to the UNFCCC in 1997 in Kyoto, Japan (see Annex G to this book for the list of COPs to the UNFCCC and MOPs to the Kyoto Protocol).

It came into force on 16 February 2005, the requisite 90 days after meeting the conditions specified in Art. 25 of the KP: being ratified by 55 parties (a target reached with the ratification of Iceland on 23 May 2002), accounting for at least 55% of the total carbon dioxide emissions of Annex I countries in 1990 (a condition fulfilled with the cooperation of Russia on 18 November 2004). Thus, "in gaining entry into force without the US, [it] has become a symbol of hope for the lead coalition favoring a regime to mitigate climate change" (Chasek et al. 2006: 127).

In Article 2 of the Convention, the ultimate objective of the Kyoto Protocol is clearly stated as being to provide "stabilization of greenhouse gas concentrations in the atmosphere at a level that would prevent dangerous anthropogenic interference with the climate system" (Art. 2, UNFCCC). In the subsequent article, the Convention sets out the basic principles with which the parties to the convention are guided to fulfil its obligations and accomplish its objective. These are: equity (Art. 3.1), common but differentiated responsibility (Arts. 3.1 and 3.2), the precautionary principle (Art. 3.3), sustainable development (Art. 3.4), and an open international economic system (Art. 3.5).

On the basis of the principle of common but differentiated responsibility, which is also emphasized in Principle 7 of the Rio Declaration, the Convention categorizes the parties into different groups.

© Springer Nature Switzerland AG 2019
Z. Savaşan, *Paris Climate Agreement: A Deal for Better Compliance?*
The Anthropocene: Politik—Economics—Society—Science 11,
https://doi.org/10.1007/978-3-030-14313-8_6

In its Annex I, it lists the developed country parties, including Western countries and ex-Soviet bloc countries, i.e. parties with economies in transition. So developing countries are defined as 'non-annex I' countries.

In its Annex II, it lists Annex I Parties, including OECD countries, which are required to provide financial and technological assistance to developing countries.

Developed countries are also distinguished according to the percentage of base year or period of the quantified emission limitation or reduction commitment. In fact, country-specific quantified emission limitation and reduction commitments, determined according to the developed country's own characteristics, are included in Annex B and Article 3.7 of the Kyoto Protocol.

Based on this categorization, the Convention contains general obligations which are binding for all parties and specific ones which only Annex I parties are required to meet.

The obligations with which all parties have to comply are counted in Article 4.1, such as to develop national inventories of greenhouse gases (4.1.a) and to implement measures to mitigate climate change and facilitate adequate adaptation (4.1b etc. (Art. 4.1(a–j)). On the other hand, specific commitments imposed on Annex I parties are indicated in the other paragraphs of the same Article (Arts. 4.2–5). Under Art. 4.2(a–g), it is stated that the developed country parties and other parties included in Annex I have to adopt policies and take corresponding measures for the mitigation of climate change by reducing their greenhouse gas emissions, protecting and enhancing their greenhouse gas sinks and reservoirs, "taking the lead in modifying longer-term trends in anthropogenic emissions consistent with the objective of the Convention" (Art. 4.2a) and "with the aim of returning individually or jointly to their 1990 levels these anthropogenic emissions of carbon dioxide and other greenhouse gases not controlled by the Montreal Protocol" (Art. 4.2b). Thus, the Convention only obliges developed country parties to undertake to return to their 1990 levels of greenhouse gas emissions.

Additionally, according to the Convention, the developed country parties and other developed parties included in Annex II have to provide "new and additional financial resources" to developing parties for complying with their obligations under Art. 12.1. and for implementing measures that are covered by Art. 4.1. and that are agreed between a developing country party and the international entity or entities referred to in Art. 11 (Art. 4.3). They are also under the obligation to help developing countries meet the costs of adaptation to the adverse effects of climate change (Art. 4.4). Furthermore, they have to promote, facilitate and finance the transfer of environmentally sound technologies and know-how to developing country parties (Arts. 4.5; 4.7).

In accordance with UNFCCC, the Kyoto Protocol also contains no specific commitments for developing countries relating to emission limitations and reduction commitments of greenhouse gases, the "so-called "six gas basket" – carbon dioxide (CO_2), methane (CH_4), nitrous oxide (N_2O), hydrofluorocarbons (HFCs), perfluorocarbons (PFCs) and sulphurhexafluoride (SF_6) – listed in Annex A of the Protocol (Oberthür/Ott 1999: 95). In accordance with Art. 4.1 of the Convention, their general obligations are mostly specified in Art. 10 of the Protocol. They are also entitled to be provided with financial resources by developed country parties through Art. 11 of the Protocol, in parallel with Arts. 4.3 and 11 of the Convention.

Like the UNFCCC, the Protocol only prescribes quantified limits and reduction obligations for Annex I parties to reduce their overall greenhouse gas emissions by an average of at least 5% below their 1990 levels in the first commitment period 2008–2012 (Art. 3.1, KP); the Doha Amendment (2012) also includes new commitments which should be carried out by the end of the second commitment period (2013–2020). For this purpose, each Annex I party is assigned an individual target amount – Assigned Amounts (AAs) – of greenhouse gas emissions listed in Annex B to the Protocol.

These commitments by developed states (in the Convention generally and in the Protocol more elaborately) were supplemented by a compliance mechanism (developed primarily by the contributions of the Joint Working Group [JWG]) through a decision of the COP (not by an amendment to the Protocol) in Marrakesh, Morocco (called the 'Marrakesh Accords') in 2001 (COP 7, Decision 24 2001: 64).

In the process of negotiations,[1] developed countries took a leading role in designing the Kyoto Protocol's compliance mechanism. To illustrate, the US succeeded in achieving agreement on clearly defined consequences: it suggested making the compliance procedure part of the Protocol prior to its entry into force to provide for binding consequences. Yet, paradoxically, the USA did not ratify the Protocol. The debate on the legal status of the consequences to non-compliance, on the other hand, still cannot be resolved. The more reluctant Annex I parties like Australia, Japan and the Russian Federation, succeeded in preventing a clear definition of the legal status of the response measures/consequences. Developing countries, on the other hand, supported a strong enforcement mechanism under the Protocol, contingent on the application of its provisions only to Annex I parties (Oberthür/Lefeber 2010).

Through Decision 1 (COP 4 1998), the COP adopted the Buenos Aires Plan of Action in 1998, which is specified in its separate decisions. Initial negotiation of the Kyoto compliance mechanism can be traced back to this Action Plan, which sets out a programme of work on issues under the Protocol for the future entry into force of the KP (Werksman 2005). At COP 4, the deadline for completion was established as COP 6 (2000). However, at COP 6 (2001) in the Hague (Part I), an agreement on the decisions under the Buenos Aires Action Plan could not be reached. With respect to the topic of compliance, key issues such as the consequences of non-compliance and the membership of the Compliance Committee could not be resolved, so they were resumed in Bonn (COP 6 2001, Part II). At COP 6 (2001) (Part II), parties adopted the Bonn Agreements on the Implementation of the Buenos Aires Plan of Action (COP 6, Decision 5 2001) on key fields including compliance (COP 6 2001: 48–49). However, since the main points regarding the compliance mechanism under the Kyoto Protocol could not be resolved fully, the draft decision proposed by the co-chairmen of the negotiating

---

[1]See Werksman (2005) for details of the negotiation of the Kyoto compliance system.

Table 6.1 Development of CM in Kyoto Protocol

Cop 4, Decision 1	Negotiation of the Kyoto compliance mechanism begins through the Buenos Aires Plan of Action in 1998
COP 6, Decision 5	Parties adopted the Bonn Agreements on the Implementation of the Buenos Aires Plan of Action on key issues, including compliance (COP 6 2001: 48–49). However, the main points regarding the compliance mechanism under the Kyoto Protocol could not be fully resolved
COP 7, Decision 24	Adoption of the CM in Marrakesh, Morocco (called the 'Marrakesh Accords') in 2001
MOP 1, Decision 27	The confirmation of COP 7, Decision 24 in Montreal, Canada in 2005, at which many issues regarding the operation of the CM – except the legal status of enforcement consequences – were resolved

Source Prepared by the author on the basis of information/documentation given on the official website of the UNFCCC. See: https://unfccc.int/decisions

group on these procedures and mechanisms was forwarded to COP 7 for further elaboration and adoption.

After their adoption by COP 7 (2001), the MOP – the Conference of the Parties serves as the meeting of the Parties to the Kyoto Protocol – also approved them to determine and address cases of non-compliance with the Protocol on the basis of Art. 18 of the Kyoto Protocol, calling on the MOP to approve them. Thus, through the confirmation of Decision 24 (COP 7 2001) in Decision 27 (MOP 1 2005: 92), held in Montreal, Canada in 2005, many of the outstanding issues necessary to bring the Protocol into operation were resolved (see Table 6.1). The exception was the legal status of enforcement consequences. MOP 1 decided to consider the issue of an amendment to the Kyoto Protocol in respect of CMs and to make a decision at the MOP 3. It also entrusted the Subsidiary Body for Implementation (SBI) to study it and report on the outcome to the MOP 3.

When the related parts regarding this issue in the reports of the SBI are examined, it is apparent that from the 24th session to the 33rd session (SBI, 24–33, except the 31st session), the SBI reports incorporate a separate section on the amendment of the Kyoto Protocol in respect of procedures and mechanisms relating to compliance. It was originally decided that its consideration of the matter should be completed at its 27th session (2007); however, at the start of that session, it was decided that compliance should continue to be discussed at subsequent sessions, and should therefore be included on the provisional agendas of subsequent sessions. Finally, in the draft conclusions of the 37th session, it was decided that no further discussion was required on this issue, and the SBI (SBI 37 2012) recommended that the MOP should complete its consideration of the proposal from Saudi Arabia (MOP 1 2005d, paras. 54–59) to amend the KP in respect of procedures and mechanisms relating to compliance. Thus, the proposal by Saudi Arabia to incorporate the compliance procedures into the Protocol through an amendment remained on the agenda of the MOP till the 37th session of the SBI.

In conclusion, it should be emphasized that this compliance mechanism has been the third attempted by negotiators. The previous two efforts were abandoned due to

Art. 13 of the Convention, which calls upon the COP to consider the establishment of a multilateral consultative process for the resolution of questions regarding the implementation of the Convention.

6.1.1   Main Components of the Kyoto Protocol Compliance Mechanism

6.1.1.1   Gathering Information on the Parties' Performance[2]

The UNFCCC and the Kyoto Protocol each contain provisions for gathering information to review the parties' performance (see Table 6.2).

The word 'reporting' is not used throughout the Convention; instead, the word 'communication' is preferred. However, the parties' communications concerning their implementation of the Convention (Art. 12, KP) have the same character as reporting (Faure/Lefevere 1999).

There are four types of report submitted under the Convention:

1. *Annex I parties' communications*: Submissions made by regular intervals by developed countries (Annex I parties) covering all aspects of implementation and compliance.
2. *Non-Annex I parties' communications*: Submissions made at regular intervals by non-Annex I parties on all aspects of implementation and compliance.
3. *Inventories*: Annual submissions by developed countries on greenhouse gas emissions and removals.
4. *National adaptation programmes of action (NAPAs)*: Submissions by least-developed countries on their needs for adaptation.

Only the first three reports are fundamental to the identification of non-compliance, so they are the only ones which will be covered in this section.

The Convention clearly states that all parties to it should communicate to the COP in line with Art. 4.1 (Art. 12.1), which instructs all parties to design national inventories on the basis of the common and differentiated principles enshrined in the Convention (Art. 4.1, UNFCCC). That is, the obligations to develop national inventories can vary according to whether the country is developed, developing or least-developed and also to the "content, timing and the availability of financial resources" (Werksman 1996: 89).

These inventories, of emissions by sources and removals by sinks of all GHGs not controlled by the Montreal Protocol, should be periodically updated (Art. 4.1a, UNFCCC). The inventories and also the measures taken by the parties to implement the Convention should be conveyed to the COP through the UNFCCC Secretariat

---

[2]See Savaşan (2012) for a comparative analysis of reporting under the Montreal and Kyoto Protocols.

Table 6.2 Gathering information under the UNFCCC and the Kyoto Protocol

Under the UNFCCC	
Annex I parties' communications	Submissions made at regular intervals by developed countries (Annex I parties) covering all aspects of implementation and compliance
Non-Annex I parties' communications	Submissions made at regular intervals by non-Annex I parties on all aspects of implementation and compliance
Inventories	Annual submissions by developed countries on greenhouse gas emissions and removals
National Adaptation Programmes of Action (NAPAs)	Submissions by least-developed countries on their needs for adaptation
Under the Kyoto Protocol	
Communications provided by Annex I and non-Annex I country parties	These basically build on the system of the Convention
Inventories by developed countries	
Initial report	Report including the party's calculation of its assigned amount submitted by Annex I parties at the beginning of the commitment period
Report on the demonstrable progress	Report which should be prepared in the middle of the commitment period by Annex I parties showing their demonstrable progress in achieving their commitments under the Protocol
Report on true-up period	Report involving final information on the party's assigned amount at the end of the commitment period

Source Prepared by the author on the basis of the provisions of the UNFCCC and the Kyoto Protocol

(Art. 4.1j, Art. 12.1, UNFCCC). The secretariat located in Bonn makes these communications publicly available when they are submitted to the COP (Art. 12.10, UNFCCC).

While all parties from both developed and developing countries are obliged under the Convention to provide information on emissions by sources and removals by sinks, only developed country parties are required to provide a detailed description of the policies and measures that they have adopted to implement their obligations (Art. 4.2b, UNFCCC). In addition, through Articles 4.3, 4.4 and 4.5 of the UNFCCC, developed parties included in Annexes I and II are required to provide new and additional financial resources to developing country parties, to help them meet the costs of adaptation to the adverse effects of climate change and to promote and finance the transfer of environmentally sound technologies and know-how to aid them in complying with their obligations under the Convention. More importantly, the ability of developing country parties to implement their commitments under the Convention is bound to the developed country parties' effective implementation of the provisions related to providing financial resources and transfer of technology to the developing country parties (Art. 4.7, UNFCCC). That is, financial assistance and technology transfer from developed country parties are regarded as prerequisites for the effective compliance of developing countries. However, regarding the assessment of compliance with the financial resource and technology transfer commitments of developed country parties, it is generally argued that while developed country parties are required to incorporate details of the measures that they have adopted under Art. 4, there is no agreed methodology for explaining "the nature and extent of these resources" (Werksman 1996: 99), and this can make the assessment more difficult.

The Convention does not only oblige developed country parties to assist developing country parties, but also instructs them not to ignore the specific needs and special situations of the least developed country parties while providing financial resources and technology transfer to other parties (Art. 4.9, UNFCCC).

Regarding least developed country parties and their specific situations concerning funding and technology transfer, there are also important COP decisions. To illustrate, while Decisions 5 and 7 (COP 7 2001) established a Least Developed Countries Fund, Decision 27 (COP 7 2001) adopted guidance for its operation. Thus, the Fund is designed to support the work programme of the least developed countries, including their preparation and implementation of the National Adaptation Programmes of Action (NAPAs) referred to in para. 11 of Decision 5 (COP 7 2001). In order to provide guidance on the preparation and implementation of NAPAs, Decision 28 (COP 7 2001) set out guidelines and Decision 29 (COP 7 2001) established an expert group to work solely on this issue.

As seen in the analysis of the Convention, this instrument does not entrust any institution with reviewing the parties' compliance with their obligations under the Convention. The Convention's review mechanism, which contains guidelines[3] for the preparation of national communications and procedures, was established by COP decisions (Decisions 2, 3, 4) in 1995 (COP 1 1995).

Through these developments, communications from the parties to the Convention have been subjected to an in-depth review (IDR) process which should undertake the six tasks described in Annex 2 to Decision 2, which involve reviewing key qualitative information and quantitative data points, policies and measures contained in national communications; assessing the progress on reaching the objectives of the Convention; revealing the expected progress in limiting emissions; and gathering together data across national communications with respect to inventories, projections, effects of measures and financial transfers (COP 1, Decision 2, Annex 2 1995).

Such a review process is conducted by expert review teams under the authority of the subsidiary bodies (COP 1, Decision 2 1995, para. 2a). These review teams carry out their work basically through parties' submitted communications (COP 1, Decision 2 1995, para. 2b). However, it is also possible to make "in-country visits" (Goldberg et al. 1998: 10; Ulfstein/Werksman 2005: 44) for these teams with the approval of the party concerned (COP 1, Decision 2 1995, para. 2c).

---

[3]See Decisions 9, 10 (COP 2 1996) and Decisions 3, 4 (COP 5 1999) for more extensive guidelines relating to country reports.

After in-depth review of a communication, each review team produces a report "written in non-confrontational language" (COP 1, Decision 2 1995: 8, para. 2d), in line with the purpose of the review process outlined in Annex I to Decision 2 (COP 1 1995) of providing an assessment "in a facilitative, non-confrontational, open and transparent manner" (COP 1, Decision 2, Annex I 1995: 9). The same Annex also indicated that this comprehensive technical assessment is made on the implementation of the Convention commitments by not only individual Annex I Parties, but also Annex I Parties as a whole. This means that there have been two kinds of review made by the expert team: one is to evaluate the overall effect of the compliance by all parties, and the other is to assess the individual party's commitments.

A draft of the review report should be presented to the party being reviewed for its comments. Then, the summary of the review report is circulated to all parties and accredited observers to the COP by the secretariat (COP 1, Decision 2 1995, para. 2d).

Under the Kyoto Protocol, on the other hand, there are five types of report. Two of them, communications provided by Annex I and non-Annex I country parties and GHG inventories by developed countries, are basically built on the system of the Convention. Three others, named "one-time reports" and submitted by Annex I parties have different characteristics (Yamin/Depledge 2004)[4]:

1. *Initial report*: This report includes information on GHG inventory time-series, the party's calculation of its assigned amount and of its commitment period reserve (CPR), and a description of its national registry and national system etc. Its chief purpose is to facilitate the calculation of the relevant party's assigned amount and demonstrating its capacity to account for its emissions and assigned amount. These reports were to be submitted by Annex I parties by 31 December 2006 or one year after the Protocol's coming into force for that party, i.e. at the beginning of the commitment period.
2. *Report on the demonstrable progress*: This report should be prepared by Annex I parties in the middle of the commitment period in the context of Art. 3.2, and should detail the demonstrable progress they have made in achieving their commitments under the Protocol.
3. *Report on the true-up period*: This report should provide final information on the party's assigned amount, and on all units retired for compliance purposes. It used to determine whether the party is in compliance with Art. 3.1 at the end of the commitment period.

The Protocol contains mainly three articles related to reporting and review: Art. 7 on communications and on implementation review by expert review teams (Art. 8) which actually "provide, without explicitly referring to non-compliance, for the first and the last step of a multilateral procedure for the implementation and

---

[4]For details of national reports, see: https://unfccc.int/documents?search2=&search3=national% 20reports&f%5B0%5D=document_type%3A3525.

treatment of non-compliance under the Protocol", and Article 5 on methodologies for inventories (Oberthür/Ott 1999: 211).[5]

Art. 5 addresses the issues regarding the estimation of greenhouse gas emissions by sources and removals by sinks. It states that Annex I parties should establish national systems for the estimation of emissions (Art. 5.1, KP). Guidelines for national systems should be set out by the MOP at its first session (Art. 5.1, KP). Methodologies employed for estimating the emissions in the reports should be provided in a uniform form by the expertise of the Intergovernmental Panel on Climate Change (IPCC) and agreed upon by Decision 2 (COP 3 1997).

Where such methodologies are not used, "appropriate adjustments" should be applied, according to methodologies agreed upon by the MOP 1 (Art. 5.2, KP). MOP should regularly review, and if it finds necessary, may revise such methodologies and adjustments, but only to assess compliance with commitments under Art. 3 (Art. 5.3, KP).

Regarding the submission of supplementary information in greenhouse gas inventories (Art. 7.1) as well as communications (Art. 7.2), Art. 7 requires Annex I parties to submit them to demonstrate their compliance situation with Art. 3 of the Protocol. The supplementary information required for greenhouse gas inventories should be submitted annually, while that required for national communications can be submitted periodically according to the dates set by the decision of the MOP based on any timetable that the COP has decided for the submission of national communications (Art. 7.3).

Annex I parties also have to report on their programmes and activities undertaken under Art. 10 (Art. 10b(ii), KP). This obligation is also valid for also non-Annex I parties (Art. 10f, KP).

Article 8, on the other hand, sets out that expert review teams should review the information submitted under Art. 7 (communications) by Annex I parties as part of the annual compilation and accounting of emissions inventories and assigned amounts if it is submitted under Art. 7.1, or as a part of the review of communications if it is submitted under Art. 7.2 (Art. 8.1, KP). The report of the team includes an assessment of the implementation of the commitments and identification of any potential problems in the fulfilment of them (Art. 8.3), and thus helps the MOP to make decisions on any matter on the related party's compliance situation (Art. 8.6).

The elements of the additional reporting that are required for Annex I Parties by articles 5, 7 and 8 under the Protocol are further elaborated by Decisions 20–23 (COP 7 2001) and the remaining work with respect to the guidelines under these articles is figured out by the decisions of COP 8 (Decision 22), COP 9 (Decisions 20, 21), COP 10 (Decisions 16, 18) and COP 11 (Decision 15). The Protocol also states that guidelines for national systems (Art. 5.1), the preparation of inventories (Art. 7.4) – see also COP 7 (Decision 19) – and the review of implementation of the

---

[5]See Herold (2012) for a detailed analysis of the experiences with Articles 5, 7 and 8 defining the monitoring, reporting and verification system under the Kyoto Protocol.

Protocol by expert review teams (8.4) should be adopted by the MOP 1 and reviewed periodically thereafter.[6]

From the above analysis of the reporting and review process of the Kyoto Protocol, it is clear that, under the Protocol, the rules on the reporting and review process are not very different from the system operating under the Convention. As "overlap and double-work" can arise otherwise, this seems a good way to ensure these procedures function effectively (Oberthür/Ott 1999: 211). However, it should also be underlined that although the Protocol builds on the Convention's reporting procedures, it contains more detailed rules than the Convention and, through these rules, promises a system leaning towards a faster and more accurate reporting and review process. That is, in brief:

- It widens the scope of in-depth review (IDR) to cover submissions of both national communications and annual inventories from Annex I parties.
- The Expert Review Teams (ERTs) prepare a report containing a "thorough and comprehensive technical assessment of all aspects of the implementation by a party of [the] Protocol" (Art. 8.3, KP).
- The ERTs' powers have been enhanced "creating the groundwork for vesting review teams with genuine investigative powers" (Goldberg et al. 1998: 11). The teams are instructed to assess the party's implementation efforts and identify any potential problems or factors influencing the fulfilment of its commitments. Their reports, together with a list of the implementation questions contained in them, are then circulated by the secretariat to all parties to the Convention.
- The MOPs take any decisions on the implementation of the party after assisting the SBSTA and, if appropriate, the SBI and considering the original submission of the party, the review team's report, and any questions of implementation listed by the secretariat.
- Whereas under the UNFCCC the inventory information is only published and reviewed periodically, under the Protocol, Art. 8. para. 1 annual inventories are reviewed on a yearly basis.
- The periodic review of the guidelines for reporting under the KP "opens up the possibility of further tightening the reporting and review mechanisms, thereby adjusting them to future developments" (Wolfrum/Friedrich 2006: 58).
- The most important difference is specifically related to the compliance issue. Under the Kyoto Protocol, any problem relating to the emission targets, identified through the reporting and review system, should be considered by the EB of the ComplCom.

Overall, with these features, it can be argued that "[t]he reporting system under the Kyoto Protocol has been transformed from an information collecting device into a true monitoring system" (Wolfrum 1999: 43).

---

[6]e.g. the MOP 1 (2005a, b) adopted the following decisions relating to articles 5, 7 and 8, KP (Decisions 13–15, 19–25, 27).

6.1.1.2 Non-Compliance Procedure (NCP)

To deal with possible difficulties Annex I parties may have in complying with the Protocol's obligations, three basic procedures should be distinguished under the Protocol:

1. *Multilateral consultative process (MCP)*: The Protocol requests the MOP to consider and modify the multilateral consultative process (MCP) elaborated under the UNFCCC for the resolution of implementation questions (Art. 16, KP).
2. *Dispute settlement procedures (DSPs)*: It also incorporates the DSPs of Art. 14, UNFCCC (Art. 19, KP).

Under the UNFCCC, the establishment of a multilateral consultative process (MCP) is left to the decision of the COP (Art. 13, UNFCCC). After the study of an Ad Hoc Group on Art. 13 (AG-13), in the last session of AG-13, the multilateral consultative process was founded in the form of a set of procedures to be served by a standing Multilateral Consultative Committee through Decision 10 (COP 4 1998: 42–46).

The Committee aims to improve the understanding of the Convention and prevent potential disputes (COP 4, Decision 10 1998: para. 2). That is, it "rel[ies] on the willingness of the parties to respect their duties and to actively promote a 'physiological' operation of the UNFCCC before any 'pathological' situation arises" (Eritja et al. 2004: 68). It also aims to provide advice on the procurement of assistance to parties to overcome their difficulties in complying with the Convention (COP 4, Decision 10, para. 2). In brief, it aims to provide resolutions to compliance problems in a "facilitative, cooperative, non-confrontational, transparent and timely manner and be non-judicial" (COP 4, Decision 10, para. 3).

Therefore, it is generally defined as an advisory mechanism. However, even though it is true that it contains elements of an advisory mechanism, and is therefore closer to being an advisory mechanism than a supervisory mechanism (Ehrmann 2002), it also contains elements of a supervisory mechanism: triggering by the party itself, by another party and by the COP (COP 4, Decision 10 1998: para. 5); the process with respect to the implementation of another party; and the option of taking measures (COP 4, Decision 10 1998: para. 12b). Yet, since the NCP created under the KP provides for a facilitation branch which has the same functions of facilitation and prevention, the importance of the MCP has been undermined. This situation has also raised the question of the overlapping which can arise between the MCP, foreseen by the Convention (Art. 13), and the NCP, foreseen by the Protocol (Art. 18), since, "the possibility of the different mechanisms operating in parallel does not exclude eventual cases of duplicity or overlapping" (Eritja et al. 2004: 85).

Regarding the overlapping problem, Art. 16, KP reveals what should be done in such a case, stating that "Any multilateral consultative process that may be applied to this Protocol shall operate *without prejudice* to the procedures and mechanisms established in accordance with Article18" (emphasis added). This means that both procedures can be exercised together in parallel; there is no reason for one to have primacy over the other.

Regarding its relationship to the DSP, there is again usage of the provision of "without prejudice". The multilateral consultative process should be separate from, and also without prejudice to, the provisions of DSP (Art. 14, VC; COP 4, Decision 10, para. 4).

3. *Non-compliance procedure (NCP)*: The Protocol also calls for the approval of procedures and mechanisms to determine and address cases of non-compliance with the provisions of the Protocol (Art. 18, KP). Hence, through the Marrakesh Accords adopted at COP 7, a set of NCPs was adopted to enforce the CM's rules, to address any compliance difficulties and to prevent calculation errors regarding emissions data and accounting of transactions under the three Kyoto mechanisms (JI, ET and CDM).

In this chapter, this set of non-compliance procedures will be analysed in detail, focusing on the institution created under these NCPs (the ComplCom), procedural phases and safeguards functioning under it, in parallel with the methods pursued under the NCP part of the CM under the Montreal Protocol.

Institution created under the NCP: Compliance Committee

The NCP adopted by Decision 24 (COP 7 2001) on the basis of Art. 18 of the Kyoto Protocol is fundamentally built on its functional body, the ComplCom and its two important branches, the Facilitative Branch (FB) and Enforcement Branch (EB). This is particularly because the Committee considers questions of implementation, pursuing a "double track system" through these two branches (Montini 2009: 409).

In addition to them, its structure is also based on a plenary and a bureau in line with Section II(2) of the Annexes to Decision 24 (COP 7 2001: 65) and Decision 27 (MOP 1 2005a: 93); and some subsidiary bodies, such as the Subsidiary Body for Implementation (SBI) and the Subsidiary Body for Scientific and Technological Advice (SBSTA), which carry out specific, delegated tasks to consider specific issues necessary to promote effective implementation of the Protocol on the basis of Art. 13.4h, KP (also Art. 7.2i, UNFCCC).

The MOP, with the assistance of the SBI and, as appropriate, the SBSTA, considers the information submitted by parties under Art. 7, the reports of the expert reviews and the questions of implementation listed by the secretariat as well as any questions raised by the parties (Art. 8(5), KP). Based on its consideration of submitted information, it makes decisions on any matter required for the implementation of the Protocol (Art. 8(6), KP).

The COP/MOP may establish subsidiary bodies to carry out specific, delegated tasks to consider specific issues necessary to promote effective implementation of the Protocol on the basis of the Art. 13.4h, KP (also Art. 7.2i, UNFCCC).

In accordance with these provisions, the SBSTA provides the COP "with timely information and advice" on scientific, technological and methodological matters relating to the Convention. It is entrusted with a number of tasks which can be further elaborated by the COP (Art. 9.3, UNFCCC), such as: providing assessments of scientific knowledge relating to climate change; preparing scientific assessments of the effects of measures taken for the implementation of the Convention; identifying innovative technologies and know-how on the ways of promoting such technologies; providing advice on the development of scientific programmes and research into climate change, as well as on ways of supporting capacity-building in developing countries and providing answers to the COP's questions on the relevant matters (Art. 9.2, UNFCCC). Within most of its tasks, it acts as an 'intermediary' between the primarily scientific body, Intergovernmental Panel on Climate Change (IPCC) and the political body, COP/MOP (Oberthür/Ott 1999: 250).

The SBI, on the other hand, helps the COP assess and review how effectively the Convention is being implemented (Art. 10.1, UNFCCC). Its tasks consist of assisting the COP in the preparation and implementation of its decisions (Art. 10.2, UNFCCC) and examining the information communicated to the COP on: national inventories of emissions; the policies, measures and other steps taken by the party to implement its commitment to the Convention under Art. 4.2(a, b) (Art. 12.2, UNFCCC); the effects of such steps; and any other issues relevant to achieving the objective of the Convention (Art. 12.1, UNFCCC).

The subsidiary bodies established by Arts. 9 and 10 of the Convention serve as subsidiary bodies of the Protocol; so the provisions relating to these two bodies under the Convention apply *mutatis mutandis* to the Protocol as well. Therefore, sessions of the meetings of the bodies of the Protocol are arranged in conjunction with the subsidiary bodies of the Convention (Art. 15(1), KP). The parties to the Convention, but not the parties to the Protocol, can participate as observers in any sessions of the subsidiary bodies. However, when the subsidiary bodies serve as the subsidiary bodies of the Protocol, decisions under the Protocol can be taken only by those that are parties to the Protocol (Art. 15(2), KP). In addition, when these bodies exercise their functions with regard to matters concerning the Protocol, any member of the Bureaux of those subsidiary bodies who is a party to the Convention, but not a party to the Protocol, should be replaced by an additional member to be elected by and from among the parties to the Protocol (Art. 15(3), KP).

The Committee is composed of twenty members elected by the MOP: ten serve in the FB, while the other ten serve in the EB (NCP, Section II(3)). Thus, both branches are composed of ten members, including one representative from each of the five official UN regions (Africa, Asia, Latin America and the Caribbean, Central and Eastern Europe, and Western Europe and others), one from the small island developing states, and two each from Annex I and non-Annex I Parties (NCP, Section IV(1); NCP, Section V(1)). For each member, an alternate member is also elected by the MOP (NCP, Section II(5)). Members and their alternates, who should be qualified in climate change and in relevant fields (the scientific, technical, socio-economic or legal fields) (for the EB, members also have legal experience (NCP, Section V(3)) should serve in their personal capacity (NCP, Section II(6);

RoP, 4).[7] Alternate members can participate in the proceedings of the plenary or the respective branch to which they belong, but they can not vote on the decision (RoP, 3.2). They can exercise voting rights only if they serve as a member during the absence of a member or when a member resigns or is unable to complete the assigned term or the functions of a member (RoP, 3.3, 3.4). The Committee meets at least twice each year in line with the desirability of holding such meetings together with the meetings of the subsidiary bodies (NCP, Section II(10)).

The Committee adopts decisions by at least a 75% majority of its members (NCP, Section II(8)). It should try to decide by consensus, but when all attempts to reach a consensus are exhausted, as a last resort it can adopt decisions approved by a 75%majority of the members present and voting (NCP, Section II(9)). While deciding, it should take into account any degree of flexibility allowed by the MOP for Annex I parties undergoing the process of transition to a market economy (NCP, Section II(11)).

The functions and powers of the Committee will now be examined according to its bodies, the plenary, the Bureau and its two branches, the EB and the FB, each of which has different functions and powers.

The Plenary. The Committee also meets in a plenary made up of members of both branches (NCP, Section III(1)). The functions of the plenary are to report on the activities of the Committee to each ordinary session of the MOP, to apply the general policy guidance received from the MOP pursuant to Section XII(c), to submit proposals on administrative and budgetary matters to the MOP, to develop any further rules of procedure that may be needed for adoption by the MOP, and to perform such other tasks which may be requested by the MOP for effective operation of the ComplCom. The rules of procedure may be amended by a decision of the MOP after the plenary has approved the proposed amendment and reported on the matter to the MOP (RoP, 26.1). Any amendment by the plenary should provisionally be applied pending its adoption by the MOP (RoP, 26.2). Its meetings are held in public, unless it decides otherwise because of very important reasons, of its own accord or at the request of the party concerned (RoP, 9).

The Bureau. The Bureau consists of four members, a chairperson and a vice-chairperson of each branch, two of whom are from Annex I parties and two from non-Annex I parties, elected for a term of two years (NCP, Section II(4)). It allocates questions of implementation (QoI) to the appropriate branch based on its mandates (NCP, Section VII(1); RoP, 19.1). It requires a majority of its members to decide on allocating a QoI to the EB, rather than the FB. The provisional agenda for each meeting of the plenary is drafted by the secretariat in agreement with the Bureau (RoP, 7.1), and the Bureau may entrust one or more members of one branch to contribute to the other branch's work on a non-voting basis (NCP, Section II(7)).

---

[7]The rules of procedure of the Compliance Committee of the Kyoto Protocol (RoP) were adopted by Decision 4 (MOP 2, 2006), and then amended by Decision 4 (MOP 4, 2008).

The Facilitative Branch (FB). The FB aims to provide advice and assistance to parties (particularly developing countries and Annex I economies in transition countries) to promote their compliance with their commitments under the Protocol on the basis of the principle of common but differentiated responsibilities and the circumstances pertaining to the questions before it (NCP, Section IV, 4).

On the basis of this purpose, it has several tasks. From its inception, it was responsible for helping Annex I parties comply with the provisions which stipulated that the overall emissions of GHGs should be reduced to at least 5 per cent below 1990 levels during the commitment period 2008–2012 and for ensuring they did not exceed the assigned amounts (Art. 3.1, KP). Its other responsibilities are to establish national systems for the estimation of emissions (Art. 5.1, KP); to invoke methodologies and appropriate adjustments for estimating emissions (Art. 5.2, KP); to incorporate the necessary supplementary information in its annual inventory of emissions (Art. 7.1, KP); to adopt guidelines for the preparation of the supplementary information; and to decide on modalities for the accounting of assigned amounts (Art. 7.4, KP; NCP, Section IV, 6). Moreover, it addresses QoIs to determine whether Annex I parties are complying with their commitments on qualified emissions limitation or reduction commitments, reporting requirements and methodologies, and eligibility requirements to access the flexible mechanisms (Arts. 6, 12, 17, Art. 3.2, KP; NCP, Section IV, 5). In order to mitigate climate change and minimize its adverse impacts on developing countries, after conducting its examination the FB is also entrusted with taking any necessary action (Art. 3.14, KP) set out in Section XIV (NCP, Section VI, 7), such as advice and facilitation of financial and technical assistance, including technology transfer and capacity-building, or the formulation of recommendations.

The Enforcement Branch (EB). The EB is a quasi-judicial body which has discretionary power to impose strict consequences. Therefore it is competent to decide on questions of implementation relating to Annex I parties' reduction commitments under Art. 3.1, methodological and reporting requirements for greenhouse gas inventories under Arts. 5 and 7 (Art. 5.1.2, Art. 7.1.4, KP) and eligibility requirements for the Kyoto mechanisms under Arts. 6, 12 and 17, KP (NCP, Section V(4)). In addition, in the case of a disagreement between a party involved and an expert review team, it can also decide on whether to apply adjustments to greenhouse gas inventories or to correct the compilation and accounting database for assigned amounts (NCP, Section V(5)). When it finds Annex I parties in non-compliance with their emission target-related commitments mentioned above, it is responsible for resolving that compliance question by applying the consequences set out in Section XV (NCP, Section V(6)). These consequences applied by the EB differ depending on the type of each case of non-compliance.

The adoption of decisions by the EB requires a 75% majority of the members present and voting. In addition, it requires a majority of present and voting members of Annex I and Non-Annex I parties (NCP, Section II(9)). That is, for the adoption of a decision on non-compliance of the parties, a 75% majority of the members present and voting is insufficient; a majority of present and voting members of Annex I and Non-Annex I parties is also necessary. So, both Annex I and

Non-Annex I parties can block EB decisions. It has been argued that this provision is designed to prevent developing countries that do not have any reduction commitments controlling Annex I parties (developed countries) which do have reduction commitments (Brunnée 2003; Wolfrum/Friedrich 2006).

In conclusion, it should be noted that there is no hierarchical relationship between the EB and the FB. They should cooperate in their operation (NCP, Section II(7)). In fact, a question of implementation submitted to the FB for examination can subsequently reach the EB, and the EB, if necessary, can refer a question of implementation to the FB for consideration (NCP, Section IX(12); RoP, 23).

Procedural Mechanism: Phases and Safeguards

Phases of the CM

The procedure within the two branches can be scrutinized in four phases: the submission and allocation of the question of implementation (QoI) phase; the preliminary examination phase; the substantive examination phase, involving the procedure before the Facilitative Branch (FB) (NCP, Section VIII; RoP, 24) and the procedure before the Enforcement Branch (EB) (NCP, Section IX–X); and the appeal phase (NCP, Section XI).

Submission Phase. The NCP may be triggered by a submission raising a question of implementation (QoI) in three different ways: by any party with respect to itself (self-triggering or 'self-denunciation' (Urbinati 2009: 72), (NCP, Section VI(1a); RoP, 14), or by any party with respect to another party (NCP, Section VI(1b); RoP, 15), or by the reports of expert review teams under Art. 8, KP (NCP, Section VI(3)).

However, the Protocol does not allow IGOs or NGOs to trigger the NCP. Also, in contrast to the MCP referred by the Convention (Art. 13), which contains triggering by the COP as well as by the party itself or by another party (COP 4, Decision 10 1998: para. 5), the NCP of the Protocol does not include initiation of the procedure by the COP/MOP.

Additionally, the secretariat is not entitled to trigger the procedure, thus not endowed "with a stronger role regarding implementation supervision" (Oberthür/ Ott 1999: 214). Due to the fact that the secretariat established under the Convention serves as the secretariat of the Protocol (Art. 14.1, KP), its functions comprise those stipulated in Art. 8, UNFCCC and other functions assigned to it under the Kyoto Protocol (Art. 14.2, KP, see also RoP, 12, 16, 19.2–3, 20.2–3, 22.2 and 23.2). According to these functions, it is only entitled to notify the members and alternate members of the branches (NCP, Section VI(1); RoP, 19.2–3) and also the party in question (NCP, Section VI(2)) in respect of which the QoI is raised, and to convey the reports of expert review teams to the Committee.

The function defined under Arts. 8.2–3, KP can be seen as one of its most important functions, since it allows the secretariat to coordinate expert review teams and circulate their reports, involving technical assessment of all aspects of the implementation by a party to the Protocol, to all parties to the Convention. It is also competent to list those questions of implementation indicated in such reports for further consideration by the MOP. This provision of the Protocol had, in fact, been applied in practice under the Convention, despite the non-existence of a specific provision about it (Oberthür/Ott 1999). This provision formally incorporates this practice into the Protocol. Thus it has further improved the position of the secretariat in evaluating the parties' implementation of the Protocol through the expert review process.

The Preliminary Phase. After the allocation of questions of implementation to the relevant branch by the bureau within seven days of receipt of the QoI (NCP, Section VII, 1; RoP 19.1), the relevant branch undertakes a preliminary examination to identify whether the questions are supported by sufficient information, are not *de minimis* or ill-founded, and are based on the requirements of the Protocol (NCP, Section VII, 2). This examination should be completed within three weeks of the date of receipt of the questions by the branch (NCP, Section VII, 3).

The Substantive Phase

The Procedure before the FB (NCP, Section VIII; RoP, 24). The general procedures set up in Section VIII are applied from this stage onwards. The party in question, which is not allowed to be present during the adoption of the decision, should designate one or more persons to represent it during the examination (NCP, Section VIII, 2). The deliberations of the relevant branch should be based on any relevant information provided by the reports of numerous bodies, including the expert review teams (Art. 8, KP), the relevant party, the party submitting the question, the MOP, the subsidiary bodies under the Convention, and the Protocol and the other branch (NCP, Section VIII, 3). Relevant factual and technical information provided by qualified IGOs and NGOs (NCP, Section VIII, 4; RoP, 20) and expert advice (NCP, Section VIII, 5; RoP, 21) can also be used.

The Procedure before the Enforcement Branch (EB) (NCP, Section IX–X). If the party does not provide a written submission within four weeks of the date of receipt of the written submission of the party concerned, or of the date of any hearing, or of the notification made by the secretariat, the EB can adopt a preliminary finding that the party concerned is not in compliance with commitments under the related articles of the Protocol (NCP, Section V, 4), or can determine not to proceed further (NCP, Section IX, 4–5). The party concerned is then notified of the preliminary finding or decision not to proceed. The secretariat makes the decision not to proceed available to the other parties and to the public (NCP, Section IX, 6).

   If the party concerned does not provide a further written submission to the EB within ten weeks of the date of receipt of the notification of the preliminary finding, the EB makes a decision – which will be final unless overturned on appeal – confirming its preliminary finding (NCP, Section IX, 7). If the party concerned does provide a further written submission, the EB considers it and adopts a final decision within four weeks of the date it received the further submission (NCP, Section IX, 8). The final decision is made available to the other parties and to the public and notified to the party concerned by the secretariat (NCP, Section IX, 10).

The Appeal Phase (NCP, Section XI). As a general rule, it is not possible to appeal the decisions of the Committee. However, there is one exception to this rule which assures the right of appeal to the MOP. It is therefore beneficial to briefly discuss the MOP and its role in compliance issues.

The Conference of the Parties (COP), the supreme body of the Convention, also serves as the MOP to the Protocol (Art. 13.1, KP); so, the rules of procedure of the COP and financial procedures applied under the Convention are applied *mutatis mutandis* under the Protocol (Art. 13(5), KP). However, because membership is different and only parties to the Protocol exercise voting rights, this body is 'independent'[8] from that of Convention (Oberthür/Ott 1999: 240). Indeed, if the MOP decides differently by consensus, this decision becomes valid for the parties; being party to the Convention is not sufficient to receive a role in the MOP. In order to have voting rights and make decisions, it is necessary to be a party to the Protocol. This is because, as in Art. 17.5 of the Convention, Art. 13.2 of the Protocol states that decisions under the Protocol should be taken only by those that are parties to it. So, if the parties to the Convention are not parties to the Protocol, they can merely participate as observers in the proceedings of any session of the MOP (Art. 13.2, KP). In addition, if any member of the Bureau of the COP representing a party to the Convention is not a party to the Protocol, it should be replaced by an additional member which is a party to the Protocol (Art. 13.3, KP).

Sessions of the MOP should be held annually in conjunction with ordinary sessions of the COP, unless otherwise decided by the MOP (Art. 13.6, KP). Extraordinary sessions, on the other hand, should be held at such other times as the MOP deems necessary or at the written request of any party, provided that within six months of the request being communicated to the parties by the secretariat, it is supported by at least one third of the parties (Art. 13.7, KP).

The participation of MOP observers from the UN, its specialized agencies and the International Atomic Energy Agency (IAEA), or any member not party to the Convention, is permitted through Art. 13.8, KP. Furthermore, any body or agency qualified in matters covered by the protocol can also be admitted, provided that at least one third of the parties does not object to its wish to be represented as an observer at a session of the MOP (Art. 13.8, KP).

The functions of MOP on reviewing and promoting the implementation of the Protocol are listed in Art. 13.4, KP and in NCP, Section XII, and include: evaluating the implementation by the parties and the overall effects of the measures taken (13.4a), periodically examining the obligations of the parties under the Protocol (13.4b), promoting the exchange of information on measures adopted by the parties (13.4c), facilitating the coordination of measures adopted by them (13.4d), providing guidance on the development and periodic refinement of comparable methodologies (13.4e), advising on any necessary matters related to the

---

[8]The same approach adopted for the relationship between the COP and MOP is used for the subsidiary bodies as well. As in the case of the relationship between the COP and the MOP, the subsidiary bodies of the Protocol are independent of those of the Convention.

implementation (13.4f), mobilizing additional financial resources (13.4g), establishing subsidiary bodies necessary for the implementation (13.4h), utilizing the services and cooperation of competent IOs, IGOs and NGOs (13.4i), considering the reports of the expert review teams and the reports of the plenary on the progress of its work, providing general policy guidance including on any issues regarding implementation, adopting decisions on proposals on administrative and budgetary matters, and deciding on appeals (NCP, Section XII(a–e)).

It is also possible to find additional functions given to the MOP in various articles of the Protocol, as in Arts. 2.3, 2.4, 3.4, 3.5, 3.6, 6.2, 12, 16 etc. Through Art. 13.4j, a "blanket clause" is also provided which gives power to the MOP to exercise such other functions which may be required for the implementation of the Protocol (Oberthür/Ott 1999: 239).

Specifically regarding the compliance issue, it should be stressed that the MOP does not have the power to decide on non-compliance or to adopt responses to it. It only has the power to decide, in the case of an appeal against the Enforcement Branch's decision, whether due process has been denied. That is, a party can appeal to the MOP against a final decision of the EB if it believes it has been denied due process, but not about the substantive content of the matter (NCP, Section XI, 1).

Any appeal has to be made within 45 days of the party being informed of the decision of the EB (NCP, Section XI, 2). If there has been no appeal within 45 days of the decision, it becomes definitive (NCP, Section XI, 4).

As a result of its examination, the MOP can override the EB's decision and refer the matter back to the EB on a 75% majority vote of the parties present and voting at the meeting (NCP, Section XII(e); NCP, Section XI, 3). This implies that the MOP can only refer the matter back to be re-examined by the EB; it cannot make a decision on the substantive content of the matter, but only on the rules of due process, so on a very narrow scope.

Thus, with this exception, as a political body, the MOP has been endowed with the power to examine the decision of a quasi-judicial body, the EB, even though this examination is particularly focused on a legal question, i.e. whether the rules of the due process have been respected with regard to the party concerned.

Procedural Safeguards

The procedure applied by the ComplCom for the resolution of non-compliance by the parties set out in NCP, Sections VIII (general procedures), IX (procedures for the EB), and X (expedited procedures for the EB) have to guarantee due process (Hovi 2005; Ulfstein/Werksman 2005; Urbinati 2009). Therefore, it should next be asked what should be done to guarantee 'due process' for the parties concerned.

The Rights of the Party Concerned. First of all, the party in question should have the right to be represented during the consideration of the QoI before the relevant branch, except during the elaboration and adoption of a decision of the branch (NCP, Section VIII, 2; RoP, 9.2).

That party, after the preliminary examination, is notified by the secretariat in writing of the decision to proceed (or, in the event of the review of eligibility requirements, of the decision not to proceed) (NCP, Section VII(4, 5)). Until the decision to proceed with the question has been made, the party concerned cannot intervene in the process, but it can comment on all relevant information and on the decision to proceed once it has been made (NCP, Section VII, 7). In addition, any decision not to proceed is made available to other parties and to the public by the secretariat (NCP, Section VII(6)). The party concerned can also comment in writing on any relevant information provided by the reports of different bodies or qualified IGOs and NGOs. The information considered by the branch is made available to the public, unless the branch decides that it should not be made available to the public until its decision has become final (NCP, Section VIII(6)). The final decision on the content of a preliminary finding or final decision (see RoP 22.1) is notified to the party concerned and made available to other parties and to the public by the secretariat (NCP, Section VIII, 7). The party concerned can also comment on this decision and any other decision of the branch (NCP, Section VIII, 8). Comments on a final decision should be submitted in writing within 45 days of receipt of that decision by the party in question (RoP, 22.2).

As for the procedures for the EB, within ten weeks of the date of receipt of the notification, the party concerned may contact the EB to disprove information submitted to the branch (NCP, Section IX, 1). Within four weeks of the date of receipt of the application, the EB holds a hearing at which the party concerned can present its views, expert testimony or opinion (RoP, 25.1). Such a hearing is held in public, unless the EB decides otherwise of its own accord or at the request of the party concerned (NCP, Section IX, 2). The EB can ask for further clarification from the party concerned, either in such a hearing or at any time in writing, in which case, the party has to provide a response within six weeks thereafter (NCP, Section IX, 3).

In brief, the party concerned has the right to access any information considered by the relevant branch (NCP, Section VIII, 6), and any information upon which decisions are made, and any decisions given by the relevant branch (NCP, Section VII, 4–5; NCP, Section VIII, 6; NCP, Section IX, 10); to comment on such information (NCP, Section VII, 7; Section VIII, 6) and on any decision of the relevant branch (NCP, Section VII, 7; NCP, Section VIII, 8; NCP, Section IX, 6; Section IX, 1); and the opportunity to present its views or expert testimony (NCP, Section IX, 2).

Predetermined Deadlines. The procedure applied by the ComplCom for the resolution of an alleged situation of non-compliance involves precise time limitations. To illustrate, the allocation of QoI by the bureau should be within seven days of receipt of the QoI (NCP, Section VII, 1; RoP 19.1); the preliminary examination should be completed within three weeks of the date of receipt of the questions by the branch (NCP, Section VII, 3); and comments in writing on a final decision should be submitted within 45 days of the receipt of that decision by the party in question (RoP, 22.2). If the EB asks for further clarification from the party concerned, the latter has to provide a response within six weeks (NCP, Section IX, 3);

if the party does not provide a written submission within four weeks of the date of receipt of the written submission of the party concerned, or of the date of any hearing, or of the notification made by the secretariat, the EB can adopt a preliminary finding (Section V, 4), or can determine not to proceed further (NCP, Section IX, 4–5); if the party concerned does not provide a further written submission to the EB within ten weeks of the date of receipt of the notification of the preliminary finding, the EB adopts a final decision (NCP, Section IX, 7). The EB may also consider adopting a final decision within four weeks of the date it received the further submission (NCP, Section IX, 8). The appeal should be made within 45 days of the party concerned being informed of the EB's decision (Section XI, 2)). Through its expedited procedure (NCP, Section X), which is provided for questions of implementation relating to eligibility requirements under Articles 6, 12 and 17 of the Protocol, both for proceedings to suspend eligibility and proceedings to have eligibility reinstated, the NCP prevents time-consuming delays and ensures a faster-operating process.

Impartiality and Independence. Other important requirements are stipulated to ensure due process. Each branch of the ComplCom's is composed of ten members and alternate members for each member. These members and their alternates should be qualified in climate change and in relevant fields (the scientific, technical, socio-economic or legal fields). The EB members also have legal experience (NCP, Section V(3)) and should serve in their personal capacities (NCP, Section II(6); RoP, 4).

The very limited role of the political organ (the MOP) should be viewed as a very effective factor to ensure due process (Ulfstein/Werksman 2005). As explained in detail above, the MOP in the CM of the Kyoto Protocol does not have the power to decide on non-compliance or to adopt responses about it, except in the case of an appeal against the denial of due process in the EB's decision (NCP, Section XII(e); NCP, Section XI).

Predetermined Consequences. NCP, Section XIV, defines which consequences should be applied by the FB, and XV defines which ones should be applied by the EB in response to three circumstances of non-compliance with different commitments: non-compliance with reporting requirements, non-compliance with eligibility requirements, and non-compliance with emission commitments. In fact, the response measures which should be applied to the non-compliant party are pre-designated certain consequences. They are imposed by the branches of the ComplCom as either positive or negative measures when they are needed to induce compliance by parties. In general, if the party has difficulty complying with the provisions of the Protocol because of its lack of material, institutional or financial resources, positive measures are applied as appropriate means to induce compliance. On the other hand, if there is deliberate non-compliance determined by the EB, negative measures involving far-reaching economic implications can be imposed on the non-compliant party to bring it back to compliance.

These "fixed consequences" that may be adopted in the NCPs of the Protocol are "well designed to give credibility and legitimacy" and to enhance the predictability of the consequences to be applied to non-compliant parties (Ulfstein/Werksman 2005: 59).

Proportionality. Response measures should not be disproportionate to the nature of the obligation or the nature of the breach. As stated in Art. 18, the Protocol requires "the development of an indicative list of consequences, taking into account the cause, type, degree and frequency of non-compliance." Therefore, it stresses, through this statement, respect for the principle of proportionality between the nature of the non-compliance and the type of measures applied in the event of that non-compliance.

Transparency. The procedure should provide access to information to both parties and non-parties (IGOs-NGOs) with a high level of transparency, in line with the "right to know" required by important binding instruments (see e.g. the 1998 Aarhus Convention[9]). This is particularly because transparency may induce the parties to comply with their commitments, even before the initiation of the CM, through providing information to NGOs which initiates a debate on the relevant issue in the domestic sphere, thereby establishing "shaming infrastructures" (Hovi et al. 2005: 8).

There are also several opportunities for NGOs to participate in the processes of the CM (Andresen/Gulbrandsen 2005). To illustrate, the NCP of the Protocol allows the competent intergovernmental (IGOs) and non-governmental organizations (NGOs) to submit relevant factual and technical information to the relevant branch, and also allows each branch to seek expert advice (NCP, Section VIII, paras. 4–5). They are also allowed to participate in the EB deliberations and hearings with 'observer status', in accordance with Art. 7(6), Convention. They can monitor and evaluate certified CDM and sink projects, and attend the Executive Board CDM meetings. They can convince the parties (using the shame-effect) to refrain from buying GHG emission allowances from the countries within a hot air loophole, such as Russia and Central and Eastern European Countries (CEECs). They may also exercise a potential influence not only on Annex I parties, but also Non-Annex I parties, by ensuring "the quality of technology transferred from Annex I to Non-Annex I parties as well as its appropriateness to local circumstances" (Andresen/Gulbrandsen 2005: 180). Besides these internal instruments provided for in the CM, NGOs can also apply external strategies, like reputation of the parties in the international community, to influence the parties' compliance.

In summation, although the NCP of the Protocol has eventually become "somewhat less transparent and open than the NGOs had advocated" (Andresen/Gulbrandsen 2005: 176); it can be argued that they have been notably successful in "overall goal attainment on the CM" (Andresen/Gulbrandsen 2005: 182).[10]

---

[9]See: https://www.unece.org/fileadmin/DAM/env/pp/documents/cep43e.pdf.

[10]See also Dannenmaier (2012) for a detailed analysis of the role of non-state actors in climate compliance.

Finally, in order to ensure publicity, the information considered by the EB (NCP, Section VIII, 6), the decisions not to proceed (NCP, Section VII, 6; NCP, Section IX, 6), final decisions (NCP, Section VIII, 7; NCP, Section IX, 10) including the consequences (RoP 22f), hearings of the EB (NCP, Section IX, 2), and also of all meetings of plenary and branches – except the adoption of decisions (RoP, 9.1) – are made available to the public in the procedure.

Appeal. As an exception, a party can appeal to the MOP against a final decision of the EB relating to emissions targets (Art. 3(1), KP) if it believes it has been denied due process (Section XI, 1). Thus, even though the exception is very narrowly framed – only if the rules on the due process have been disregarded, not if the substantive content of the matter has been infringed – the appeal results in giving power a political body, the MOP, to examine the decision of a quasi-judicial body, the EB. Indeed, through this appeal procedure, the parties attain the opportunity to send the EB's decision back to it for re-examination, but, the MOP cannot give a final decision on the QoI; it can only refer it back to the EB.

6.1.1.3 Non-Compliance Response Measures

When response measures aim to ensure compliance with the Kyoto Protocol, the consequences (Art. 18, KP) designated by the COP/MOP (NCP, Section XIV–XV) according to the circumstances of non-compliance with different commitments (which will be detailed in the subsequent sections), should be taken into account.

In order to ensure the compliance of the parties, these consequences may consist of positive measures, such as cooperation, assistance through financial means and capacity-building, monitoring and verification and assistance through procedural means (the consequences particularly applied by the FB), or negative measures,, such as the deduction of tonnes of a party's assigned amount at a penalty rate equal to 1.3 times the amount in tonnes of excess emissions, or suspension of eligibility to participate in the flexible mechanisms. In principle, these consequences are particularly applied by the EB.

Positive Response Measures

Positive measures may be adopted by the FB after its examination based on the ERT's assessment and identification of 'any potential problems' in the fulfilment of the relevant party's commitments. If it finds that this party has difficulty meeting its Protocol targets and is therefore in non-compliance with the provisions of the Protocol, it can apply one or more of the consequences which it is entitled to apply.

With regard to those consequences listed within NCP, Section XIV(a–d), the FB can provide advice and assistance to the non-compliant parties, and can facilitate financial and technical assistance (Art. 11, VC), including technology transfer[11] and capacity-building, from sources other than those established under the Convention and the Protocol for developing countries or having regard to Arts. 4.3, 4.4, 4.5, VC.[12] It can also adopt recommendations which take into account Art. 4.7, Convention, according to which the extent of the developing parties' implementation depends on the effective implementation by developed country parties of their commitments related to financial resources and the transfer of technology.

As seen from the consequences applied by the FB, the FB strongly requires financial resources to help the developing countries that need technical and financial assistance in order to facilitate compliance. It can meet the costs of their compliance, including sometimes the operation of relevant projects, sometimes training of national officials, or the enhancement of scientific or technological facilities or data systems etc. Hence the effectiveness of the FB, i.e. the facilitation of compliance, ultimately depends on the availability of funds to provide the resources necessary to meet all these costs. As the funds are "likely to fall considerably short of the needs", strengthening the funds becomes a particularly significant aspect of the activities of the FB (Hovi et al. 2005: 9).

On the other hand, when the positive measures employed by the EB are analysed, two circumstances of non-compliance with different commitments can be identified:

- non-compliance with reporting requirements under Arts. 5.1, KP (national system), 5.2, KP (methodologies), Arts. 7, KP (inventories), 7.4, KP (guidelines for the preparation of the information-modalities for the accounting of assigned amounts) (Section XV, 1–3), and
- non-compliance with emission commitments under Art. 3.1, KP (not to exceed assigned amount) (Section XV, 5–8).

As a response to these circumstances of non-compliance, three different consequences can be applied by the EB:

---

[11]Recalling its COP 1/Decisions 11, COP 2/Decision 7, COP 3/Decision 9, COP 4/Decision 4, COP 5/Decision 9 and the relevant provisions of its COP 4/Decision 1 on the Buenos Aires Plan of Action, and its COP 6/Decision 5, containing the Bonn Agreements on the implementation of the Buenos Aires Plan of Action, in COP 7/Decision 4, the COP establishes an Expert Group on Technology Transfer (EGTT) to facilitate and advance technology transfer activities (Decision 4/ COP 7 2001: 22, para. 2). It also sets out a number of technology transfer activities involving five key themes for meaningful actions: technology needs and needs assessments, technology information, enabling environments, capacity-building, and mechanisms for technology transfer (Annex, COP 7, Decision 4 2001: 24–30). Further, it requests the secretariat to develop an information clearing house, including a network of technology information centres, to facilitate the flow of, and access to, information on developing and transferring safe technologies (COP 7, Decision 4 2001: 26, para. 10(c, d)).

[12]Article 4.3, in line with Art. 11.2(a, b), KP: developed country parties have to provide financial resources and the transfer of technology to developing countries. Art. 4.4: developed country parties have to assist developing countries. Art. 4.5: developed country parties have to promote, facilitate and finance the transfer of, or access to, environmentally sound technologies and know-how to other parties, particularly developing country parties.

- determination of non-compliance;
- adoption of a compliance action plan;
- submission of progress reports.

Determination of Non-compliance. If a party fails to submit its national reports, annual inventories, or related information to the EB, or fails to comply with the requirements regarding them – "procedural non-compliance" – or with its reduction commitments – "actual (substantive) non-compliance" (Goldberg et al. 1998: 3) – the EB can declare that the party is in non-compliance.

Adoption of a Compliance Action Plan. After declaring that the party is in non-compliance, the EB can also ask the party to develop a compliance action plan (NCP, Section XV, 1(a, b); NCP, Section XV, 5(b)). In such a case, the party has to submit its plan within three months of the determination of non-compliance or of a longer period that the EB considers appropriate. Its plan has to include an analysis of the causes of the non-compliance, measures planned to be implemented to remedy the non-compliance and a timetable for implementing such measures, which must not exceed twelve months for the assessment of the progress made in the implementation (NCP, Section XV, 2(a–c)). In the case of non-compliance of emission commitments, the plan has to contain domestic policies and measures that the party plans to implement to meet its quantified emission limitation or reduction commitment in the subsequent commitment period and a timetable for implementing its plans that does not exceed three years or the end of the subsequent commitment period (NCP, Section XV, 6(a–c)).

Submission of Progress Reports. The non-compliant party is also requested to submit its progress on the plan's implementation, reporting to the EB regularly (NCP, Section XV, 3; NCP, Section XV, 7).

Negative Response Measures

With regard to the negative measures employed by the EB, two circumstances of non-compliance with different commitments can be identified:

- non-compliance with eligibility requirements under Arts. 6, 12 and 17, KP (flexibility mechanisms) (Section XV, 4);
- non-compliance with emission commitments under Art. 3.1, KP (not to exceed assigned amount) (Section XV, 5–8).

As a response to these circumstances of non-compliance, three different consequences can be applied by the EB:

- suspension of eligibility to participate in the flexible mechanisms;
- deduction of tonnes of a party's assigned amount at a penalty rate;
- suspension of the eligibility to make transfers.

Suspension of Eligibility to Participate in the Flexible Mechanisms. The EB can withdraw the eligibility of a party which has failed to meet the criteria for participating in the flexibility mechanisms, i.e. can suspend the party's accession to the flexibility mechanisms, in line with the relevant provisions under those articles.

Eligibility may only be restored in accordance with the procedure in Section X, 2 (NCP, Section XV, 4). The party can ask the EB to reinstate its eligibility either directly or through an expert review team (ERT). In response to a request submitted through an ERT, if the ERT confirms that a QoI no longer exists regarding the eligibility of the party concerned, the EB should restore the party's eligibility, unless it decides that such a QoI continues to exist. In response to a request submitted directly, if the EB decides on the non-existence of a QoI, it reinstates that party's eligibility. If it decides otherwise, it has to apply the procedure set out under Section X, 1 which is regulated specifically for QoIs relating to eligibility requirements under Arts. 6, 12 and 17 of the Protocol.

This response measure is generally criticised, as suspending a party's ability to use the flexibility mechanisms can prevent it bringing itself into compliance through these mechanisms (Crossen 2004).

Deduction of Tonnes of a Party's Assigned Amount. If the party concerned is found in breach of its emissions target, the EB can deduct tonnes of a party's assigned amount at a penalty rate (equal to 1.3 times the amount in tonnes of excess emissions) in a subsequent commitment period from the party's assigned amount for the second commitment period (NCP, Section XV, 5(a)). That is, parties not in compliance with their commitments in the first commitment period (2008–2012) have to deduct 1.3 tonnes from their second period assigned amount for every tonne of gas they emitted in excess of their assigned targets in the first period. For subsequent commitment periods, the rate can be determined by an amendment (NCP, Section XV, 8). The concern relating to the deduction penalty is that the parties can inflate their assigned amount for the second commitment period to accommodate the deduction (Crossen 2004; Wang/Wiser 2002). In addition, this response raises the problem of unpaid emissions debts, as it theoretically allows parties to borrow from the subsequent period unlimitedly, leading to excessive leniency in the calculation of the deduction rate (Wang/Wiser 2002).

Suspension of the Eligibility to Make Transfers. After declaring that the party is in non-compliance with its emission commitments under Art. 3.1, KP, the EB can suspend the party's eligibility to make transfers on its surplus – not emission credits for its own compliance – under emissions trading (Art 17, KP).

The party's eligibility to make transfers under emissions trading is suspended until it is restored by the EB in accordance with the procedure developed particularly on the eligibility requirements for emissions trading under Section X, 3–4 (NCP, Section XV, 5c).

It should be emphasized here that the consequences applicable to Annex I parties in breach of their emissions targets cannot be fully tested until the first commitment period is expired. That is to say, after the completion of the expert review process, Annex I parties have 100 days to fulfil their commitments under Art. 3(1), KP, regarding compliance with emission targets by acquiring – or by the transfers of other parties to such party – Emission Reduction Units (ERUs) from JI projects,

Certified Emission Reductions (CERs) from the CDM, Assigned Amount Units (AAU)s from emissions trading and Reduction Units (RUs) from afforestation, reforestation and deforestation-related activities under Arts. 6, 12 and 17, KP, provided the eligibility of any such party has not been suspended in accordance with Section XV, 4 (NCP, Section XIII).

In short, the parties are allowed an additional period (true-up period (Goldberg et al. 1998; Mitchell 2005) or grace period (Ulfstein/Werksman 2005) to acquire and transfer emission units to meet their targets after the end of the commitment period running from 2008 to 2012. If the party's total emissions of regulated greenhouse gases exceed its assigned amount at the end of the commitment period, and following the additional period during which the parties can bring themselves within their assigned amounts, the EB can declare the non-compliance of the concerned party and can apply the consequence.

Binding Effect of Response Measures

The legal status of the response measures of the branches is till controversial. Pursuant to Art. 18, KP, any compliance procedures and mechanisms entailing binding consequences should be adopted by means of an amendment to the Protocol (see also MOP 1, Decision 2 2005c, para. 5), either as "separate amendments" (Goldberg et al. 1998: 64, footnote 31) adopted for individual forms of consequences, or as "omnibus amendments" (Goldberg et al. 1998: 64, footnote 31) adopted in a collective manner. This suggests that response measures would not be binding unless they become part of such an amendment, even if there is a decision by a branch of the Committee using "a prescriptive or mandatory language" (Eritja et al. 2004: 101).

However, it is likely that any amendment to the Protocol, which only binds the parties which have accepted it, would be very difficult to enforce effectively in practice (Eritja et al. 2004; Oberthür/Ott 1999; Wang/Wiser 2002). This is particularly because amendments enter into force only after ratification by at least three-quarters of the parties to the Protocol (Art. 20.4, KP). Ratification may therefore take a long time, and during this period some parties may prefer to stay out of it, and not be bound by the consequences. It also blurs the status of those parties wishing to sign up to the Protocol, but not to an amendment that reflects the binding nature of these consequences (Barrett 2003; Werksman 2005). That implies that when some parties to the Protocol are not also parties to the amendment, the unamended Protocol would govern the relationship between parties just to the Protocol and parties to both the Protocol and the amendment. Meanwhile the amended Protocol would be applied merely to parties who had ratified both the Protocol and the amendment.

It was proposed to adopt "some form of supplementary legal instrument, which all parties would agree to ratify at the same time that they ratified or acceded to the Protocol" (Wang/Wiser 2002: 198). However, the dispute on the legal status of the response measures is unresolved at present. So it is now necessary to discuss how the term 'binding' should be interpreted in response to those response measures.

Ulfstein/Werksman (2005) argue that the implications regarding whether or not a response measure should be considered binding under Art. 18 of the Protocol can be different for each of the consequences.

In particular they argue that soft consequences, such as determination of non-compliance, adoption of an action plan, and submission of progress reports, and also consequences which prevent the concerned party enjoying a 'privilege' given by the organs of the Protocol but not a 'right' stemming from the agreement itself, like suspension of the eligibility to use the flexible mechanisms (Ulfstein/ Werksman 2005: 58),[13] can be regarded as part of the "implied powers" of the organs of the Protocol. Therefore, they can be admitted as within the competence of the EB. So to become a binding measure, they should not be evaluated as requiring an amendment of the Protocol under its Art. 18.

On the other hand, in a case of non-compliance with the emission commitments under Art. 3.1, obliging the concerned party to deduct tonnes from its assigned amount at a penalty rate, the same authors assert that this response measure requires amendment. In this case, it can be claimed that parties which have withheld their consent to response measures adopted by the decisions of the COP/MOP (COP 7, Decision 24; MOP 1, Decision 27) have a legal basis for arguing that they are not bound by these deductions.

Thus, except the measure requiring deductions of assigned amounts at a penalty rate, other measures can be considered to be adopted by the EB in the exercise of its implied powers. However, it should not be forgotten that insistence on adoption of a measure requiring deductions in legally binding form by an amendment to the Protocol can result in the unwillingness of some parties to ratify it. This directly affects the compliance attitudes of the parties as well as the functioning of the whole regime.

6.1.2   Kyoto Protocol's Compliance Mechanism in Practice

How effectively the CM of the Protocol operates in practice largely depends on the ability of its provisions to induce compliance. In order to clarify to what extent its provisions are influential in practice, this part will focus on the impacts of those provisions on the functioning of the mechanism in practice.

When the implementation of the CM in practice is examined, it is apparent that, to date, there has been one question of implementation (QoI) to the FB by South Africa, as Chairman of the Group of 77 and China and eight QoIs to the EB by ERTs.

---

[13]See also Goldberg et al. (1998) for the same view on the binding status of the measure requiring the suspension of eligibility.

6.1.2.1 The Facilitative Branch (FB) Cases

The QoI by South Africa to the FB was raised on 26 May 2006 against the 15 Annex I parties which had not provided their reports demonstrating progress, even though the deadline (1 January 2006) for the submission of reports had passed (MOP 1, Decision 22 2005a; COP 7; Art. 3.2, KP).

This submission was made on the grounds that failure to submit timely information about national communication and non-compliance can constitute an "early warning of potential non-compliance" by the parties with commitments under Art. 3.1, KP (Ulfstein/Werksman 2005: 45). Raising the QoI indicated that the FB was expected to encourage the parties to submit the requisite information as well as provide advice and facilitate timely submission and early warning of potential non-compliance.

The FB began a preliminary examination of the QoI on 31 May 2006. However, it managed to adopt only two decisions: on Latvia and Slovenia. The Branch decided not to proceed against either of these parties after considering their cases. This was because Latvia's fourth national communication and its progress report had been received by the secretariat on 25 May 2006, before the FB's consideration of its case started (CC-2006-8-3/Latvia/FB; CC-2006-8-4/FB), and those of Slovenia had been received on 12 June 2006, just after its consideration proceeded (CC-2006-14-2/Slovenia/FB; CC-2006-14-3/FB).

The other thirteen parties' cases were not considered despite a number of attempts by the branch to arrive at a consensus. Thus, due to the lack of consensus and also the lack of a 75% majority of the members present and voting to adopt either a decision to proceed or a decision not to proceed (NCP, Section II, para. 9; NCP, Section VII, paras. 4, 6), no decision could be taken about them (CC-2006-1-2/FB; CC-2006-2-3/FB; CC-2006-3-3/FB; CC-2006-4-3/FB; CC-2006-5-2/FB; CC-2006-6-2/FB; CC-2006-7-2/FB; CC-2006-9-2/FB;

Table 6.3 Kyoto Protocol CM in practice/Facilitative Branch (FB)

Question of Implementation by South Africa, as Chairman of the Group of 77 and China(2006)		
Parties concerned	Subject of the Question	Results
Austria, Bulgaria, Canada, France, Germany, Ireland, Italy, Liechtenstein, Luxembourg, Poland, Portugal, Russian Federation, Ukraine	Non-compliance with the emission commitments under Art. 3.1, KP (failure in the submission of national communications and reports)	No decision due to the lack of consensus and majority of 75% (21 June 2006)
Latvia, Slovenia	Non-compliance with the emission commitments under Art. 3.1, KP (failure in the submission of national communications and reports)	Decision not to proceed against either party

Source Prepared by the author on the basis of information/documentation, given on the official website of the UNFCCC. See: https://unfccc.int/process/bodies/constituted-bodies/compliance-committee-cc/questions-of-implementation/question-of-implementation-numerous-annex-1-party

CC-2006-10-2/FB;   CC-2006-11-3/FB;   CC-2006-12-3/FB;   CC-2006-13-2/FB;
CC-2006-15-2/FB).

6.1.2.2   The Enforcement Branch (EB) Cases

Nine QoIs by ERTs to the EB have been sent to the Committee for consideration
about eight parties to date. The cases on these parties – Bulgaria, Canada, Croatia,
Greece, Lithuania, Romania, Slovakia, Ukraine – will be analysed in the next
paragraphs.

The Bulgaria Case

Bulgaria, which first became eligible to participate in the mechanisms on 25
November 2008, submitted its 2009 annual inventory submission on 13 April 2009,
in order to provide information to maintain its eligibility to participate in the
flexibility mechanisms.

On 9 March 2010 the ERT finalized its Annual Report Review (ARR) 2009,
which contained a QoI relating to the national system of Bulgaria (CC-2010-1-1/
Bulgaria/EB; FCCC/ARR/2009/BGR). Specifically, the QoI related to compliance
with the guidelines for national systems, as functions of the national system of
Bulgaria did not ensure that Bulgaria's 2009 annual submission was sufficiently
complete and accurate, as required by the guidelines for national systems. In
addition, the ERT found Bulgaria's institutional arrangements and arrangements for
the technical competence of staff within the national system insufficient to enable
the adequate planning, preparation and management of the party's annual submission in accordance with the guidelines for national systems and other related
guidelines, such as guidelines for UNFCCC reporting, IPCC good practice, and
IPCC good practice for LULUCF. So it found Bulgaria ineligible to participate in
the flexibility mechanisms.

After considering the matter, the EB issued a preliminary finding of non-
compliance with respect to Bulgaria (CC-2010-1-6/Bulgaria/EB) and confirmed this
finding with its final decision on 28 June 2010 (CC-2010-1-8/Bulgaria/EB).

The consequences applied by the branch to Bulgaria were again the same as
those applied in the previous QoIs considered by the EB. In accordance with these
consequences, on 12 August 2010, Bulgaria submitted an updated improvement
plan to address its non-compliance (CC-2010-1-11/Bulgaria/EB).

As the EB 11 decided that this plan did not fully meet the necessary requirements and did not include an analysis of the causes of non-compliance (CC/EB/11/
2010/2), Bulgaria submitted a revised compliance action plan on 1 October 2010
(CC-2010-1-12/Bulgaria/EB) and a progress report on the implementation of its
revised plan on 27 January 2011 (CC-2010-1-15/Bulgaria/EB).

In order to benefit from the ERT's report of the review of the 2010 annual
submission of Bulgaria, the EB postponed its review and assessment of Bulgaria's
revised plan until after the publication of the ERT's report on 29 November 2010.

Table 6.4 The Bulgaria Case/Actions in Chronological Order

Date	Action
13 April 2009	Bulgaria submitted its 2009 annual inventory submission to the secretariat
28 September–3 October 2009	The ERT conducted an in-country review
9 March 2010	ERT finalized its ARR 2009, containing a QoI (CC-2010-1-1/Bulgaria/EB; FCCC/ARR/2009/BGR)
9 March 2010	Bulgaria was given official notification of the question of implementation
16 March 2010	The question was allocated to the EB
17 March 2010	The secretariat notified the members of the EB of the QoI
31 March 2010	The EB decided to proceed with the question (CC-2010-1-2/Bulgaria/EB)
8 April 2010	The EB received a request for a hearing from Bulgaria (CC-2010-1-3/Bulgaria/EB)
15 April 2010	The EB agreed to invite three experts drawn from the UNFCCC experts (CC-2010-1-4/Bulgaria/EB)
5 May 2010	A written submission was made by Bulgaria (CC-2010-1-5/Bulgaria/EB)
10–12 May 2010	The EB 9 meeting (CC/EB/9/2010/2)
10 May 2010	The EB 9 held a hearing
12 May 2010	The EB 9 resulted in a preliminary finding of non-compliance (CC-2010-1-6/Bulgaria/EB)
15 June 2010	Bulgaria made a further written submission (CC-2010-1-7/Bulgaria/EB)
28 June 2010	The EB 10 (CC/EB/10/2010/2) considered further written submission made by Bulgaria and gave its Final Decision (CC-2010-1-8/Bulgaria/EB)
21 July 2010	Comments were submitted by Bulgaria on the Final Decision (CC-2010-1-9/Bulgaria/EB)
3 August 2010	Bulgaria resubmitted comments on the Final Decision (CC-2010-1-10/Bulgaria/EB)
12 August 2010	Bulgaria submitted an updated improvement plan to address its non-compliance (CC-2010-1-11/Bulgaria/EB)
16 September 2010	The EB 11 (CC/EB/11/2010/2) noted that the plan did not fully meet the necessary requirements, and encouraged Bulgaria to submit a complete plan no later than 1 October 2010
1 October 2010	Bulgaria submitted a revised compliance action plan (CC-2010-1-12/Bulgaria/EB)
4–9 October 2010	In-country review was conducted
25 October 2010	The EB decided to defer the completion of the review and assessment of the revised plan until after the publication of the ERT's ARR 2010 (CC-2010-1-13/Bulgaria/EB)
29 November 2010	The ERT's ARR 2010 was published (CC/ERT/ARR/2010/18)
2 December 2010	Bulgaria submitted a request for the reinstatement of its eligibility (CC-2010-1-14/Bulgaria/EB)

(continued)

Table 6.4   (continued)

Date	Action
27 January 2011	Bulgaria submitted a progress report on the implementation of its revised plan (CC-2010-1-15/Bulgaria/EB)
3–4 February 2011	The EB 12 meeting was held (CC/EB/12/2011/2)
4 February 2011	The EB 12 decided that there was no longer a QoI, and that Bulgaria was fully eligible to participate in the Kyoto mechanisms

Source Prepared by the author on the basis of information/documentation on the official website of the UNFCCC. See: http://unfccc.int/kyoto_protocol/compliance/questions_of_implementation/items/5451.php

Then, at its twelfth meeting, held from 3 to 4 February 2011, taking into account the ERT's report and Bulgaria's revised plan, the EB 12 decided that there was no longer the need for a QoI with respect to Bulgaria's eligibility, as it had again become eligible to participate in the flexibility mechanisms (Table 6.4).

The Canada Case

Canada submitted its initial report to the secretariat on 15 March 2007. The ERT's report for Canada, published on 11 April 2008, involved a QoI in relation to Canada's compliance with the guidelines for the preparation of the information required under Art. 7, KP (MOP 1, Decision 15) and the modalities for the accounting of assigned amounts under Art. 7, para. 4, KP (MOP 1, Decision 13). In particular, the ERT asserted that Canada's national registry was not in accordance with the guidelines and modalities, and so it was not eligible to use market mechanisms.

After a preliminary examination, the EB decided to proceed with the submission made against Canada (CC-2008-1-2/Canada/EB), and then to apply the same consequences as applied to Greece (see section on Greece, below).

On 15 June 2008, after considering views and advice given by independent experts at the hearing held on 14–15 June 2008, it decided not to proceed further (CC-2008-1-6/Canada/EB).

Canada's recent declaration of its intention not to meet its emissions target for the first commitment period has raised the possibility of applying the FB process for Annex I parties which may be at risk of missing their emission reduction targets based on the demonstrable progress reports. Because the FB had never previously been used by the parties or the ERT for the purpose of facilitation on compliance with the emission reduction targets of Annex I parties, the Canada case indicated the need for reforming the FB into a system in which the FB's early warning function, in accordance with MOP 1, Decision 27, Section IV, paras. 4 and 6(a), can be used by Annex I parties in need of facilitation to meet their targets (FB 12 2012: Annex). The FB, however, stated that: "its early warning function is based on information contained in review reports and, as such, its ability to engage in an early warning exercise depends heavily on the timeliness and accuracy of such information" (FB 12 2012: para. 18). It also stressed the importance of timely information to carry out its early warning exercise, drawing attention to the "time gap between the year for which a Party reports and the year in which a review report is published" (FB 16 2014: para. 11).

After the withdrawal of Canada from the KP in December 2011 – in accordance with Art. 27 of the KP, allowing any party to withdraw by giving written notification to the Depository at any time after three years from the date on which the Protocol entered into force for that Party – the situation of Canada and its reporting obligations under the KP became controversial, particularly in light of Art. 70, VCLT. On this issue, the secretariat prepared a note focusing on the following essential points: Because Canada officially ceased to be a party to the KP from 15 December 2012, it did not have to further implement the KP in relation to the CP 2. In other words, from 15 December 2012, Canada no longer has any reporting obligations under the Protocol pursuant to Art. 70, paras. 1(a) and 2, VCLT. However, Canada remains bound by the mitigation and reporting requirements under the UNFCCC (Secretariat 2014: para. 15).

With regard to the effects of the withdrawal of Canada from its obligations in relation to the CP 1, if there is any obligation that was established before the date of

Table 6.5 The Canada Case/Actions in Chronological Order

Date	Action
15 March 2007	Canada submitted its initial report to the secretariat
5–10 November 2007	In-country review by ERT took place
11 April 2008	The Initial Review Report (IRR) for Canada was published (CC-2008-1-1/Canada/EB; FCCC/IRR/2007/CAN)
11 April 2008	The secretariat received a QoI in the report of the ERT
14 April 2008	The QoI was deemed to have been received by the ComplCom
14 April 2008	Canada was given official notification of the QoI
16 April 2008	The Bureau of the ComplCom allocated the QoI to the EB
17 April 2008	The secretariat notified the members of the EB of the QoI
2 May 2008	The EB decided, after a preliminary examination, to proceed with the QoI (CC-2008-1-2/Canada/EB)
5 May 2008	Canada was given official notification of the decision
20 May 2008	Canada requested a hearing (CC-2008-1-4/Canada/EB)
21 May 2008	The EB agreed to seek expert advice (four experts on national registries) (CC-2008-1-3/Canada/EB)
5 June 2008	Canada's written submission (CC-2008-1-5/Canada/EB)
14–15 June 2008	The hearing was held
14–16 June 2008	The hearing formed part of the EB 5 meeting (CC/EB/5/2008/2)
15 June 2008	The EB decided not to proceed further (CC-2008-1-6/Canada/EB)
16 June 2008	Canada was given official notification of the decision
14 July 2008	Further written submission of Canada (CC-2008-1-7/Canada/EB)

Source Prepared by the author on the basis of information/documentation on the official website of the UNFCCC. See: http://unfccc.int/kyoto_protocol/compliance/questions_of_implementation/items/5451.php the formal withdrawal of Canada and not affected by that withdrawal (Art. 70, para. 1(b), VCLT), or there is an obligation of Canada to include supplementary information for the period before 15 December 2012 when it was no longer a party to the Protocol, the situation becomes more blurred. Therefore a profound discussion was necessary to decide on its implications for the review process and the ComplCom (Secretariat 2014: para. 15). To illustrate, the question arose as to whether Canada's 2013 and 2014 annual submissions should be reviewed, as Canada remained a party to the KP until its withdrawal took effect on 15 December 2012. The EB, while noting that Canada's national inventory report and common reporting format tables submitted on 11 April 2014 were reviewed, also stressed that its mandate was limited to consideration of QoIs. So it completed its consideration of this item (EB 27 2015: para. 6) (Table 6.5).

The Croatia Case

The Croatia case plays a crucial role in other cases brought before the EB, as it was the first in which a party found non-compliant by the EB appealed to the MOP.

The process began with the ERT's report (CC-2009-1-1/Croatia/EB; FCCC/IRR/2008/HRV), which included two QoIs and was finalized on 26 August 2009, nearly one year after the submission of Croatia's initial report.

The QoIs were related to non-compliance with Croatia's assigned amount and its commitment period reserve. They were questioning whether a decision taken under the Convention would allow Croatia to issue more credits under the Protocol and thereby increase its commitment period reserve.

The EB adopted its final decision confirming the preliminary finding of non-compliance (CC-2009-1-6/Croatia/EB), thus confirming the consequences contained in para. 23 of the preliminary finding on 26 November 2009 (CC-2009-1-8/Croatia/EB). As a result of this final decision, the same consequences involving declaration of non-compliance, preparing a compliance action plan and being suspended from participation in market mechanisms, which were applied previously to both Greece and Canada, were applied to Croatia as well.

As it was found to be non-compliant with its assigned amount and commitment period reserve, it was advised that its plan should address the calculation of the assigned amount and the commitment period reserve of Croatia in line with Art. 3, paras. 7 and 8, KP, and the modalities for the accounting of assigned amounts contained in Decision 13 (MOP 1 2005b).

Croatia did not submit its plan, which should have been drafted to address its non-compliance within three months, expressing its appeal as a basis for its non-submission. Instead, on 14 January 2010, Croatia lodged an appeal against the final decision of the EB. This application to the MOP for appeal initiated a debate on the conditions of appeal. This is because although appeal is only possible for issues relating to the parties' 2012 targets and is additionally only permitted in the case of denial of due process under the CM of the KP, Croatia's application for appeal was not based on denial of due process, but on some other issues.

In its appeal, Croatia referred to the following eight reasons:

1. The final decision of the EB was not in line with Art. 31, paras. 1 and 2, VCLT, requiring that a treaty should be interpreted in good faith and in light of its object and purpose. Therefore both the Annex and the Preamble of the treaty should be taken into consideration for interpretation.

   On the basis of this reason, Croatia argued that it should be granted the same flexibility in the process of the implementation of its commitments under the KP as five other countries undergoing a similar process of transition to a market economy, namely Bulgaria, Hungary, Poland, Romania and Slovenia.

   In order to support its argument, it referred to Art. 4, para. 6, UNFCCC, COP 12, Decision 7 and COP 2, Decision 9; and it also asked to be allowed to add 3.5 Mt CO_2 equivalent to its 1990 level of GHG emissions (amounting to 31.7 Mt CO_2) not controlled by the MP, for the purpose of establishing the level of emissions for the base year for implementation of its commitments under Art. 4, para. 2, UNFCCC.

2. Secondly, Croatia claimed there had been violation of Art. 31, para. 3(b), VCLT, stating that regarding the treaty's interpretation, any subsequent practice in the application of the treaty should be taken into consideration. It put forward that, contrary to this provision, the EB did not take into account the flexibility for application of the KP target allowed under the Convention in the identical cases of Bulgaria, Hungary, Iceland, Poland, Romania and Slovenia.

3. Thirdly, Croatia asserted that the EB's final decision of non-compliance was manifestly absurd and unreasonable for several different reasons, particularly on the grounds of Croatia's historical circumstances. This resulted in the violation of Art. 32, VCLT, stipulating that when interpretation of the treaty is inadequate, and results in ambiguous, obscure, or manifestly absurd or unreasonable meanings, supplementary means of interpretation, such as the preparatory work of the treaty and the circumstances of its conclusion, should also be taken into account.

4. Moreover, Croatia alleged that the EB interpreted Art. 3, para. 5, KP in a very strict, inflexible and purely grammatical way, and this interpretation concluded in the application of flexibility only in the use of a base year or period other than 1990, and only to four EIT Parties explicitly identified in COP 2, Decision 9. So it decided that Croatia could not invoke Art. 3, para. 5, KP.

5. Another argument of Croatia, leaning on paragraph 3c of the EB's final decision, was about the violation of COP and MOP decisions and provisions of KP. In that paragraph, the EB stated that because the COP and the MOP are two different decision-making bodies, the fact that all parties to the Protocol are also parties to the UNFCCC does not provide a basis for establishing the application of COP decisions, such as Decision 7/COP 12, under the KP.

   Croatia, relying on Art. 7, 8, KP, MOP 1, Decision 27 (2005a), Annex, Section II, and MOP 1, Decision 13 (2005b), claimed that the EB's final decision directly contradicted all these provisions/decisions and also, instead of applying both Decision 9/COP 2 and Decision 7/COP 12, disregarded them.

6. By comparing the Iceland and Slovenian cases to its own case, Croatia also argued that it had not been accorded equal treatment. As these countries gained flexibility under the Convention and have been allowed to use it for the implementation of their KP commitments, Croatia contended that, under the KP, COP 12, Decision 7 should be accepted as directly applicable to its own case.

7. Furthermore, Croatia stressed the EB's reference to the EU delegation's remark expressed at COP 12, and underlined that the EB, both in its preliminary finding and final decision, did not provide an explanation for the EU delegation's remark and its implications for Croatia's case, and also did not grant Croatia an opportunity to respond to the EB on the matter in writing. Based on these reasons, it claimed the EB had violated its obligation to provide information relevant to its decision and to grant Croatia the right to respond.

8. Croatia's final argument was about the violation of independence, impartiality and conflict-of-interest principles (Rule 4, RoP). Here, it stressed that one of the alternate members of the EB who participated in the consideration and elaboration of the preliminary finding with respect to Croatia was also a member of the EU delegation at COP 12 and there advocated that flexibility for Croatia could not be applied for the purpose of implementation of the KP, and this clearly formed an apparent conflict of interest and also a violation of the principles of independence and impartiality.

In short, Croatia's appeal included the following arguments: violation of Art. 31 (paras. 1, 2 and 3(b)), and Art. 32 of the VCLT; strict and inflexible application of Art. 3 (para. 5) of the KP; violation of: COP and MOP decisions and provisions of the KP; the equal-treatment principle; the granting explanation/information about the decision in question and the right to respond in writing; and the independence, impartiality, and conflict-of-interest principles.

On the basis of its arguments, at the end of its appeal, Croatia requested a MOP decision confirming the application of COP 12, Decision 7 and any subsequent commitment period; or referring the matter back to the EB to decide not to proceed with QoIs questions, thus, following COP 12, Decision 7, to allow Croatia to add 3.5 Mt CO_2 equal to its 1990 GHG emissions not controlled by the MP for the purpose of establishing the level of emissions for the base year for implementation of its commitments under the KP.

This appeal was considered by the MOP 6 held in Cancun, Mexico, from 29 November to 10 December 2010. However, the MOP 6 (2010: 16, paras. 62–68) was unable to complete its consideration, so it was put into the agenda of MOP 7 (2011). Because Croatia withdrew its appeal on 4 August 2011 (FCCC/KP/CMP/2011/2), MOP 7 solely took note of Croatia's withdrawal of its appeal and terminated its consideration (COP 12, Decision 7; MOP 7, Decision 14).

As already stated, the MOP does not have the power to decide on non-compliance or to adopt responses to it under the KP compliance mechanism. It can only decide on non-compliance in the case of an appeal against the EB's decision for denial of due process, hence not about the substantive content of the matter (NCP, Section XI, 1). Thus, appeals to the MOP are only considered on a very narrow matter – whether the rules of the due process have been properly

Table 6.6   The Croatia Case/Actions in chronological order

Date	Action
27 August 2008	Croatia submitted its initial report to the secretariat
26 August 2009	The ERT finalized its IRR of Croatia, which included QoIs (CC-2009-1-1/Croatia/EB; FCCC/IRR/2008/HRV)
27 August 2009	Croatia was given an official notification of the QoIs
28 August 2009	The questions were allocated to the EB
7 September 2009	The EB decided to proceed with the questions (CC-2009-1-2/Croatia/EB)
24 September 2009	The EB agreed to seek expert advice (CC-2009-1-3/Croatia/EB)
25 September 2009	A request for a hearing was submitted by Croatia (CC-2009-1-4/Croatia/EB)
9 October 2009	A written submission was made by Croatia (CC-2009-1-5-Croatia/EB)
11–13 October 2009	The EB 7 (CC/EB/7/2009/2) held a hearing
13 October 2009	Preliminary finding (CC-2009-1-6/Croatia/EB)
12 November 2009	Further written submission from Croatia, opposing the arguments and the conclusion of the preliminary finding (CC-2009-1-7/Croatia/EB)
23–24 November 2009	The EB 8 (CC/EB/8/2009/2) meeting
26 November 2009	Final decision (CC-2009-1-8/Croatia/EB)
28 December 2009	Comments from Croatia opposing the final decision (CC-2009-1-9/Croatia/EB)
14 January 2010	Croatia lodged an appeal against the final decision to the MOP (FCCC/KP/CMP/2010/2)
2 March 2010	The plan to address Croatia's non-compliance was due
8 March 2010	Croatia indicated that it did not intend to submit such a plan in view of its submission of an appeal against the final decision
16 September 2010	The EB 11 (CC/EB/11/2010/2) noted that its decisions stand pending when appealed and agreed to bring the matter of Croatia's non-submission of a plan to the attention of the MOP
29 November–10 December 2010	The MOP 6 could not complete its consideration of the appeal and decided to include it on the provisional agenda for MOP 7 (FCCC/KP/CMP/2010/12, para. 67)
4 August 2011	Croatia withdrew its appeal (FCCC/KP/CMP/2011/2)
11 November 2011	Croatia submitted its plan (CC-2009-1-10/Croatia/EB)
14 November 2011	Croatia's plan was deemed to have been received by the EB
14–18 November 2011	The EB 16 (CC/EB/16/2011/2) reviewed and assessed Croatia's plan
18 November 2011	The EB 16 reviewed and assessed the plan and concluded that the measure reflected in the plan was expected to remedy Croatia's non-compliance (CC-2009-1-11/Croatia/EB)
28 November–9 December 2011	MOP 7 noted Croatia's withdrawal of its appeal and terminated its consideration (COP 12, Decision 7; MOP 7, Decision 14)
21 December 2011	Croatia submitted its revised plan (CC-2009-1-12/Croatia/EB)
30 December 2011	The secretariat sent a letter to Croatia confirming that the compilation and accounting database had been updated accordingly (CC-2009-1-13/Croatia)

(continued)

Table 6.6 (continued)

Date	Action
7, 8–10 February 2012	The EB 18 (CC/EB/18/2012/3) considered the request to reinstate Croatia's eligibility
8 February 2012	The EB decided that there was no longer a QoI with respect to Croatia's eligibility, and that Croatia was fully eligible to participate in the Kyoto mechanisms (CC-2009-1-14/Croatia/EB)

Source Prepared by the author on the basis of information/documentation on the official website of the UNFCCC. See: http://unfccc.int/kyoto_protocol/compliance/questions_of_implementation/items/5451.php

applied with regard to the party concerned – and cannot be made about the substantial issues. Of the grounds for appeal advanced by Croatia, only the last ones could be assessed under the scope of appeal. If Croatia had not withdrawn its appeal, it would have been possible to experience how the MOP would act against those arguments about the substantial issues, rather than just the denial of due process; whether it would restrict itself to a narrow interpretation of the scope of the appeal, or would widen the scope of its interpretation. However, this opportunity was lost through Croatia's decision to withdraw its appeal.

After its withdrawal, Croatia submitted its plans (both its original plan and a revised one) and request for the reinstatement of its eligibility to participate in the Kyoto mechanisms (CC-2009-1-10/Croatia/EB; CC-2009-1-12/Croatia/EB). After the EB's consideration (CC/EB/18/2012/3) of its plans and request to reinstate its eligibility, on 8 February 2012 it was decided that Croatia was fully eligible to participate in the market mechanisms (CC-2009-1-14/Croatia/EB) (Table 6.6).

The Greece Case

After the submission of Greece's initial report on 29 December 2006, the ERT's report for Greece was published on 28 December 2007 (CC-2007-1-1/Greece/EB; FCCC/IRR/2007/GRC). This report concerned a QoI related to Greece's non-compliance with the guidelines for national systems under Art. 5, para. 1, KP, the guidelines for the preparation of the information required under Art. 7, KP, and the eligibility requirement under Arts. 6, 12 and 17, KP.

During its third meeting (CC/EB/3/2008/2), the EB held a hearing regarding this QoI. As a result, it found that, due to the fact that Greece was not in compliance with the guidelines for national systems under Art. 5, para. 1, KP (MOP 1, Decision 19 2005) or the guidelines for the preparation of the information required under Art. 7 KP (MOP 1, Decision 15 2005b), it could not be regarded as a party meeting the eligibility requirement under Arts. 6, 12 and 17, KP to have a national system proper to the conditions of Art. 5, para. 1, KP or the requirements in the related guidelines. So the EB adopted a preliminary finding of non-compliance (CC-2007-1-6/Greece/EB) with national system requirements for countries with 2012 targets (Annex B Parties).

Table 6.7 The Greece Case/Actions in chronological order

Date	Action
29 December 2006	Greece submitted its initial report to the secretariat
28 December 2007	The Initial Review Report (IRR) for Greece was published (CC-2007-1-1/Greece/EB; FCCC/IRR/2007/GRC)
28 December 2007	The secretariat received a QoI indicated in the report of the ERT
31 December 2007	The QoI was deemed to have been received by the ComplCom
7 January 2008	The Bureau of the ComplCom allocated the QoI to the EB
8 January 2008	The secretariat notified the members of the EB of the QoI
22 January 2008	The EB decided to proceed with the QoI (CC-2007-1-2/Greece/EB)
8 February 2008	The EB agreed to invite four experts on national systems drawn from the UNFCCC experts (CC-2007-1-3/Greece/EB)
11 February 2008	The EB received a request for a hearing from Greece (CC-2007-1-4/Greece/EB)
26 February 2008	The EB received a written submission from Greece (CC-2007-1-5/Greece/EB)
4–5 March 2008	As requested by Greece, a hearing was held
4–6 March 2008	The hearing formed part of the EB 3 meeting (CC/EB/3/2008/2)
6 March 2008	Preliminary finding (CC-2007-1-6/Greece/EB)
9 April 2008	The EB received a further written submission from Greece (CC-2007-1-7/Greece/EB)
14 April 2008	Greece was given an official notification of the QoI
16–17 April 2008	The EB 4 (CC/EB/4/2008/2) evaluated further written submission
17 April 2008	Final decision (CC-2007-1-8/Greece/EB)
16 July 2008	Greece submitted its plan (CC-2007-1-9/Greece/EB)
September 2008	The in-country review of the greenhouse gas inventories of Greece submitted in 2007 and 2008 was completed in September
6–7 October 2008	The EB 6 (CC/EB/6/2008/3) reviewed and assessed the document submitted by Greece
7 October 2008	The EB asked Greece to submit a revised plan (CC-2007-1-10/Greece/EB)
17 October 2008	The annual review report (ARR) of Greece submitted in 2007 and 2008 was published (FCCC/ARR/2008/GRC; CC-2007-1-12/Greece/EB)
20 October 2008	The secretariat forwarded the review report to the ComplCom
27 October 2008	Greece submitted a revised plan in response to the request of the EB (CC-2007-1-11/Greece/EB)
7 October 2008	Greece also submitted a request to reinstate its eligibility to participate in the Kyoto mechanisms

(continued)

Table 6.7 (continued)

Date	Action
28 October 2008	The request was deemed to have been received by the EB
13 November 2008	The EB decided that there was no longer a QoI with respect to Greece, making Greece fully eligible to participate in the Kyoto mechanisms (CC-2007-1-13/Greece/EB)

Source Prepared by the author on the basis of information/documentation on the official website of the UNFCCC. See: http://unfccc.int/kyoto_protocol/compliance/questions_of_implementation/items/5451.php

In its fourth meeting (CC/EB/4/2008/2), the EB then adopted a final decision confirming its preliminary finding. Thus, for the first time, a State had been officially found in non-compliance with a KP requirement (Table 6.7).

In accordance with NCP, Section XV, it was decided to apply the following consequences to Greece for its non-compliance:

a. declaration of non-compliance;
b. Greece to be required to submit a plan, including measures to ensure the maintenance of the national system through transitions and appropriate administrative arrangements to support an in-country review by the ERT of the new national system of Greece, within three months of the determination of non-compliance;
c. Greece to be declared ineligible to participate in the flexibility mechanisms. This implied that Greece was suspended from trading in the Kyoto carbon market – selling and transferring credits, and gaining credits – set up by these mechanisms until the decision stating that a QoI with respect to Greece's eligibility no longer existed, permitting full reinstatement by the EB.

After the final decision, Greece submitted a compliance action plan (CC-2007-1-9/Greece/EB), but the EB found it insufficient to enable the branch to complete the required assessment, as it did not meet the requirements set out in para. 2, Section XV – causes of non-compliance, measures to remedy the non-compliance and a timetable for implementing such measures – or the issues set out in para. 18b of the Annex to the final decision clarifying the requirements set out in para. 2, Section XV, to enable the branch to complete the required assessment (CC-2007-1-10/Greece/EB).

Nearly three months later, Greece submitted a revised plan to comply with para. 2, Section XV and issues set out explicitly in para. 18(b) of the Annex to the final decision of the EB (CC-2007-1-8/Greece/EB), pursuant to the request of EB 6 (CC-2007-1-10/Greece/EB).

After considering the most recent ERT report (FCCC/ARR/2008/GRC; CC-2007-1-12/Greece/EB) with respect to Greece and Greece's revised plan, on 13 November 2008 the EB decided that Greece was fully eligible to participate in the Kyoto market mechanisms (CC-2007-1-13/Greece/EB).

The severe impact of the global financial crisis on Greece can be summarized in three main dimensions: the increase in the price of fuel oil, resulting in the use of fireplaces and stoves; the emissions released from cars; and the emissions released from Greek industries, which were more apparent at national level because the biggest industries are located outside the capital. All these factors increased air pollution (Nanou 2015: 56). Therefore, in order to address this challenge, environmental policies implemented during the period 2008–2014 related "to the reduction of the use of fireplaces and stoves, the 'green' ring, the withdrawal of old cars, the enhancement of the monitoring system for air pollution, the reinforcement of public means of transport, and the use of diesel for the cars" (Nanou 2015: 57). Although these measures were not found sufficient to address air pollution in Greece, over the period 2008–2011, the EU-15 met its target through of nine countries, including Greece (Portugal, Greece, Ireland, Finland, France, Belgium, Sweden, the UK and Germany) which achieved their targets in the burden-sharing agreement (Haita 2012).

The Lithuania Case

A QoI on Lithuania was brought before the EB by the ERT with regard to the guidelines for national systems as well as Art. 7, KP guidelines on reporting commitments (CC-2011-3-1/Lithuania/EB; CC/ERT/ARR/2011/33); like the former cases dealt with by the EB it hence related to the eligibility requirement.

After considering a further written submission by Lithuania, on 21 December 2011 the branch adopted a final decision to confirm its preliminary finding (CC-2011-3-6/Lithuania/EB). As a result of the final decision (CC-2011-3-8/Lithuania/EB), the same consequences were applied by the EB as in the previous cases.

The compliance action plan prepared by Lithuania in response to the EB's decision and its first progress report on its implementation (CC-2011-3-9/Lithuania/EB) was reviewed by the EB. It decided that the plan adequately addressed each of the elements specified in para. 2, Section XV, and that Lithuania was expected to remedy the non-compliance, if implemented in accordance with this decision (CC-2011-3-11/Lithuania/EB).

However, the EB 20, held between 9–14 July 2012, decided that, on the basis of expert advice, it was not possible to reinstate Lithuania's eligibility to participate in the Kyoto mechanisms, and that another in-country review was required (CC-2011-3-14/Lithuania/EB).

Since this meeting of the EB, there have been two reviews concerning Lithuania: the expedited in-country review for Lithuania conducted from 28 to 29 September 2012 (expedited review report) assessed whether the QoI had been resolved; and the in-country review of the annual report submitted by Lithuania's 2012 annual submission conducted from 1 to 6 October 2012 involved an assessment of Lithuania's 2012 annual submission.

The expedited in-country review for Lithuania (expedited review report) (CC/ERT/EXP/2012/1) stated that Lithuania's archiving system was fully in line with the guidelines for national systems (para. 24a); it had also compiled all the necessary data to identify the lands subject to activities under Art. 3, paras. 3–4, KP and to enable accurate estimates of greenhouse gas emissions and removals (para. 24b). It had also fully addressed the relevant issues for improvement raised in the review reports of Lithuania's 2010 and 2011 annual submissions (para. 25).

The in-country review of the annual report submitted by Lithuania's 2012 annual submission had involved a review of the land use, land-use change and forestry data in the 2012 annual submission. It concluded that Lithuania was capable of providing the information on activities under Art. 3, paras. 3–4, KP necessary for it to meet the reporting requirements defined in the Article 7 guidelines.

Relying on the information provided by the expedited review report, the expert advice received, and also on the information submitted by Lithuania indicating that most measures in its plan had been implemented, the EB concluded that the QoI regarding Lithuania had been resolved.

In addition, although it underlined that some measures described in Lithuania's plan were yet to be fully implemented and so encouraged Lithuania to further strengthen its national system by implementing these measures, it found Lithuania fully eligible to participate in the flexibility mechanisms (CC-2011-3-18/Lithuania/ EB) at the end of the EB 21 (22–24 October 2012) (Table 6.8).

The Romania Case

The QoI relating to Romania was raised within the ERT ARR 2010 (CC-2011-1-1/ Romania/EB; CC/ERT/ARR/2011/21; FCCC/ARR/2010/ROU) because in its review the ERT found that Romania's national system had failed to perform some specific functions on inventory preparation required by the guidelines for national systems, and it did not comply with the requirements for the preparation of information under Art. 7, para. 1, KP specifically for the land use, land-use change and forestry (LULUCF) activities.

After its preliminary examination, the EB decided to proceed with the question (CC-2011-1-2/Romania/EB). Then Romania requested a hearing, which was held as part of the EB 13 meeting on 7 July 2011. At the end of the EB 13 meeting and the hearing CC-2011-1-6/Romania/EB), the EB decided on the consequences to be applied to Romania, contained in para. 24 of the preliminary finding of the EB, and also confirmed by and annexed to the final decision (CC2011-1-8/Romania/EB; CC-2011-1-9/Romania/EB; CC-2011-19/Romania/EB/Add.1).

The consequences applied by the branch were that Romania was:

a. declared to be in non-compliance;
b. required to submit a plan to address its non-compliance within three months;
c. declared ineligible to participate in the market-based mechanisms (ET, Art. 17, CDM, Art. 12, JI, Art. 6). The exception was that if there were emission reduction units generated from JI projects hosted by Romania and verified under a special track II JI procedure, Romania could trade credits, and if forwarded by a host developing country, it could acquire credits.

So, the consequences applied were the same as those applied in the four previous cases brought before to the EB.

Table 6.8 The Lithuania Case/Actions in chronological order

Date	Action
14 April 2010	Lithuania submitted its annual inventory submission to the secretariat
20–25 September 2010	The ERT conducted a centralized review of Lithuania's submission
7 September 2011	The ERT finalized its 2010 ARR (FCCC/ARR/2010/LTU) containing a QoI relating to its national system as well its reporting commitments (CC-2011-3-1/Lithuania/EB; CC/ERT/ARR/2011/33)
8 September 2011	The QoI was received by the ComplCom and Lithuania was given official notification
15 September 2011	The question was allocated to the EB by the Bureau of the ComplCom
16 September 2011	The secretariat notified the members and alternate members of the EB of the QoI
4 October 2011	Preliminary Examination; the EB decided to proceed (CC-2011-3-2/Lithuania/EB)
11 October 2011	The EB agreed to invite three experts drawn from the UNFCCC experts (CC-2011-3-3/Lithuania/EB)
19 October 2011	The EB received a request for a hearing from Lithuania (CC-2011-3-4/Lithuania/EB)
9 November 2011	A written submission made by Lithuania (CC-2011-3-5/Lithuania/EB)
14–18 November 2011	A hearing was held as part of the EB 16 meeting (CC/EB/16/2011/2)
15–16 November 2011	At the request of Lithuania, the EB held a hearing
16 November 2011	Further questions from the ERT on the 2011 review of the GHG inventories of Lithuania (CC-2011-3-5/Lithuania/EB/Add.1)
17 November 2011	Preliminary finding of the EB (CC-2011-3-6/Lithuania/EB)
19 December 2011	Lithuania made a further written submission (CC-2011-3-7/Lithuania/EB)
21 December 2011	Final Decision (CC-2011-3-8/Lithuania/EB)
26 March 2012	Lithuania submitted a plan to address its non-compliance, which also included a first progress report on its implementation (CC-2011-3-9/Lithuania/EB)
13 April 2012	Lithuania requested a review and assessment of its plan (CC-2011-3-10/Lithuania/EB)
2 May 2012	The EB reviewed and assessed the plan (CC-2011-3-11/Lithuania/EB)
14 June 2012	Lithuania submitted the second progress report on the implementation of the plan and a request to reinstate its eligibility to participate in the mechanisms (CC-2011-3-12/Lithuania/EB)
27 June 2012	The EB decided to seek expert advice on the 2011 ARR (CC/ERT/ARR/2011/33) and the implementation of the plan by Lithuania (CC-2011-3-13/Lithuania/EB)
9–14 July 2012	The EB 20 considered the request for reinstatement (CC/EB/20/2012/2)

(continued)

Table 6.8 (continued)

Date	Action
14 July 2012	The EB decided that there continued to be a QoI with respect to Lithuania's eligibility (CC-2011-3-14/Lithuania/EB)
18 July 2012	The secretariat received a request from Lithuania for the EB not to initiate the expedited procedure before the review report of the National Greenhouse Gas Inventory (CC-2011-3-15/Lithuania/EB)
31 July 2012	The EB decided not to initiate the expedited procedure (CC-2011-3-16/Lithuania/EB)
28–29 September 2012	The expedited review of Lithuania was conducted
11 October 2012	The report on the expedited review of Lithuania, which concluded that the QoI for Lithuania had been fully resolved, was published
12 October 2012	The secretariat forwarded the expedited review report to the ComplCom
22–24 October 2012	The EB 21 (CC/EB/21/2012/2) concluded, after receipt of expert advice, that there was no longer a QoI
23 October 2012	The EB decided to seek expert advice (CC-2011-3-17/Lithuania/EB)
24 October 2012	Lithuania became fully eligible to participate in the mechanisms (CC-2011-3-18/Lithuania/EB)

Source Prepared by the author on the basis of information/documentation on the official website of the UNFCCC. See: http://unfccc.int/kyoto_protocol/compliance/questions_of_implementation/items/5451.php

Nearly three months after the final decision, Romania submitted its plan and first progress report on 2 November 2011 (CC-2011-1-9/Romania/EB), (CC-2011-1-9/Romania/EB/Add.1). It submitted its second progress report on the implementation of the plan (CC-2011-1-12/Romania/EB) on 2 February 2012, and its third progress report and a request to reinstate its eligibility to participate in the mechanisms (CC-2011-1-13/Romania/EB) on 26 March 2012.

The EB 20 (CC/EB/20/2012/2–9 August 2012) considered the request for reinstatement together with the 2011 ARR (FCCC/ARR/2011/ROU), which was published (CC/ERT/ARR/2012/4) on 27 February 2012. As a consequence of its analysis, on 13 July 2012 it decided that the QoI had been resolved, making Romania eligible to participate in the mechanisms (Table 6.9).

The Slovakia Case

Slovakia was involved in a question of implementation on the ERT's 2011 ARR (CC/ERT/ARR/2012/17; FCCC/ARR/2011/SVK). The questions was about the national system of Slovakia and its calculation of several estimates of greenhouse gas emissions.

The EB 20 meeting, also holding a hearing, discussed the issue, which led to a preliminary finding (CC-2012-1-7/Slovakia/EB) which was confirmed by the final decision of the EB (CC-2012-1-9/Slovakia/EB) on 17 August 2012.

Table 6.9 The Romania Case/Actions in chronological order

Date	Action
15 April 2010	Romania submitted its annual inventory submission to the secretariat
20–25 September 2010	The ERT conducted a centralized review of Romania's submission
11 May 2011	The ERT finalized its 2010 ARR (FCCC/ARR/2010/ROU), which contained a QoI (CC-2011-1-1/Romania; CC/ERT/ARR/2011/21)
12 May 2011	The QoI was received by the ComplCom and Romania was given official notification
16 May 2011	The question was allocated to the EB by the Bureau of the ComplCom
17 May 2011	The secretariat notified the members and alternate members of the EB of the QoI
27 May 2011	Preliminary Examination; the EB decided to proceed (CC-2011-1-2/Romania/EB)
3 June 2011	The EB agreed to invite four experts drawn from the UNFCCC experts (CC-2011-1-3/Romania/EB)
14 June 2011	The EB received a request for a hearing from Romania (CC-2011-1-4/Romania/EB)
29 June 2011	A written submission was made by Romania(CC-2011-1-5/Romania/EB)
6–8 July 2011	EB 13 meeting (CC/EB/13/2011/2–18 July 2011)
7 July 2011	At the request of Romania, a hearing was held as part of the EB 13 meeting
8 July 2011	Preliminary finding of the EB (CC-2011-1-6/Romania/EB)
11 August 2011	Romania made a further written submission (CC-2011-1-7/Romania/EB)
27 August 2011	Final decision of the EB (CC-2011-1-8/Romania/EB)
26 September–1 October 2011	In-country review
2 November 2011	Romania submitted its plan and first progress report (CC-2011-1-9/Romania/EB), (CC-2011-1-9/Romania/EB/Add.1)
14 November 2011	Expert Advice on Plan (CC-2011-1-10/Romania/EB)
14–18 November 2011	The EB 16 meeting (CC/EB/16/2011/2–26 November 2011) provided inputs to Romania
15 November 2011	The EB adopted a decision on the review and assessment of its plan (CC-2011-1-11/Romania/EB)
2 February 2012	Romania submitted its second progress report on the implementation of the plan (CC-2011-1-12/Romania/EB)
7, 8–10 February 2012	The EB 18 (CC/EB/18/2012/3–24 February 2012) made a number of recommendations (CC/EB/18/2012/3)
27 February 2012	The 2011 ARR (FCCC/ARR/2011/ROU) was published (CC/ERT/ARR/2012/4)
1 March 2012	The secretariat forwarded the 2011 ARR to the ComplCom
26 March 2012	Romania submitted its third progress report and a request to reinstate its eligibility to participate in the mechanisms (CC-2011-1-13/Romania/EB)

(continued)

Table 6.9 (continued)

Date	Action
27 June 2012	Decision on Expert Advice (CC-2011-1-14/Romania/EB)
9–14 July 2012	The EB 20 (CC/EB/20/2012/2–9 August 2012) considered the request for reinstatement
13 July 2012	The QoI having been resolved, the EB decided that Romania was eligible to participate in the Kyoto mechanisms (CC-2011-1-15/ Romania/EB)

Source Prepared by the author on the basis of information/documentation on the official website of the UNFCCC. See: http://unfccc.int/kyoto_protocol/compliance/questions_of_implementation/ items/5451.php

Firstly, according to the decision, during the review of Slovakia's 2011 annual submission, there was a partial operational impairment of the performance of some of the specific functions of Slovakia's national system, causing non-compliance with Art. 5, para. 1, KP and the related guidelines, but not resulting in non-compliance with the eligibility requirements under Arts. 6, 12 and 17, KP. Secondly, the QoI relating to the disagreement over whether to apply adjustments was resolved by the decision. It was decided not to apply the adjustment with respect to the estimates of emissions from road transportation, but to apply the adjustments with respect to the estimates of emissions from the consumption of halocarbons and SF_6 (CC-2012-1-6/Slovakia/EB). Finally, the EB emphasized the need for an in-country review of Slovakia's national system to assess whether the measures developed and implemented by Slovakia would prevent the operational impairment.

This decision was taken on the basis of the 2011 ARR, the advice of the experts invited by the branch, Slovakia's written submission and its updated information on its national system (with regard to its institutional structure and the roles of the institutions involved in the preparation of the inventory).

In the opinion of the 2011 ARR (CC/ERT/ARR/2012/17; FCCC/ARR/2011/ SVK), the national system of Slovakia did not fully perform the functions required for national systems, as it basically relied on individual external expertise rather than institutional expertise, even in the case of those managing the data sources. Furthermore, its estimates of emissions from road transportation and the consumption of halocarbons and SF_6 were incomplete or had been prepared inconsistently with the methodological and reporting requirements of the Revised 1996 IPCC Guidelines and the IPCC Good Practice Guidance.

Expert advice also stressed the adjustments calculated and recommended by the ERT, and the national system's problems with performing some of its functions, particularly those relating to the collection of sufficient activity data, process information and emission factors (CC-2012-1-7/Slovakia/EB).

On the other hand, in its written submission (CC-2012-1-5/Slovakia/EB), while admitting that there were some problems in the functioning of its national system, Slovakia defended itself by stating that those were either addressed during the review of its 2011 annual submission, or were not serious disabilities which could result in significant distortions from the requirements for national systems. It also expressed its disagreement with the expert advice which had cited the adjustments as an indicator of a structural problem of the national system. However, at the hearing, it accepted adjustments with respect to estimates of emissions from the consumption of halocarbons and SF_6, and provided additional information on the estimates of emissions from road transportation (CC-2012-1-6/Slovakia/EB, paras. 13–14). Based on this additional information, expert advice considered that the recommended adjustments with respect to estimates of emissions from road transportation were no longer necessary. In return for its acceptance of the recommended adjustments with respect to estimates of emissions from the consumption of halocarbons and SF_6, the experts stated that the QoI relating to the disagreement over whether to apply adjustments had been resolved.

In accordance with the above-mentioned findings, the EB applied the following consequences to Slovakia:

a. declared it to be in non-compliance;
b. required it to develop a compliance action plan, to submit it within 3 months to the EB – before the in-country review of its 2012 annual submission – as part of its plan to inform the EB of its preparations for this in-country review, and to report on its implementation progress (Table 6.10).

In line with the decision and its consequences, Slovakia submitted its plan and first progress report (CC-2012-1-10/Slovakia/EB) on 21 September 2012, before the in-country review of its 2012 annual submission conducted between 1–6 October 2012. In its plan and progress report (CC-2012-1-10/Slovakia/EB) , Slovakia rendered an overview of the measures undertaken to improve its national system. It explained in detail the measures it had taken prior to the in-country review of its 2012 annual submission, and how they had contributed to the preparation of the in-country review. It also argued, during the in-country review of its 2012 annual submission, that there had been no partial operational impairment of the specific functions of the national system. The expert advice, on the basis of its plan and progress report as well as the 2012 national inventory report (NIR), also acknowledged that the national system's functions performance has been strengthened, particularly through the implementation of the measures addressed in the recommendations of previous Expert Review Teams (CC-2012-1-12/Slovakia/EB).

Based on the above facts, the review and assessment of the plan by the EB in its 21st meeting concluded with the following decisions (CC-2012-1-12/Slovakia/EB):

a. the plan adequately addressed each of the elements specified in Section XV, para. 2;
b. if implemented properly, it was expected to remedy the non-compliance;
c. ARR 2012 was also required in order to reach a decision on whether all the QoIs had been resolved.

Table 6.10 The Slovakia Case/Actions in chronological order

Date	Action
15 April 2011	Slovakia submitted its 2011 annual inventory submission to the secretariat
22–27 August 2011	In-country review of Slovakia's 2011 annual submission
8 May 2012	The ERT finalized its 2011 ARR (FCCC/ARR/2011/SVK) containing QoIs (CC/ERT/ARR/2012/17)
9 May 2012	Slovakia was given official notification of the QoIs
16 May 2012	The questions were allocated to the EB by the Bureau of the ComplCom
17 May 2012	The secretariat notified the members and alternate members of the EB of the QoIs (CC-2012-1-1/Slovakia/EB)
1 June 2012	Preliminary Examination: the EB decided to proceed (CC-2012-1-2/Slovakia/EB)
27 June 2012	The EB agreed to invite two experts drawn from the UNFCCC experts (CC-2012-1-4/Slovakia/EB)
8 June 2012	The EB received a request for a hearing from Slovakia (CC-2012-1-3/Slovakia/EB)
4 July 2012	A written submission was made by Slovakia (CC-2012-1-5/Slovakia/EB)
9–14 July 2012	A hearing was held as part of the EB 20 meeting (CC/EB/20/2012/2)
10–11 July 2012	At the request of Slovakia, the EB held a hearing
14 July 2012	Decision, with respect to the estimates of emissions from road transportation, not to apply the adjustment; with respect to the estimates of emissions from the consumption of halocarbons and SF6, to apply the adjustments (CC-2012-1-6/Slovakia/EB)
14 July 2012	Preliminary finding of the EB (CC-2012-1-7/Slovakia/EB)
18 July 2012	The EB received a written notification from Slovakia (CC-2012-1-8) which indicated that it did not intend to make a further written submission under NCP, Section X, para. 1e
16 August 2012	The secretariat confirmed that no further written submission had been received from Slovakia by 15 August 2012, the due date for any further written submission
17 August 2012	Final decision of the EB (CC-2012-1-9/Slovakia/EB)
21 September 2012	Slovakia submitted its plan and first progress report (CC-2012-1-10/Slovakia/EB)
1–6 October 2012	In-country review of Slovakia's 2012 annual submission
15 October 2012	The EB agreed to invite an expert drawn from the UNFCCC experts to provide advice (CC-2012-1-11/Slovakia/EB)
22–24 October 2012	The EB 21(CC/EB/21/2012/2) reviewed and assessed the plan submitted by Slovakia
23 October 2012	The EB decides that if implemented, the plan was expected to remedy the non-compliance, but, it was also necessary to examine 2012 ARR to find out whether all the QoIs had been resolved (CC-2012-1-12/Slovakia/EB)

(continued)

Table 6.10   (continued)

Date	Action
15 March 2013	Second progress report on the plan submitted by Slovakia (CC-2012-1-13/Slovakia/EB)
19 June 2013	EB decision on expert advice (CC-2012-1-14/Slovakia/EB)
4 July 2013	Decision on resolution of the QoIs with respect to Slovakia (CC-2012-1-15/Slovakia/EB)

Source Prepared by the author on the basis of information/documentation on the official website of the UNFCCC. See: http://unfccc.int/kyoto_protocol/compliance/questions_of_implementation/items/5451.php

After its submission of the second progress report on the plan (CC-2012-1-13/Slovakia/EB) and the receipt of the ERT's review report on Slovakia's annual submission in 2012, provided in document FCCC/ARR/2012/SVK (2012 ARR) by the secretariat, the EB addressed the QoIs concerned at its 23rd meeting on 3–4 July 2013. It decided that although not all the measures prescribed in Slovakia's plan had been implemented, and that Slovakia therefore needed to fully implement those measures, there was no QoI with respect to Slovakia anymore (CC-2012-1-15/Slovakia/EB).

The Ukraine Case I

There have been two cases regarding Ukraine to date. One is regarding the eligibility requirements. The other is on Ukraine's true-up period report.

Similar to the processes of the previous cases, in the first Ukraine case, after Ukraine's annual inventory submission to the secretariat and the ERT's review of this submission in 2010, on 3 June 2011 the ERT prepared its 2010 ARR (FCCC/ARR/2010/UKR) containing a QoI (CC-2011-2-1/Ukraine/EB; CC/ERT/ARR/2011/28).

As in the earlier cases, the QoI did not relate to whether Ukraine was in compliance with its 2012 emissions target, but to its compliance with the guidelines for national systems and therefore its eligibility requirements. In fact, the ERT found that the national system of Ukraine failed:

- to perform the functions required by the guidelines for national systems;
- to ensure that its 2010 annual submission was under the conditions required by the related guidelines, such as the guidelines for national systems, for the preparation of the information required under Art. 7, KP, for UNFCCC reporting, for IPCC good practice, and for IPCC good practice for LULUCF;
- to ensure that areas of land subject to KP LULUCF activities under Art. 3, paras. 3 and 4, KP were identifiable in accordance with para. 20 of the guidelines relating to land use (MOP 1, Decision 16 2005a: Annex).

The EB's examination of the issue resulted in the decision that Ukraine was not in compliance with its national systems requirements; hence, it did not meet the eligibility requirements under Arts. 6, 12 and 17, KP (CC-2011-2-6/Ukraine/EB).

Table 6.11 Ukraine Case I/Actions in chronological order

Date	Action
12 April 2010	Ukraine submitted its annual inventory submission to the secretariat
30 August–4 September 2010	The ERT conducted a centralized review of Ukraine's submission
3 June 2011	The ERT finalized its 2010 ARR (FCCC/ARR/2010/UKR) containing a QoI (CC-2011-2-1/Ukraine; CC/ERT/ARR/2011/28)
6 June 2011	The QoI was received by the secretariat and ComplCom; Ukraine was given a official notification
13 June 2011	The question was allocated to the EB by the Bureau of the ComplCom
14 June 2011	The secretariat notified the members and alternate members of the EB of the QoI
29 June 2011	Preliminary Examination: the EB decided to proceed (CC-2011-2-2/Ukraine/EB)
6 July 2011	The EB agreed to invite four experts drawn from the UNFCCC experts (CC-2011-2-3/Ukraine/EB)
19 July 2011	The EB received a request for a hearing from Ukraine (CC-2011-2-4/Ukraine/EB)
2 August 2011	A written submission was made by Ukraine (CC-2011-2-5/Ukraine/EB)
24 August 2011	At the request of Ukraine, the EB held a hearing
22–27 August 2011	A hearing was held as part of the EB 14 meeting (CC/EB/14/2011/2)
25 August 2011	Preliminary finding of the EB (CC-2011-2-6/Ukraine/EB)
2 September 2011	Ukraine submitted a request to postpone the EB 15 (CC-2011-2-7/Ukraine/EB)
28 September 2011	The EB received a further written submission from Ukraine (CC-2011-2-8/Ukraine/EB)
11–12 October 2011	The EB 15 considered this further written submission in elaborating a final decision
12 October 2011	Final decision of the EB (CC-2011-2-9/Ukraine/EB)
10–15 October 2011	In-country review
7 December 2011	Ukraine submitted its plan (CC-2011-2-10/Ukraine/EB)
20–21 December 2011	The EB 17 (CC/EB/17/2011/2) reviewed and assessed Ukraine's plan
21 December 2011	The EB concluded that the plan met the relevant requirements (CC-2011-2-11/Ukraine/EB)
13 January 2012	The 2011 ARR (FCCC/ARR/2011/UKR) was published
18 January 2012	The secretariat forwarded the 2011 ARR to the ComplCom (CC/ERT/ARR/2012/2)
23 January 2012	Ukraine submitted a request for reinstatement of its eligibility to participate in the mechanisms (CC-2011-2-12/Ukraine/EB)
7 February 2012	Ukraine submitted its first progress report on the implementation of its plan (CC-2011-2-13/Ukraine/EB)

(continued)

Table 6.11 (continued)

Date	Action
7, 8–10 February 2012	The EB 18 (CC/EB/18/2012/3) decided to defer the adoption of a decision on the request, pending the receipt of expert advice (CC-2011-2-14/Ukraine/EB)
21 February 2012	A corrigendum to the ARR 2011 on paras. 163c and 189c (FCCC/ARR/2011/UKR/Corr.1) was submitted (CC/ERT/ARR/2012/2/Corr.1)
6 March 2012	The EB decided to invite two experts to provide advice to the branch (CC-2011-2-15/Ukraine/EB)
8–9 March 2012	The EB 19 (CC/EB/19/2012/2) considered the request to reinstate Ukraine's eligibility
9 March 2012	The QoI having been resolved, the EB decided that Ukraine was eligible to participate in the mechanisms (CC-2011-2-16/Ukraine/EB)

Source Prepared by the author on the basis of information/documentation on the official website of the UNFCCC. See: http://unfccc.int/kyoto_protocol/compliance/questions_of_implementation/items/5451.php

As in the previous cases, and in accordance with Section XV, NCP, KP, the EB also decided to apply three consequences – a declaration of non-compliance; mandatory submission of a plan; and suspension of trading in market mechanisms.

On 2 September 2011 Ukraine submitted a request to postpone the EB 15, the consideration of the further written submission and the adoption of a final decision (CC-2011-2-7/Ukraine/EB). In that request, it noted that it would make a further submission no later than 26 September 2011. It also asked for the meeting to take place on 15 October 2011, since the persons who were to attend the EB 15 meeting (11–12 October 2011) on behalf of Ukraine also needed to attend the in-country review on 10–15 October 2011.

Despite this request, Ukraine sent the EB its further written submission, including a list of participants on behalf of Ukraine, on 27 September 2011 (CC-2011-2-8/Ukraine/EB). At the end of its written submission, it requested the EB to postpone the final decision and reassess it on the basis of its 2011 ARR (FCCC/ARR/2011/UKR), or to confirm the preliminary finding but defer its coming into effect until the outcome of the in-country review of the annual inventory of Ukraine (Table 6.11).

In line with the EB's decision, Ukraine submitted firstly its plan (CC-2011-2-10/Ukraine/EB), and then a request for reinstatement of its eligibility to participate in the mechanisms (CC-2011-2-12/Ukraine/EB).

After its first progress report on the implementation of its plan (CC-2011-2-13/Ukraine/EB), the EB 18 (CC/EB/18/2012/3) considered its request, but decided to wait for expert advice before making a decision about the request (CC-2011-2-14/Ukraine/EB).

The EB 19 (CC/EB/19/2012/2) reconsidered the request to reinstate Ukraine's eligibility on the basis of expert advice, the 2011 ARR, the first progress report on the plan and also the additional information presented by Ukraine at the EB 18 and 19.

Based on its findings, on 9 March 2012 the EB concluded that all the measures described in Ukraine's plan had been implemented, or were expected to be fully implemented in the 2013 annual submission, and that all information submitted was sufficient to decide that the QoI had been resolved, making Ukraine eligible to participate in the mechanisms (CC-2011-2-16/Ukraine/EB).

Ukraine Case II

According to MOP 1, Decision 13 and MOP 10, Decision 3, each Annex B Party, after the first commitment period of the KP ended on 31 December 2012, was required to transfer and acquire KP units from the first commitment period to ensure that it would meet its target; and to submit its true-up period report to the secretariat prior to 2 January 2016.

In accordance with Art. 8, KP, the review of the initial true-up period report submission of Ukraine was finalized by an ERT on 8 April 2016; the report contained two QoIs, triggering the KP's CM to be subsequently considered by the ComplCom. Of those QoIs, the first one was about the late submission of the true-up period report, and the inconsistencies between information submitted by Ukraine and the international transaction log (ITL) maintained by the secretariat. The second was about Ukraine's emission reduction target, which was found insufficient to cover its total greenhouse gas emissions for the first commitment period (CP 1).

Confirming its preliminary finding, in its final decision the EB stated that Ukraine was in non-compliance with Art. 7, para. 1, in conjunction with para. 4, KP. However, it did not determine whether Ukraine was in non-compliance with its qualified emission reduction commitment under Art. 3, para. 1. This was due to the fact that there was no accepted procedure in the relevant MOP decisions that would allow Ukraine to demonstrate its formal compliance with Art. 3, para. 1, KP for CP 1. The EB also decided to apply the following measures to Ukraine: 1. Declaration of being in non-compliance; 2. Requirement of the submission of a plan to address its non-compliance within three months; 3. Suspension of eligibility of Ukraine to participate in Kyoto mechanisms, in accordance with the relevant provisions under Arts. 6, 12 and 17, KP (CC-2016-1-6/Ukraine/EB).

In response to these measures, on 5 December 2016 Ukraine submitted a plan to address its non-compliance (CC-2016-1-7/Ukraine/EB). In its review of the plan, the EB concluded that Ukraine had met the relevant requirements. In addition, since Ukraine's full implementation of the measures set out in the plan was required to consider whether QoI has been resolved, Ukraine was requested to submit periodic reports – at least every three months – on the implementation of the plan (CC-2016-1-8/Ukraine/EB).

After Ukraine's resubmission of its true-up period report – including the standard electronic format tables for the period 1 January to 18 November 2015 and for the years 2015, 2016 and 2017, the list of serial numbers for the KP units in the retirement account at the end of the true-up period, and the list of serial numbers for the emission reduction units (ERUs) and assigned amount units (AAUs) that the

Table 6.12 Ukraine Case II/Actions in chronological order

Date	Action
23 February 2016	ERT draft review report was sent to Ukraine
9 March 2016	In response to the draft review report, Ukraine made its true-up period report submission
8 April 2016	ERT Report on Ukraine's true-up period report (CC-2016-1-1/Ukraine/EB)
8 April 2016	Secretariat received QoI
11 April 2016	ComplCom received QoI
18 April 2016	The Bureau allocated the QoI to the EB
On 19 April 2016	The secretariat notified the members and alternate members of the EB of the QoI
3 May 2016	EB Decision on Preliminary Examination (CC-2016-1-2/Ukraine/EB)
30 May 2016	EB Decision on expert advice (CC-2016-1-3/Ukraine/EB)
21 June 2016	EB Preliminary Finding (CC-2016-1-4/Ukraine/EB)
22 July 2016	In response to the preliminary findings of the EB, Ukraine's written submission (CC-2016-1-5/Ukraine/EB)
30 August 2016	Letter from Ukraine containing additional information on reconciliation and time synchronization
7 September 2016	EB Final Decision (CC-2016-1-6/Ukraine/EB)
5 December 2016	Plan submitted by Ukraine pursuant to the final decision (CC-2016-1-7/Ukraine/EB)
21 December 2016	Decision on the Review and Assessment of the Plan Submitted Under Para. 2 of Section XV (CC-2016-1-8/Ukraine/EB)
30 March 2017	Ukraine's First progress report (CC-2016-1-9/Ukraine/EB)
3 July 2017	Ukraine's Second Progress Report (CC-2016-1-10/Ukraine/EB)
10 August 2017	ERT Report on the individual review of the resubmitted report upon expiration of the additional period for fulfilling commitments (FCCC/KP/CMP/2017/TPR/UKR/CC-2016-1-12/Ukraine/EB)
15 August 2017	Report upon expiration of the additional period for fulfilling commitments by Ukraine (CC-2016-1-11/Ukraine/EB)

Source Prepared by the author on the basis of information/documentation on the official website of the UNFCCC. http://unfccc.int/kyoto_protocol/compliance/questions_of_implementation/items/9575.php

party was requesting to be carried over to the second commitment period (CP 2) – on 3 July 2017 the ERT concluded its reassessment that the information provided by Ukraine in its resubmitted true-up period report covered all the points required by relevant decisions, such as MOP 1/Decision 13, MOP 1/Decision 15 and MOP 10/Decision 3. It therefore decided that Ukraine's total anthropogenic GHG emissions CP 1 did not exceed the quantities of ERUs, certified emission reductions (CERs), temporary CERs, long-term CERs, AAUs and removal units. It also found that the quantities of AAUs and ERUs to be carried over by Ukraine to CP 2 were consistent with the requirements set out in MOP 1, Decision 13, Annex, para. 15. Accordingly, no QoIs were identified by the ERT (CC-2016-1-12/Ukraine/EB; FCCC/KP/CMP/2017/TPR/UKR) (Table 6.12).

6.1.3   Lessons Learned: Some Tentative Conclusions

The experience attained in the context of the CM under the Kyoto Protocol shows that FB has been applied only in the case of the QoI raised against the 15 Annex I parties by South Africa, due to the fact that they did not submit their progress reports on time.[14] However, no decision could be taken about these submissions; in the cases of Latvia and Slovenia, the FB decided not to proceed, but it could not give a decision on their substance (see Table 6.3).

This failure of the FB to come to an agreement on this submission has been ascribed to the "limitations of the Committee's voting rules and the danger of politicization of its proceedings", and is seen as "a serious threat to the credibility of the Committee" (Oberthür/Lefeber 2010). As it has not been used to provide early warning of non-compliance by the parties with commitments under Art. 3.1, KP, it is usually argued that the FB has not got a crucial place in the practical application of the process. Therefore, it needs to be further developed to address potential non-compliance through its "early warning function" (Oberthür/Lefeber 2010: 137, 155). In order to overcome that failure, it is also proposed to engage the FB earlier than at the present stage, giving the FB an opportunity to review ERT reports and to initiate the process on its own (Doelle 2010). Thus, the FB can infer the non-compliance problems of the parties from the ERT reports, and then intervene in the process to solve the problems.

Despite the existence of the opportunity for the parties to trigger the mechanism, there has been no submission to the EB by the parties concerned. The submissions made, up to now, have been triggered by the ERTs (see Table 6.13).

The role of the reports of expert review teams under Art. 8, KP (NCP, Section VI (3)) forms the most outstanding issue that should be underlined here, since their role "has the potential to de-politicise the process and lead to more acceptance of the procedures" (Wolfrum/Friedrich 2006: 59). When it is thought that the parties are generally reluctant to trigger the procedure with regard to the compliance of any other party with its target, their role on triggering the procedure becomes more significant. The survey in practice acknowledges this significance, as the procedure in the EB has always been triggered by a question raised by an ERT up to now (see Table 6.13). Even if a party ventures to raise a QoI directly against another party, the reports of the ERT again play an effective role in determining the decision and the consequences applied to the party. However, even though the reports play a

---

[14]For a general analysis of the weaknesses and strengths of the current system, see Savaşan (2015c).

Table 6.13 QoIs brought before the Enforcement Branch (EB) in brief

Parties concerned	Preliminary finding	Final decision	Decision under Section X, para. 2	Appeal
Greece (CC-2007-1/ Greece/EB)	Non-compliance with the guidelines of national systems and Art. 7 guidelines, hence, Greece does not yet meet the eligibility requirement (6 March 2008)	Confirmation of the preliminary decision (17 April 2008)	Found eligible to participate in the Kyoto mechanisms (13 November 2008)	No Appeal
Canada (CC-2008-1/ Canada/EB)	No preliminary finding, but, at the end of preliminary examination, the EB decided to proceed; the QoI is not *de minimis* or ill-founded (2 May 2008)	No final decision-but the EB decided to not to proceed further (15 June 2008)	No decision required under Section X, para. 2	No Appeal
Croatia (CC-2009-1/ Croatia/EB)	Croatia does not have its assigned amount pursuant to Art. 3, paras. 7–8, calculated and recorded in accordance with MOP 1, Decision 13 and therefore does not yet meet the eligibility requirements (13 October 2009)	Confirmation of the preliminary finding (26 November 2009)	EB decided that Croatia is fully eligible to participate in the mechanisms (8 February 2012)	Croatia lodged an appeal against the final decision of the EB (14 January 2010) and then withdrew its appeal (4 August 2011)
Bulgaria (CC-2010-1/ Bulgaria/EB)	Non-compliance with national system requirements, hence not yet meeting the eligibility requirement (12 May 2010)	Confirmation of the preliminary finding (28 June 2010)	Bulgaria is fully eligible to participate in the mechanisms (4 February 2011)	No Appeal
Romania (CC-2011-1/ Romania/EB)	Non-compliance with national system requirements, hence not yet meeting the eligibility requirement (8 July 2011)	Confirmation of the preliminary finding (27 August 2011)	There is no longer a QoI with respect to Romania's eligibility (13 July 2012)	No Appeal
Ukraine I	Non-compliance with national system	Confirmation of the	There is no longer a QoI with respect	No Appeal

(continued)

Table 6.13   (continued)

Parties concerned	Preliminary finding	Final decision	Decision under Section X, para. 2	Appeal
(CC-2011-2/ Ukraine/EB)	requirements, hence not yet meeting the eligibility requirement (25 August 2011)	preliminary finding (12 October 2011)	to Romania's eligibility (9 March 2012)	
Ukraine II (CC-2016-1-1/ Ukraine/EB)	Non-compliance with eligibility requirements and with Art. 3, para. 1, KP (21 June 2016)	Confirmation of the preliminary finding (7 September 2016)	No QoIs were identified by the ERT (15 August 2017)	No Appeal
Lithuania (CC-2011-3/ Lithuania/EB)	Non-compliance with national system requirements, hence not yet meeting the eligibility requirement (17 November 2011)	Confirmation of the preliminary finding (21 December 2011)	Continuing QoI with respect to Lithuania's eligibility (14 July 2012) Lithuania subsequently fully eligible to participate in the mechanisms (24 October 2012)	No Appeal
Slovakia (CC-2012-1/ Slovakia/EB)	Non-compliance with national system requirements (14 July 2012)	Confirmation of the preliminary finding (17 August 2012)	There is no longer a QoI with respect to Slovakia (4 July 2013)	No Appeal

Source Prepared by the author on the basis of information/documentation on the official website of the UNFCCC.   See: http://unfccc.int/kyoto_protocol/compliance/questions_of_implementation/items/5451.php

remarkable role in the operation and the results of the procedure, they are not authorized to make decisions on non-compliance.

The ERTs are composed of experts selected from those nominated by the parties to the Convention and, as appropriate, by IGOs, and are coordinated by the secretariat (Art. 8(2), KP). They review the information submitted under Art. 7, KP by each party included in Annex I pursuant to the relevant decisions of the COP and in accordance with guidelines adopted for this purpose by the MOP. Indeed, they review the information submitted under Art. 7(1), KP as part of the annual compilation and accounting of emissions inventories and assigned amounts, and the information submitted under Art. 7(2) as part of the review of communications (Art. 8(1), KP).

The coordination of expert review teams, selected from those nominated by parties and by IGOs, is conducted by the secretariat (Art. 8.2). On the basis of the

Table 6.14 Subjects of the QoIs under the EB

Greece	Non-compliance with the guidelines for national systems (MOP 1, Decision 19 2005a) and the guidelines for the preparation of the information (Decision 15/MOP 1 2005b), hence found insufficient to meet the eligibility requirement
Canada	Non-compliance with national registry[b] requirements, with the Art. 7 guidelines for the preparation of the information (MOP 1, Decision 15 2005b) and the modalities for the accounting of assigned amounts under Art. 7, para. 4, KP (MOP 1, Decision 13 2005b)
Croatia	Non-compliance with the assigned amount[c] (it does not have its assigned amount pursuant to Art. 3, paras. 7–8, KP, calculated and recorded in accordance with MOP 1, Decision 13 2005b), and with its commitment period reserve[d]
Bulgaria	Non-compliance with the guidelines for national systems (MOP 1, Decision 19 2005a), the UNFCCC reporting guidelines, the IPCC Good Practice Guidance and the IPCC Good Practice Guidance for LULUCF
Romania	Non-compliance with the guidelines for national systems[a] (MOP 1, Decision 19 2005a), with the Art. 7 guidelines for the preparation of the information (Annex to MOP 1, Decision 15 2005b) and the definitions, modalities, rules and guidelines relating to LULUCF activities (MOP 1, Decision 15 2005b, and MOP 1, Decision 16 2005a)
Ukraine I	Non-compliance with the guidelines for national systems (MOP 1, Decision 19 2005a), Art. 7 guidelines for the preparation of the information (MOP 1, Decision 15), the UNFCCC reporting guidelines, the IPCC Good Practice Guidance and the IPCC Good Practice Guidance for LULUCF, and definitions, modalities, rules and guidelines relating to LULUCF activities (MOP 1, Decision 16 2005a)
Ukraine II	Non-compliance with Art. 7, para. 1, in conjunction with para. 4, and the mandatory requirements set out in the "Modalities for the accounting of assigned amounts under Art. 7, para. 4, of the Kyoto Protocol" (Annex to MOP 1, Decision 13) and the "Guidelines for the preparation of the information required under Art. 7 of the Kyoto Protocol" (Annex to MOP 1, Decision 15) Non-compliance with Art. 3, para. 1, KP. Been unable to formally demonstrate its compliance with its commitment under Art. 3, para. 1, KP in accordance with the relevant procedures set out in MOP 1, Decision 13
Lithuania	Non-compliance with the guidelines for national systems (MOP 1, Decision 19 2005a) and the Art. 7 guidelines (MOP 1, Decision 15 2005b), and the definitions, modalities, rules and guidelines relating to KP-LULUCF activities (MOP 1, Decision 16 2005a), the UNFCCC reporting guidelines, the IPCC Guidance and the IPCC Good Practice Guidance for LULUCF
Slovakia	Non-compliance with the guidelines for national systems (MOP 1, Decision 19 2005a), with the methodological and reporting requirements of the Revised 1996 IPCC Guidelines, and the IPCC Good Practice Guidance

(continued)

Table 6.14 (continued)

- None of the QoIs have been related directly to whether the concerned party is in compliance with its 2012 emissions target
- Yet Ukraine's II Case is about the fact that the total GHG of Ukraine for the CP 1 exceed the quantities of emission reduction units (ERUs), certified emission reductions (CERs), temporary CERs (tCERs), long-term CERs (lCERs), assigned amount units (AAUs) and removal units (RUs) in the retirement account of Ukraine for the CP 1
- Only in the Slovakia case did the EB decide on the disagreement between the ERT and Slovakia over whether to apply adjustments to emission inventories (see CC-2012-1-6/Slovakia/EB)
- All are related to the eligibility requirement – referred to in para. 31c, Annex to MOP 1, Decision 3; para. 21c, Annex to MOP 1, Decision 9 (2005b); and para. 2c, Annex to MOP 1, Decision 11 (2005b)
- Ukraine's II Case is also related to compliance with Art. 3, para. 1, KP, i.e. Ukraine's emission reduction target for CP 1
- In all, the expedited procedures as contained in NCP, Section X were applied

In fact, in Ukraine's II Case as well, the EB followed the expedited procedures for both QoIs, taking into account that they were related to each other

Source Prepared by the author on the basis of information/documentation on the official website of the UNFCCC. See: http://unfccc.int/kyoto_protocol/compliance/questions_of_implementation/items/5451.php

[a] A national system includes the institutional, legal and procedural arrangements for estimating emissions and removals, sinks of GHGs covered by the Protocol, and for reporting and archiving this information. Each country is required to produce a reliable account of its emissions and demonstrate compliance with its 2012 emissions target

[b] A national registry is a computerized system with certain technical standards used to track holdings of GHG credits, obliging the countries to account for their emissions credits

[c] The assigned amount sets a party's base emission limit for its first commitment period (2008–2012)

[d] The commitment period reserve is to limit the amount of credits that a party can sell

review process, which should provide a "thorough and comprehensive technical assessment" (Art. 8(3)) of all aspects of the implementation by a party of the Protocol, they prepare a report to the MOP involving their assessment of the implementation by the party, the identification of any potential problems in meeting its commitments and questions of implementation (Art. 8(3), KP). When the report comes before the MOP, the MOP decides on any matter on the related party's compliance situation (Art. 8.6), based on the information submitted by parties (Art. 7, Art. 8.5a) and the questions of implementation listed by the secretariat (Art. 8.5b).

Accordingly, due to the fact that the parties have not wanted to encounter "the political costs of accusing other states and of undermining a sense of common purpose in the regime" (Mitchell 2005: 75–76), in practice – apart from the South African submission made to the FB – all submissions have been launched by the ERTs.

To date, the subjects of the QoIs that come before the EB have generally been about national system requirements, but have not been directly related to whether the concerned parties are in compliance with their 2012 emissions target (see Table 6.14). This is because non-compliance with emissions targets is a subject that cannot be brought before the EB until after the end of the first commitment period in 2012. In fact, the EB can decide on emission targets of the parties only after the end of the first period. Additionally, the non-compliant party has 100 days to comply with its commitment after the expert review of its final emissions inventory. The submission of inventories for emissions in the year 2012 should finish on 15 April 2014, and ERT has one year to review them (MOP 1, Decisions 15–22). Therefore, consideration of compliance for the first period could not begin until July 2015 or later (Oberthür/Lefeber 2010). If such a party still does not meet its target, it has to encounter the consequences (NCP, Section XV, 5–8) of non-compliance with emission commitments under Art. 3.1, KP. It is then expected to make up its shortfall in the second commitment period (Tables 6.15 and 6.16).

Only in the case of Slovakia, by the end of 2012, did the EB have the opportunity to decide on the disagreement between the ERT and Slovakia on whether to apply adjustments to emission inventories by determining the correct amount for the party and applying adjustments (see CC-2012-1-6/Slovakia/EB). In the previous cases, the parties resolved such disagreements by cooperating with the ERTs before the formal proceeding had been initiated by the ComplCom. In all cases, the EB sought expert advice; in particular, it asked members of the expert review team to present their report and advice, and also asked other independent experts for their advice. The concerned parties have made written submissions and presented their evidence during a hearing held as a part of the EB meetings. To date, only Croatia has applied for appeal to the MOP against an EB decision, but it withdrew its appeal.

Although the NCP of the Protocol involves various response measures (positive-negative, except financial measures or loss of credits), most of them have been very rarely applied in practice; if applied at all, they have been very limited in scope. In all eight cases of non-compliance, the EB has used the same three

Table 6.15 Consequences applied to the non-compliant parties under the EB

Greece	a. Declared to be in non-compliance
Canada	b. Required to submit a compliance action plan to address its non-compliance within three months
Croatia	c. Declared to be ineligible to participate in the market mechanisms
Bulgaria	
Romania	
Ukraine I–II	
Lithuania	
Slovakia	a. Declared to be in non-compliance b. Required to submit a compliance action plan to address its non-compliance within three months

Source Prepared by the author on the basis of information/documentation on the official website of the UNFCCC. See: http://unfccc.int/kyoto_protocol/compliance/questions_of_implementation/items/5451.php

consequences (except in the case of Slovakia, when only two of them were applied; suspension of participation in the flexibility mechanisms was not applied):

a. making a public declaration of non-compliance;
b. the obligation to submit a compliance action plan addressing their non-compliance within three months, which would be subject to review and assessment by the EB;
c. suspension of trading in the Kyoto carbon market set up by the ET, CDM and JI mechanisms (implying that the party is suspended from trading in the Kyoto carbon market set up by these mechanisms until reinstatement by the EB).

In a nutshell, like the CM of the MP, those measures have worked just for entailing reputational damage rather than immediate material costs/consequences. Additionally, regarding the response measures applied by the branches of the ComplCom, it can be argued that they have been adopted in line with "a *pro futuro* approach", and thus follow a forward-looking perspective as they do not aim to punish the non-compliant party, but to encourage it to comply with its obligations (Eritja et al. 2004: 53). Yet, it may be criticized as it has not developed mechanisms to reward "overcompliance and innovation", such as "public awards, white lists, accession to financial mechanisms or reduced project verification requirements" (Mitchell 2005: 74). Because of this, the parties are not expected to make any effort to exceed their emission targets or to undertake projects providing uncertain (but high level) reductions.

With regard to the question, which of the two different logics of behaviour – namely, the logic of consequences and the logic of appropriateness – should be regarded as fundamental to the CM of the Kyoto Protocol, it should be underlined that, like other MEAs and their compliance mechanisms, the CM of the Kyoto

Table 6.16 The road towards the Paris Climate Agreement: Key COPs/MOPs

Key COPs/ MOPs No/Date/ Place	Key documents	Key decisions
COP 13/MOP 3 3–14 December 2007 Bali, Indonesia	COP 13, Decision 1: Bali Road Map	– Ad Hoc Working Group on Long-term Cooperative Action (AWG-LCA) is established – Gateway towards long-term cooperative action beyond 2012 to reach an agreed outcome
COP 14/MOP 4 1–12 December 2008 Poznan, Poland	No available document directly/ or indirectly related to the Paris Climate Agreement[22]	No available decision directly/or indirectly related to the Paris Climate Agreement
COP 15/MOP 5 7–18 December 2009 Copenhagen, Denmark	COP 15, Decision 1: Outcome of the work of the AWG-LCA COP 15, Decision 2: Copenhagen Accord MOP 5, Decision 1: Outcome of the work of the AWG-KP	Extension of the AWG-LCA's mandate
COP 16/MOP 6 29 November– 10 December 2010 Cancun, Mexico	COP 16, Decision 1: The Cancun Agreements: Outcome of the work of the AWG-LCA MOP 6, Decision 1: Cancun Agreements: Outcome of the work of the AWG-KP 15	Continuation of the work towards a post-2012 legally-binding agreement
COP 17/MOP 7 28 November–9 December 2011 Durban, South Africa	COP 17, Decision 1: Establishment of an AWG-DP COP 17, Decision 2: Outcome of the work of the AWG-LCA MOP 7, Decision 1: Outcome of the work of the AWG-KP 16	– Agreement on a second commitment period of the KP – Recognition of the need for a universal, legal agreement – Launch of a new platform for a legal agreement with legal force by 2015 for the period beyond 2020
COP 18/MOP 8 26 November–7 December 2012 Doha, Qatar	MOP 8, Decision 1: Doha Amendment COP 18, Decision 1: Agreed outcome pursuant to the Bali Action Plan COP 18, Decision 2: Advancing the Durban Platform	– Setting out a timetable to adopt a universal agreement by 2015 to come into effect in 2020 – Completing the work under the Bali Action Plan to concentrate on the 2015 agreement – Finalization of the Works of both the AWG-LCA and the AWG-KP
COP 19/MOP 9 11–22 November 2013 Warsaw, Poland	COP 19, Decision 1: Further advancing the Durban Platform	– A timeline is put forward for the development of the 2015 agreement – Decision to initiate domestic preparations for INDCs towards the agreement

(continued)

Table 6.16 (continued)

Key COPs/ MOPs No/Date/ Place	Key documents	Key decisions
COP 20/MOP 10 1–12 December 2014 Lima, Peru	COP 20, Decision 1: Lima Call for Climate Action	– A further important step is put forward for a new 2015 agreement – Elaboration of the elements of the new agreement – Agreement on the rules on how countries can submit contributions to the new agreement during the first quarter of the next year
COP 21/MOP 11 30 November– 11 December 2015 Paris, France	COP 21, Decision 1: Adoption of the Paris Agreement	– Paris Climate Agreement is adopted – Launch of a Facilitative Dialogue between parties with the aim of understanding the parties' progress towards the long-term goal of the Agreement (Art. 4.1) and on the preparation of the NDCs (Art. 4.8) (COP 21, Decision 1, 2015: para. 20)

Source Prepared by the author on the basis of information/documentation on the official website of the UNFCCC. See: https://unfccc.int/decisions

Protocol does not always function through only one logic, but both can influence its functioning. To Mitchell (2005), in predicting the short-term influence of the Kyoto regime, it is appropriate to adopt the logic of consequences model. To this, in the short term, the incorporation of flexibility into the regime is likely to contribute to the compliance rates, encouraging further behavioural changes by committed states and further compliance by contingent states (but with not much behavioural change), and convincing resistant states to re-evaluate their position on remaining parties to the agreement and its requirements.

For the long term, Mitchell (2005) stresses the need to adopt the logic of appropriateness model with new supplements that can contribute to climate change and deeper transformations of the goals and norms. He argues that flexibility can provide these "deeper transformations" (Mitchell 2005: 66). These transformations are assured by reinforcing the committed states' views on protecting the climate, by convincing some contingent states to comply with the relevant requirements of the Protocol, and enabling some resistant states to join the regime during the second commitment period as contingent states.

This is in parallel with the MP's CM in that, even though there seems to be a system of bringing together the facilitation and enforcement functions under the same mechanism, a management approach which applies not only LoC but also LoA and which is equipped with facilitative-positive measures emerges as the more dominant approach adopted in the KP's CM as well.

Overall, it may be argued that although the CM under the KP is based on a shorter time of experience than the one under the MP, it "has proved its ability to significantly contribute to meeting the related functional demands" with its "overall design" and "individual elements" establishing an "important benchmark" for the climate regime (Lefeber/Oberthür 2012: 101). Indeed, it "offers a good starting point for any effort to combine facilitation and enforcement for an effective compliance system that deals with the range of motivations, from self-interest to lack of capacity" (Doelle 2012: 120). Through its two-branch structure – of which one branch is grounded on a quasi-judicial character, and certain time limitations for different steps in the process – it has the potential to contribute to ensuring and promoting better compliance (Savaşan 2015c).

6.2   The Paris Climate Agreement: The Path Forward on Compliance?

6.2.1   The Road Towards the Paris Climate Agreement: Key Meetings/Decisions

The first decisive step at international level regarding the issue of global warming and climate change has been the global cooperation regime established by the United Nations Framework Convention on Climate Change (UNFCCC). The ultimate goal of the regime has been determined as "to achieve stabilization of greenhouse gas concentrations in the atmosphere at a level that would prevent dangerous anthropogenic interference with the climate system" (Art. 2, UNFCCC). Identification of the basic principles for combating the climate change problem, the adoption of institutional and procedural bases and of the main obligations were all made with the input of the UNFCCC, which is a framework convention; and with the Kyoto Protocol (KP) adopted in 1997, the obligations have been made more concrete and detailed (Arts. 3.14, 5.1, 6.2, 7.4, 8.4, 12.7, 16, 17, 18, KP).

In the Kyoto Protocol, emission reduction commitments have been identified, the general framework of the mechanisms for implementation has been put forward, and arrangements have been made to monitor whether the commitments are fulfilled. As a requirement of the Common but Differentiated Responsibility Principle-Respective Capabilities (CDRP-RC), developed countries have the primary responsibility for emission reduction. As a matter of fact, the Protocol foresaw that during the first commitment period the countries included in the Annex I list would reduce their greenhouse gas emissions by at least 5% compared to 1990 levels (Art. 3, KP). However, in order to balance the situation of these countries with that of developing countries and help them meet their emission reduction targets, flexibility mechanisms such as Joint Implementation, Clean Development and Emission Trading were also established (Arts. 6, 12, 17, KP).

The cooperation regime on climate change has mostly been developed in accordance with the decisions of the COP/MOP, the first of which took place in Berlin in 1995. During the COP/MOP negotiations held annually, further decisions and procedural rules were decided; and the provisions of the UNFCCC and the KP have been implemented through the COP decisions. Significant steps were particularly taken at the Buenos Aires (COP 4) and Marrakech Meetings (COP 7).

An Ad Hoc Working Group on Further Commitments for Annex I Parties under the Kyoto Protocol (AWG-KP) was established to discuss the determination of additional obligations for the Annex I countries at the COP 11/MOP 1 convened in Montreal in 2005.[15] Another important development at the Montreal COP/MOP with regard to the obligation to adapt to climate change was the launch of the Nairobi Work Programme (NWP) established through COP 11, Decision 2 as a mechanism under the Convention to facilitate the development and dissemination of information for supporting adaptation policies and practices, whose implementation has been coordinated by the Subsidiary Body for Science and Technological Advice (SBSTA).

As the first commitment period covered only 2008–2012, the Bali Roadmap was adopted at the COP 13/MOP 3, held in Bali in 2007, in order to regulate the period after 2012. This was the most important development on the road towards the Paris Climate Agreement. Indeed, through the Bali Roadmap, a set of decisions which can be seen as the key steps towards the Agreement were accepted. According to the Bali Action Plan, it was decided to establish an Ad Hoc Working Group on Long-term Cooperative Action (AWG-LCA).[16] It was also agreed to present the results of the work to be carried out by this working group in 2009 at COP 15 (COP 13, Decision 1 2007: paras. 1–2). Therefore, the significance of the measurable, reportable, verifiable commitments is also underlined in the Plan to lead to enhanced national/international action for mitigating climate change. That is, it is emphasized that nationally appropriate mitigation actions should be considered in a measurable, reportable and verifiable manner, taking into account developed countries' quantified emission reduction targets on the one hand, and developing countries' actions in the context of sustainable development, supported by technology, financing and capacity-building activities on the other hand (COP 13, Decision 1, para. 1.b.i–ii). Within the negotiations carried out by AWG-LCA 6, taken during the sixth session held in Bonn on 1–12 June 2009, the views of the parties in the context of the Bali Action Plan have also been taken into account.[17]

---

[15]For the activities of the Ad Hoc Working Group on Further Commitments for Annex I Parties under the Kyoto Protocol (AAWG-KP), see http://unfccc.int/bodies/body/6409.php.

[16]For the activities of the Ad Hoc Working Group on Long-term Cooperative Action (AWG-LCA), see: http://unfccc.int/bodies/body/6431.php.

[17]For details of AWG-LCA 6, see: http://unfccc.int/meetings/bonn_jun_2009/session/6302.php.

In the 2009 Copenhagen Accord adopted through COP 15, Decision 2, it is emphasized that the average global temperature increase should be kept below 2 °C (paras. 1–2) and that Annex-I countries should undertake to implement quantified emission reduction targets for 2020 (para. 4). It is also recommended that Non-Annex-I countries should submit their national plans for emission reduction targets and their plans and activities to the secretariat (para. 5). Additionally, the Copenhagen Accord provided for novel and additional resources for both commitment periods of the KP (I. Commitment Period: 2010–2012; II. Commitment Period: 2012–2020) (para. 8). The establishment of a Green Climate Fund for financing the adaptation and mitigation activities of developing countries is also mentioned in the Accord (para. 10). In line with the final objective of the UNFCCC, it is stated that an assessment should have been made in 2015 on the implementation of the Accord (para. 12).

It was thought that effective decisions on the KP's second commitment period would be made at the COP 15/MOP 5 held in Copenhagen in 2009, but the expectations were not met (COP 15, Decision 2 2009). Therefore, the Copenhagen Accord has been defined as a document to cover the failure (Dimitrov 2010). At the Copenhagen summit, however, there was a paradigm shift in the struggle with climate change, which found concrete expression in the Paris Climate Agreement. Indeed, all the factors which were included in the Copenhagen Accord become concrete in the Paris Climate Agreement of 2015. The decision adopted in Cancun in 2010 at the COP 16/MOP 6, containing the final regulations on the outstanding aspects of Copenhagen Accord, is of particular importance as well.[18]

The COP 17/MOP 7 negotiations, which took place in Durban between 28 November and 9 December 2011, were also crucial for determining the climate change regime that became valid from 2012 and for implementing decisions made in Cancun. The main development recorded at the Durban meeting was the establishment of the Ad Hoc Working Group on the Durban Platform for Enhanced Action (AWG-DP), which played a crucial role in the process leading to the Paris Agreement.[19] The Durban Platform was established as a subsidiary body under the Convention with the mandate of preparing a protocol, another legal instrument or an agreed outcome with legally binding force applicable to all parties.

It was decided by the COP 17, Decision 1 (2011) that: the Platform would complete its work no later than 2015 and come into effect from 2020 (paras. 2–4); a workplan would be launched in the first half of 2012 to identify ways/means to improve mitigation ambition to close the ambition gap (paras. 6–8); and the working of AWG-LCA should be extended for one year in order to achieve an agreed outcome by the COP 18, when the working of AWG-LCA terminated (para. 1). In order to ensure the highest possible mitigation efforts by all parties, COP 17, Decision 1 foresaw that all developed and developing countries would take responsibility, thus giving the signal for the abolition of the division between

---

[18]See also Morgan (2012) for a detailed analysis on the emerging post-Cancun climate regime.

[19]For the activities of the AWG-DP, see: http://unfccc.int/bodies/body/6645.php.

developed and developing countries (Preamble, MOP 8, Decision 1 2012; COP 18, Decision 2 2012: para. 5). At the MOP 7, it was also decided that the second commitment period under the KP would begin on 1 January 2013 and end either on 31 December 2017 or 31 December 2020 with the decision of the Ad Hoc Working Group on Further Commitments for Annex I Parties under the Kyoto Protocol (AWG-FCAP) (MOP 7, Decision 1 2011: para. 1).

In the COP 18/MOP 8 negotiations held in Doha in 2012, the KP's second commitment period was set as 2013–2020 through the Doha Amendment (MOP 8, Decision 1 2012: para. 4). It was also anticipated that Clean Development, Joint Implementation and International Emissions Trading Mechanisms would continue in the second period as well (MOP 8, Decision 1 2012: paras. 12, 13, 15). Stressing the importance of the work under the AWG-DP to adopt a legal instrument no later than 2015, the continuation of preparatory work on a new agreement to come into effect and to be implemented from 2020 was also envisaged (MOP 8, Decision 1 2012; Preamble; COP 18, Decision 2 2012: para. 4). It was also decided that the AWG-KP had fulfilled its mandate by adopting MOP 8, Decision 1 (Doha Amendment). The negotiations carried out since 2007 (COP 13) within the framework of long-term cooperation actions were terminated at the COP 18 with the expiry of the mandate of the AWG-LCA (COP 18, Decision 1 2012).

The most important innovation under the COP 19/MOP 9 negotiations held in Warsaw in 2013 was the invitation of state parties to submit their intended nationally determined contributions (INDC), which form the basic elements of the climate action to be implemented in the post-2020 period, under the new climate agreement (COP 19, Decision 1 2013: para. 2b).[20]

However, the problems arising from the structural characteristics of the Protocol in the KP first commitment period (2008–2012) have brought about a paradigm shift in the future of the UNFCCC global climate regime. In fact, the USA did not become a party to the KP at the beginning of the process, and Canada took the decision to withdraw from the Protocol a day before the COP 18/MOP 8. The fact that these two economic powers, which are expected to assume a significant responsibility for greenhouse gas emissions, are out of process; that only 37 parties are bound to the KP for the second commitment period; and that some large economies, e.g. Brazil, China, still are not expected to cut back on their emissions as they have the status of developing countries, has brought about system discussions on how to make the climate change regime effective. As a result of efforts to overcome the deadlocks in climate change negotiations, a new road map for the post-2020 period of the climate change regime and the Paris Climate Agreement

---

[20]In the COP 20/MOP 10 negotiations held in Lima in 2014, invitations made to the state parties at the Warsaw meeting in 2013 for the presentation of their INDCs were repeated (COP 20, Decision 1 2014: paras. 8–10, 13–16). In addition, the draft negotiating text developed by the Durban Platform in Lima confirmed that the AWG-DP should complete its work at the COP 21 to adopt an instrument legally binding under the Convention, applicable to all parties, on the basis of the principle of CBDR in the light of different national circumstances (COP 20, Decision 1 2014: paras. 1–3, 5–7 and Annex).

was adopted as a concrete legal framework for reconciliation at the COP 21 held in Paris from 30 November to 13 December 2015 (COP 21, Decision 1 2015).

On 5 October 2016, the conditions for the entry into force of the Paris Agreement were met. It entered into force on 4 November 2016, in accordance with its Article 21, after the deposit of 55 instruments of ratification, acceptance, approval or accession, corresponding to an estimated 55 per cent of the total global greenhouse gas emissions (Art. 21, PCA).[21]

In sum, the need for a paradigm shift in the future of the UNFCCC global climate regime began to be discussed at the COP 15/MOP 5 held in Copenhagen in 2009, because the Kyoto Protocol's first commitment period (2008–2012), which began in 2008, ended on 31 December 2012; and the change in question has been tangibly expressed in the Paris Climate Agreement (PCA).

6.2.2 Paris Climate Agreement: Fundamental Features/ Outcomes

6.2.2.1 Legal Nature of the Paris Agreement and Its Relationship with the Climate Convention

According to the Bali Action Plan, in 2007 it was decided to launch a comprehensive process in order to reach an agreed outcome (COP 13, Decision 1). After three years, in 2010, this outcome was stated as a "legally binding outcome" (COP 16, Decision 1). Arguably, the statement may be regarded as a requirement that the Paris agreed outcome should have legally binding force (COP 17, Decision 1 2011: para. 2). Indeed, the Paris Agreement can be considered as an "international agreement concluded between States in written form and governed by international law," according to Art. 2.1(a), VCLT. Nevertheless, it should be stressed that its various provisions just provide guidance; they are not worded as positive obligations. The main example for this is represented by Art. 4, PCA, which does not contain any obligations for state parties in the reduction of their GHG emissions. That is, each party determines, in its domestic law, the "contributions that it intends to achieve" (Art. 4.2, PCA). This is the basic compromise: the majority of states agreed to conclude an international treaty (i.e. a legally binding instrument) provided that it did not contain binding obligations on the reduction of emissions.

At present (16 April 2018), 195 of 197 parties to the UN Framework Convention of Climate Change (UNFCCC) signed the Paris Agreement,[22] and 178 parties to the

---

[21](Information obtained on 10 June 2018). See the status of ratification of the Paris Agreement at: http://unfccc.int/paris_agreement/items/9444.php.

[22]New research reveals that all countries currently signed up to the Paris Agreement now have at least one national climate law or piece of legislation or national climate policy in place. See Grantham Research Institute on Climate Change and the Environment and Sabin Center for Climate Change Law (2018) for details.

Convention have ratified the Paris Agreement,[23] and as the Agreement (Art. 27) does not allow reservations in line with the UNFCCC (Art. 24) and the KP (Art. 26), no parties may place caveats on their ratification. Considering that the Kyoto Protocol has 192 parties (191 States and 1 regional economic integration organization)[24] and the Montreal Protocol has 197 parties achieving universal ratification,[25] the Paris Agreement is on the way towards being another universal deal like the Montreal Protocol, given the decision of accession of Nicaragua on 23 October 2017, and news on Syria's possible decision to sign the Agreement (leaving the USA as the only country outside the Agreement, even though the United States Climate Alliance, involving 16 states and Puerto Rico, committed to the goals of the PCA[26]).[27]

With its entry into force, it is expected that the Paris Agreement should be performed by the parties in good faith, i.e. in accordance with the object and purpose of the treaty, to avoid confusion between the obligations of respective treaties in good faith and the general rule of interpretation. The VCLT acknowledges the binding status of the treaties for the parties that have expressed their consent to be bound by it, and expects the parties to act in accordance with the principle of good faith (Preamble, Art. 26, 31, VCLT).

With respect to the relationship of the Paris Agreement with the UNFCCC and the Kyoto Protocol, it should first be recalled that the UNFCCC is a framework convention and that the KP is a legal instrument adopted under the same convention, which remains in force even after the entry into force of the Paris Agreement.

As a binding instrument adopted within the framework of the UNFCCC, the Paris Agreement should be consistent with the objective and principles of the Convention as explicitly pointed out in many provisions of the Agreement (Preamble, para. 3, PCA). In addition, the Paris Agreement should enhance the implementation of the Convention, including its objective to strengthen the global response to the threat of climate change (Art. 2.1, PCA). Finally, as the Paris Agreement is a part of the UNFCCC system, it also ensures the continuation of the existing obligations under the Convention; e.g. developed country parties are obliged to provide financial support to developing country parties in continuation of their existing obligations under the Convention (Art. 9.1, PCA).

The Paris Agreement's institutional structure is also modelled on the UNFCCC; it establishes a MOP (called the Conference of the Parties, serving as the Meeting of

---

[23]Paris Agreement – Status of Ratification. http://unfccc.int/paris_agreement/items/9444.php. See also: https://treaties.un.org/Pages/ViewDetails.aspx?src=TREATY&mtdsg_no=XXVII-7-d&chapter=27&clang=_en.

[24]Kyoto Protocol – Status of Ratification: http://unfccc.int/kyoto_protocol/status_of_ratification/items/2613.php.

[25]Montreal Protocol – Status of Ratification: http://ozone.unep.org/sites/ozone/modules/unep/ozone_treaties/inc/datasheet.php.

[26]For details of the US Climate Alliance see: https://www.usclimatealliance.org/.

[27]See: http://www.independent.co.uk/news/world/middle-east/syria-paris-agreement-us-climate-change-donald-trump-world-country-accord-a8041996.html?amp.

the Parties to the Paris Agreement (CMA)) which meets simultaneously with the Convention's COP, but as a separate meeting with its own rules/procedures (Art. 16.1, PCA). The parties to the Convention who are not parties to the Paris Agreement may participate in the proceedings of any session of the CMA but only as observers (Art. 16.2, PCA).

Thus, through the Paris Agreement, the Conference of the Parties serving as the Meeting of the Parties (CMA 1) started convening its sessions for the first time in conjunction with COP 22 of the UNFCCC in November 2016.

Additionally, the Technology Mechanism (Art. 10.3), the institutional arrangements regarding the capacity-building activities (Art. 11.5, PCA), the secretariat (Art. 17.1, PCA), the Subsidiary Body for Scientific and Technological Advice (SBSTA) and the Subsidiary Body for Implementation (SBI) (Art. 18.1), and even subsidiary bodies or other institutional arrangements, other than those referred to in the Paris Agreement, established by or under the Convention, serve for the Paris Agreement as well.

In parallel with these provisions allowing the Convention bodies to serve for the Paris Agreement as well, the provisions of the Convention on the functions and the functioning of the secretariat (Art. 17.2, PCA) as well as on the functioning of both subsidiary bodies (SBSTA) and (SBI) (Art. 18.1, PCA) are applied *mutatis mutandis* to the Agreement. Similarly, the rules of procedure of the COP and the financial procedures applied under the Convention apply *mutatis mutandis* to the Paris Agreement (unless otherwise decided by consensus of the CMA) (Art. 16.5, PCA).

As far as decision-making and voting procedures are concerned, it is clearly put forth that parties to the Convention who are not parties to the Paris Agreement may participate as observers in the proceedings of any session of the CMA. As a result, decisions can only be taken by parties to the Agreement (Art. 16.2, PCA).

Moreover, the provisions of Art. 14 of the Convention on dispute settlement (Art. 24, PCA), Art. 15 on the adoption of amendments to the Convention (Art. 22, PCA), and Art. 16 on the adoption and amendment of annexes to the Convention (Art. 23) apply, again *mutatis mutandis,* to the Paris Agreement. Finally, it is to be stressed that the transparency framework and necessary transparency arrangements should be built on the Convention system (Art. 13.4, PCA); the withdrawal from the Convention entails the withdrawal from the Paris Agreement as well (Art. 28.3, PCA).

The withdrawal from the Paris Agreement is possible only after three years from the date on which the Paris Agreement entered into force for that party (Art. 28, PCA). Considering that the withdrawal takes effect at the earliest one year after the Depositary received such notification, no party can withdraw from the Agreement before four years have passed after its entry into force for that party. To illustrate, in June 2017, Donald Trump, the President of the USA, declared that the USA was quitting the Paris Agreement. Afterwards, in January 2018, he announced that the USA might rejoin the Agreement, but at the time of writing it is still not clear what the decision will be.[28] However, the USA will have to abide by the exit process established under Art. 28 PCA; so the earliest date for the USA to withdraw from the Paris Agreement is November 2020, around the time of the next USA presidential election.

6.2.2.2 What is New in the Paris Climate Agreement?

A new paradigm was prescribed by the PCA in an attempt to overcome the trouble experienced in the context of the KP. Even though they were not able to agree on a binding legal text, the participants of COP 15 in Copenhagen did agree that changes to the existing paradigm of the KP should be made. Although the initiative of the Durban Platform failed to reach a consensus on most of the controversial and complex issues and the Copenhagen Accord is generally regarded as a 'disappointment' (Bodansky 2015), "[it] should be probably seen as an outcome which perfectly exemplifies the present international situation, where most of the countries are not willing to accept binding and costly obligations under international environmental law." Therefore, it "represents a very important milestone in terms of 'reality' (Montini 2011: 13).

The elements raised at the Copenhagen Summit and developed with the Durban Platform's efforts were finally made concrete by the Paris Climate Agreement. Thus, the Paris Summit, which has been regarded as a great success, yielded a document based on mutually agreed compromises that evolved as a result of the Copenhagen Accord. That is, the Paris Agreement both recapitulates the Copenhagen Accord and builds on it with a less differentiated structure, stronger transparency and a legal form (Bodansky 2016a: 317).

In the Paris Agreement, an ambitious long-term goal was also put forward to limit the increase in temperature. It was agreed that the global average temperature rise should be kept below 2 °C from the pre-industrial level and that the temperature increase should be limited to 1.5 °C above the pre-industrial period (Art. 2.1 (a), PCA). In order to achieve its main goal, the PCA also articulates two other sub-goals: 1. to reach global peaking of greenhouse gas (GHG) emissions in the short term, reserving the fact that this could take longer for developing country parties; 2. to achieve a balance between anthropogenic emissions by sources and removals by sinks in the long term in the second half of this century (Art. 4.1, PCA).

In the process of meeting these objectives, issues such as adaption to the adverse impacts of climate change and the development of climate-resilient development without threatening food security, the use of financial resources within the framework of low emissions and climate-change-resilient development objectives,

---

[28]See: https://www.theguardian.com/us-news/2018/jan/28/donald-trump-says-us-could-re-enter-paris-climate-deal-itv-interview.

sustainable development and combating poverty have been highlighted (Art. 2.1(b–c), PCA).

It is stated that the "common but differentiated responsibility and respective capabilities principle" (CBDR-RC), which is at the core of the UNFCCC (Preamble, Arts. 3, 4, UNFCCC; Art. 10, KP), will provide a basis for implementation of the PCA in a way that takes account of the fairness, national conditions and the relative capabilities of the parties (Art. 2.2, PCA). Thus, the principle of CBDR-RC, one of the fundamental principles of the UNFCCC, is not abandoned by the PCA; however, unlike the KP, a general obligation addressed to all countries has been provided for, instead of the approach of specifying countries' obligations in two separate lists.

The Paris Agreement also contains references to developed and developing countries; however, unlike the UNFCCC system, it does not contain the lists of Annex I (developed) and non-Annex I countries (developing). In particular, it provides common commitments requiring all parties to show ambitious efforts (Arts. 4, 7, 9, 10, 11, 13) and improve them over the years ahead to achieve the main purpose of the Agreement set out in Art. 2 (Art. 3, PCA). However, while establishing a common framework with those commitments for all parties; it also allows self-differentiation in accordance with their varying national capacities and circumstances (Art. 4.4), and recognizes that special circumstances apply to the least developed countries and small island developing states (Art. 4.6).

According to the Agreement, developed country parties should continue to take the lead by undertaking economy-wide absolute emission reduction targets. Developing countries should advance their mitigation efforts and progress towards economy-wide emission reduction targets over time on the basis of their national circumstances (Art. 4.4, PCA). The Paris Agreement also reaffirms the obligations of the developed countries under the UNFCCC system to ensure support to the developing countries for the implementation of their commitments (Arts. 4.5, 10, 11, PCA). That is, the developed countries continue to remain the main responsible actors for ensuring financial support for mitigation and adaptation in developing countries (Arts. 6.6, 7.6, PCA). Additionally, unlike the UNFCCC system, the Paris Agreement encourages voluntary contributions from developing countries. Similarly, while requiring biennial communications on financial resources to be made available by the developed country parties, it just encourages biennial communications prepared by the developing countries on a voluntary basis (Art. 9.5, PCA).

As there is the need for intensive investment, transfer of technology and financial resources provided by developed countries in the context of both reduction and compliance obligations provided for under Arts. 9, 10, 11, PCA for the countries which have a very limited impact on greenhouse gas emissions but are adversely affected by the increase in temperature due to climate change and so are very fragile in the face of climate change problems, the PCA explicitly refers to the Warsaw International Mechanism for Loss and Damage associated with Climate Change Impacts, established in 2013 (COP 19). Established as an interim body at COP 19 (Art. 8, PCA), it is charged with helping those countries to cope with the adverse effects of climate change and to address loss and damage stemming from it (Art. 8.1). Cooperation and facilitation mechanisms, such as early warning systems, emergency preparedness and slow onset events, are also envisaged (Art. 8.4).

A fund called the Green Climate Fund and defined as "the cornerstone of a new global climate finance architecture" has also been established as a source of climate financing to increase the fighting capacities of the countries which are most affected by climate change but have the lowest ability to combat it (COP 21, Decision 1 2015: paras. 46, 54, 58). The Fund "coexists with other related funds inside and outside the multilateral climate change regime" (Machado-Filho 2012: 238).[29]

The concept of 'carbon budget', which is not included in the KP, is also considered to be one of the important gains of the Paris Agreement (COP 21, Decision 1 2015: para. 17). The total amount of emissions that can be released to the atmosphere in order to ensure that global warming can be kept under 2 degrees is 3.650 billion tons of CO_2; since 2,000 billion tons of this have already been released, a maximum of 1.650 billion tons remains that can be released. This is identified as the carbon budget (Kazokoğlu 2015). In this context, state parties are expected to reduce their emission values as quickly as possible in order not to consume the total global carbon budget.

The system of INDCs, which form the backbone of the PCA, has been adopted as a result of the matured common will on the premise that the effectiveness of the global climate change regime will significantly increase in the case of implementation of a system that will operate from the bottom up (COP 21, Decision 1, paras. 12–21; PCA, Arts. 3–4). This is because the top-down approach predominant in the KP, which stipulates that the precise emission reductions should be met within a precise time frame within the framework of a dual understanding of obligations, has failed in its struggle against climate change. The new approach, called the INDCs system, consists of a new method of emission reduction based on national commitments which will be determined by the parties to the PCA themselves.

Under the INDCs system, all parties are required to prepare, communicate and maintain NDCs. To achieve the long-term purpose of the Agreement, they should also pursue domestic measures which reflect their highest possible ambitious efforts. Each party's NDC should represent a progression beyond its current NDC. In line with their national and special circumstances, the least developed countries and small island developing states should also prepare and communicate strategies, plans and actions for low greenhouse gas emissions development (Arts. 4.2, 4.3, 4.4, 4.6, PCA).[30]

---

[29]See Machado-Filho (2012) for a detailed analysis of the financial mechanisms under the climate regime.

[30]For guidance on the determination of INDCs, see International Law Association (2014). *Legal Principles Relating to Climate Change*, Washington Conference. https://papers.ssrn.com/sol3/papers.cfm?abstract_id=2461556. See also Oslo Principles on Global Obligations to Reduce Climate Change, adopted on 1 March 2015: https://law.yale.edu/system/files/area/center/schell/oslo_principles.pdf.

While the Paris Agreement allows autonomy/flexibility for states through NDCs on the one hand, on the other hand, it requires their possible highest ambitions for climate action and reinforces the supervision through *a transparency system, global stocktake process and implementation/compliance mechanism.*

Through Article 13,[31] the Paris Agreement establishes a new "transparency framework built-in flexibility," based on the principle of CBDR-RC (Art. 13.2, PCA). Further developments of the system of transparency were negotiated firstly by CMA 1, and then by CMA 1.2 (Bonn, 6–17 November 2017). This transparency system will "be implemented in a facilitative, non-intrusive, non-punitive manner, respectful of national sovereignty, and avoid placing undue burden on Parties" (Art. 13.3, PCA). In fact, the Agreement uses a mandatory wording ('shall') to express to duty of the parties to adopt and implement domestic mitigation measures, making them binding obligations. However, it does not provide a binding obligation with respect to the implementation of and compliance with NDCs. It only envisages a technical expert review. This expert review includes assistance in identifying capacity-building needs and areas of improvement for the relevant party to consider its implementation and achievement of NDC (Arts. 13.11, 13.12, PCA). Thus, in this review process, the information reported by parties is evaluated and the progress made by the relevant party and its respective implementation and achievement of its NDC are questioned in a facilitative/multilateral manner, providing support to developing country parties for their implementation (Arts. 13.14–15) on the basis of CBDR-RC, i.e. in the light of the respective national capabilities and circumstances of developing country parties (Arts. 13.11–12, PCA).

The purpose of the global stockage is to track the progress of parties' individual NDCs, recorded in a public registry maintained by the UNFCCC Secretariat (Art. 4.12), to inform the global stocktake, a five-yearly review of the impact of parties' actions to cope with climate change (Art. 14), to provide clarity on parties' adaptation actions (Art. 7), and to ascertain the support provided and received by relevant parties (Arts. 4, 7, 9, 10, 11), particularly the financial, technology transfer and capacity-building support provided by the developed country parties (or other volunteer parties, if there are any) to developing country parties (Arts. 13.5–10, PCA).

In order to evaluate progress towards achieving the goals set in the Agreement, the CMA is entrusted to periodically scrutinize its implementation. It will undertake its first global stocktake in 2023 and every five years thereafter, NDCs will be submitted in line with the information provided by the outcomes of the global stocktake, with the expectation that the representations will show a progression ahead of former undertakings (Arts. 14.1–2, PCA). This is because the NDCs are already expected to reflect the "highest possible ambition" (Art. 4.3, PCA), and, in line with the principle of global stocktake, the parties are required to communicate their successive NDCs every five years to assess their collective progress.

---

[31]For a detailed analysis of Article 13, see Dagnet/Levin (2017).

The NDCs' cycle is carried out through clauses which ensure a certain degree of flexibility in its implementation; for instance, parties whose INDCs contain a time frame up to 2025 are required to communicate a new NDC by 2020 (COP 21, Decision 1 2015: para. 23), and parties whose INDC contains a time frame up to 2030 are required to communicate or update by 2020 (COP 21, Decision 1, para. 24).[32]

The last feedback mechanism consists of the compliance/implementation mechanism, that is planned to "be expert-based, facilitative in nature and function in a manner that is transparent, non-adversarial and non-punitive" (Art. 15, PCA). Its elaborations were expected to be negotiated at the next CMAs.[33]

In conclusion, the evolution of the climate change regime established under the UNFCCC consists of varying directions. It is usually accepted that the KP's approach is a top-down approach, as it includes binding emissions targets for developed countries, and no binding commitments for developing countries. In line with the UNFCCC system, it is also based on the principle of CBDR-RC. However, the Kyoto experience has already shown that imposing emissions reduction commitments and making them legally binding are not very effective ways to cope with the challenges of climate change, and simply waste time and resources. Flexibility is a better way to manage these challenges (Morel/Shishlov 2014). Therefore, a more bottom-up approach is adopted, providing a paradigm shift "from centralized standard setting to a 'bottom-up' approach" under the regime through the efforts made for the 2009 Copenhagen Accord and 2010 Cancun Agreements (Brunnée 2012: 52). After the call at the Durban Platform for Enhanced Action for a legal instrument applicable to all parties under the Climate Convention, the call of the COP 19 in Warsaw to submit INDCs indicates the bottom-up approach incorporated into the Agreement. Thus, unlike the KP, the Paris Agreement brings a balance between ambitious climate action and fairness between the parties, combining both bottom-up and top-down approaches (Doelle 2017; Rajamani 2016; Voigt/Ferreira 2016; Bäckstrand et al. 2017). As a result, in the new period beginning with the Paris Climate Agreement, the idea that all countries in the negotiations of the post-2012 regime should contribute their efforts to combat global climate change to the extent of their national capacities has become tangible with the mechanism which foresees the collective responsibility of all countries where INDCs are at the centre.

---

[32]Communication of the parties' first NDCs (COP 21, Decision 1 2015: para. 22) was undertaken during the ratification process. For INDCs as communicated by Parties, see: http://www4.unfccc.int/submissions/indc/Submission%20Pages/submissions.aspx.

For an online climate data platform bringing together national climate pledges under the Paris Agreement and greenhouse gas (GHG) emissions datasets, see: http://cait.wri.org/.

For another online platform to explore and visualize the most up-to-date data on carbon fluxes, see: http://www.globalcarbonatlas.org/en/CO2-emissions.

For an up-to-date assessment of individual country assessments by the Climate Action Tracker, see: http://climateactiontracker.org/.

[33]See the forthcoming sections for further discussions on this issue.

The experience gained in the process of the KP has also made it clear that states tend to abandon the international cooperation regime in order not to be sanctioned if they fail to fulfil those obligations. It is not a matter of preference in terms of the effectiveness of the international environmental policies that the distresses stemming from their commitments cause these countries to opt out of the international cooperation system. For this reason, it can be argued that the Paris Agreement, which incorporates a controlled freedom mechanism for States which are failing to determine their own national goals, has brought about an important innovation.

In addition, the Paris Climate Agreement, on the one hand, puts forward that coping with climate change is an obligation for all states, foreseeing that this obligation is necessary for the interests of all states; on the other hand, it emphasizes the necessity of climate justice as well. Therefore, it seems that the obligations of the responsible actors in the struggle with climate change (coercion) are complemented by interests and virtues. The inclusion of these three "fundamental levers of human action" (coercion, interests, virtues) that can influence human behaviour can be used to argue that the Agreement contains more realistic obligations than the KP. Therefore, although it may be argued that "Through this structure the Paris Agreement establishes a political narrative that is clearer than its less prescriptive and precise legal backbone" (Bodle/Oberthür 2017: 103); it still has the potential to be more successful than the KP (Vinuales 2016).

However, it should be underlined that there is still no consensus on the possible success or failure of the Agreement. On the one hand, the PCA is considered to be a successful example of global cooperation "on an issue where divisions were deep and stakes were high" (Ivanova 2015); on the other hand, it may be regarded as mere 'propaganda', ensuring the legitimization of the big emitters' rejection of making the deep cuts in their GHG emissions which are necessary to keep the global average temperature rise below 2 °C.[34] Actually, as Bodansky states (2016a: 316), "the truth lies somewhere in between." Though the PCA may not be sufficient to solve all aspects of the climate change problem, it can still go further than generally considered in many areas, such as reducing emissions, expected warming, improving the national planning process in many countries, promoting transparency and progression. So "it is inevitably a high stakes experiment, as there are no guarantees of success, and no time left for second chances in case of failure" (Doelle 2017: 387). It actually depends on future practice. Yet the Paris Agreement is nevertheless a historic breakthrough, as it "creat[es] a new dynamic of cooperation and motivation for collective global action" (Doelle 2017: 387). Supportively, in the 2017 G-20 leaders' declaration, the significance and the irreversibility of the Paris Agreement was stressed by all leaders, except the USA (G-20 2017).

---

[34]See: http://www.rappler.com/thought-leaders/151482-deja-vu-strongly-against-philippines-signing-paris-climate-deal

6.2.3 The Compliance Mechanism Under the Paris Climate Agreement

Before the adoption of the Paris Agreement, various proposals were discussed and negotiated with regard to implementation and compliance under three main documents: the Lima Call for Climate Action (COP 20, Decision 1 2014: 40, para. 88), the Negotiating Text adopted in Geneva (ADP 2.8 2015a: 82–84, paras. 194–201) and the Draft Agreement adopted in Bonn (ADP 2.11 2015b: 25–27).[35]

Under the Lima Call for Climate Action (Section L, para. 88) four options were suggested to facilitate implementation and promote the compliance of the parties: leaving any elaboration on the subject to the governing body; the establishment of a compliance mechanism or committee; strengthening the provisions for transparency and reconsidering the multilateral consultative process (MCP) (Art. 13, UNFCCC; Art. 16, KP); and the non-inclusion of any specific provisions on compliance.

The Geneva Negotiating Text (Section K, paras. 194–201) involved three basic options. The first considered the proposals discussed within the Lima Call in a more detailed manner. The second left the elaboration of the modalities of the mechanism/committee to the COP. The third just mentioned the establishment of a compliance committee and gave some details of its two branches (the enforcement branch (EB) and the facilitative branch (FB) – like the KP's CM) and their duties. Contrary to the KP's CM, it did not include a distinction between positive and negative measures; it just recommended actions as measures. The establishment of an international climate justice tribunal for supervising and sanctioning non-compliance was also suggested first by Bolivia in the AWG-LCA (2010).[36]

The Bonn Draft Agreement (Article 11) contained three options as well: a set of detailed clauses under seven titles with further sub-options: establishment, objective and scope, nature, structure, triggers, consequences, relationship to the CMA; the non-inclusion of specific provisions on facilitating implementation and compliance; the establishment of an International Tribunal of Climate Justice (for details of the proposals presented by these three documents, see Savaşan 2017b).

Under the Paris Agreement, there is no reference either to the establishment of an international climate justice tribunal or to a multilateral consultative process (MCP) under the compliance mechanism (Art. 13, UNFCCC; Art. 16, KP). It contains a specific provision on compliance (Article 15), which mentions the establishment of a mechanism to "facilitate implementation" and "promote compliance" of the parties with their commitments under the Agreement (Art. 15.1).

This is in contrast to the KP and the MP, where there is no specific provision creating a compliance mechanism. Indeed, Art. 8 of the MP, the adoption of the procedures and institutional mechanisms for governing non-compliance was

---

[35]See also Oberthür (2014) for a detailed debate on the options for a compliance mechanism in the 2015 Climate Agreement.

[36]Sands (2016) also addresses the potential role of international adjudication in the form of an Advisory Opinion from the ICJ or ITLOS.

delegated to the MOP 1. Similarly, in Art. 18, KP, the MOP was mandated to approve appropriate and effective procedures and mechanisms to determine and address cases of non-compliance with the provisions of the Protocol and to develop an indicative list of consequences for the effective operation of these procedures.[37]

The establishment of a compliance procedure by the Paris Agreement through a governing body of the Agreement makes its operation faster and more flexible, but at the same time it makes its binding status controversial. However, as it specifically mentions the establishment of a mechanism under the Agreement itself, this makes its legal status stronger, even though its details will be established by the subsequent CMAs.

Under Art. 15.2, PCA, the Committee governing the non-compliance mechanism is defined as being "expert-based and facilitative in nature" and operating under the principle of differentiation in a "transparent, non-adversarial and non-punitive manner" in comparison to the system of the MP, in which there is no reference to being expert or possessing competence in relevant scientific, technical, social economic or legal fields. Unlike the MP, the members of the EB under the KP system should have legal experience (NCP, Section V(3)) and should serve in their personal capacities (NCP, Section II(6); RoP, 4).

According to Art. 15, PCA and COP 21, Decision 1 (para. 102), the parties agreed that the Committee would consist of 12 members elected by the CMA, including two members each from the five regional groups of the UN and one member each from the small island developing states and the least developed countries, on the basis of equitable geographical representation, like both the KP and MP systems (KP NCP, Section IV(1); KP NCP, Section V(1); MP NCP, para. 5), but also on the basis of 'gender balance'. As in the KP system, it is also expected that the members will have competence in relevant scientific, technical, social economic or legal fields.

The Paris Agreement does not provide any further information/reference explained above on the mechanism mentioned under Art. 15; it leaves any other elaboration to the CMA 1 which was conducted in conjunction with the COP 22 and the CMP 12 (Art. 15.3, PCA; COP 21, Decision 1 2015: para. 103).

In order to consider further issues with regard to the Paris Agreement, an Ad Hoc Working Group was established to meet for the first time (APA 1) at the Bonn Climate Change Conference on May 16–26, 2016. The parties were asked for their views and proposals on the Compliance Committee under the COP 22 meeting in November 2016.

The APA 1.2 (2016) also invited parties to submit, by 30 March 2017, their views and proposals on modalities and procedures required for the effective

---

[37]Based on these provisions, the KP's non-compliance procedure was established by a decision of the COP (COP 7, Decision 24) Marrakesh, Morocco (the so-called 'Marrakesh Accords') in 2001, and then approved by the MOP (MOP 1, Decision 27) in Montreal, Canada in 2005; and the MP's procedure was finalized by the MOP 4 (Decision 1V/5) held in Copenhagen on 23–25 December 1992. See *supra* Sect. 6.1: *The 1997 Kyoto Protocol Compliance Mechanism.*

operation of the Compliance Committee, and questions that should be addressed through such modalities and procedures (APA 1.2 2016: paras. 23, 26).

Furthermore, in the APA 1.3 (2017) that took place from 8 to 18 May 2017 in Bonn in conjunction with the SBI 46 and the SBSTA 46, it was reaffirmed that the Paris Agreement establishes a mechanism to facilitate implementation and promote compliance, as well as a Committee to function under the modalities and procedures adopted at CMA 1. Yet, it was only decided that the APA would continue to work on this item and other relevant matters related to implementation (APA 1.3 2017: Section 7, paras. 29–32; APA 1.3 2017: Section 8, paras. 33–36).

COP 23 (MOP 13/CMA 1.2/APA 1.4), held between 6–17 November 2017 in Bonn, aimed to accelerate and conclude the work programme under the Paris Agreement, including issues related to the implementation of the Paris Agreement (such as the mitigation section of COP 21, Decision 1, the transparency framework, the global stocktake, matters related to finance, adaptation, loss and damage, and implementation and compliance), to make it operative in practice by no later than the next COP in Poland (COP 24), held in December 2018 (APA 1.3 2017; COP 23, Draft Decision 1, 2018).

The preparations for the implementation of the Paris Agreement and consultations with parties on the organization of the 2018 Facilitative Dialogue were mentioned as the items taken into consideration at the COP 23 held with the Presidency by the Government of Fiji.[38] The working agenda of APA 1.4 also involved negotiations on both compliance and implementation issues and elaborated on matters related on them.[39] Therefore, for the next steps, under COP 23, the APA's report (APA 1.4 2017) on the topic, prepared under the second part of the CMA 1 (CMA 1.2), was submitted on 31 January 2018. According to the outcomes produced by that meeting, a re-examination of the subject was required, because it just reaffirmed its commitment to fulfil its mandate by COP 24 (December 2018), but did not lead to tangible outcomes.[40]

Accordingly, there is currently no concrete regulation of the compliance issue. It is just expected to get the work which is conducted by the APAs; after the completion of its work, the plan is to finalize the details of the Committee and the rules governing how it functions in practice. In short, even if a mechanism is currently under development, at the time of writing there is no concrete resolution on compliance issues, or of the Compliance Committee referred under Art. 15, Paris

---

[38]See Moosmann et al. (2017) for an overall and specific analysis, including the process leading to the COP 23 and the international developments and the challenges related to that meeting.

[39]The progress of the negotiations relating to the Paris Agreement is summarised in the Progress Tracker document updated regularly by the UNFCCC secretariat. See: http://unfccc.int/files/paris_agreement/application/pdf/pa_progress_tracker_200617.pdf.

[40]Under the report of the Ad Hoc Working Group on the Paris Agreement on the first part of its first session (APA 1.1 2016), to assist parties to develop their consideration on the elements of the compliance committee, the details of which were left to be negotiated to the CMA 1, it was also decided to ask for the preparation of guiding questions from its Co-Chairs by 30 August 2016 (APA 1.1 2016: para. 9.7–25; COP 21, Decision 1 2015: para. 103).

Agreement. Therefore, the main topics that need to be further clarified in the next CMAs are as follows:

- What will the mechanism be called? As the related provision of the Paris Agreement just mentions a mechanism, but does not provide any specific clarification, it is necessary to determine whether it is an implementation mechanism, or an implementation and compliance mechanism.
- How it will be structured? With two branches – the facilitative branch and the enforcement branch – like the KP system, or just one body like the ImplCom of the MP system? Given its composition based on the COP 21, Decision 1 (2015: para. 102), it will consist of 12 members; so it is not expected to be similar to the KP's system.
- Will it be facilitative for both developed and developing countries, or just for developing countries?
- According to the Agreement, it will be an expert-based body, and will operate in a facilitative, transparent, non-adversarial and non-punitive manner. So there is no sign that it will be a mechanism with an enforcement power involving negative response measures as well as positive ones.
- There is need for further clarification on several procedural rules/safeguards, such as the phases of the non-compliance procedure; the triggering process; deadlines for both the submission of the parties and the decisions to be taken by the relevant bodies; the rights granted to the parties under scrutiny and the rights provided to the submitting parties; the rights of confidentiality and transparency/ publicity within the procedures; the impartiality and independence of the Committee; and the consequences of non-compliance.

6.2.4   The Post-Paris Period: Searching for Ways to Improve Compliance

6.2.4.1   What Happens Next? Future Options for a Compliance Mechanism under the Paris Climate Agreement

Given the current conditions, in a discussion on probable future developments of the compliance mechanism under the Paris Climate Agreement, which existing legal sources/documents should be applied?

(a) *PCA, Art. 15:*

- Art. 15 mentions a mechanism on implementation and compliance.
- This mechanism will consist of an expert-based, facilitative committee functioning in a transparent, non-adversarial and non-punitive manner.
- The Committee will take into account the principle of CBDR-RC in its functions.

- The methods and procedures regarding the Committee will be adopted by the CMA 1.
- The Committee will submit annual reports to the CMA.

(b) *COP 21, Decision 1:*

- Gives a few details about the members of the Committee (para. 102).

(c) *APA's work:*

- To date there have been four meetings: APA 1.1, 1.2, 1.3, 1.4.
- No tangible formulation has been developed under this working group to date.

Accordingly, there is no adequate background information that can be used to design a fully-fledged compliance mechanism under the Paris Agreement. However, it is possible to draw key lessons from the experience gained under the current compliance mechanisms established under the MP, or particularly under the KP for the establishment of a compliance mechanism. In addition, it is also necessary to design modalities "carefully tailored to the unique features of the Paris Agreement, including its broad scope, the differing legal character of its provisions, the unilateral nature of some of the provisions and its nuanced treatment of differentiation" (Dagnet and Northrop 2017: 351).

Of the commitments provided for under the Paris Agreement, the NDCs are the core ones on which the Agreement's system has been built. Unlike the KP, the Agreement does not categorize parties as Annex I or non-Annex I with certain general obligations (Art. 4, UNFCCC) valid for all parties and specific ones for Annex I parties. However, on the basis of its overall objective, which requires to the increase in the global average temperature to be held well below 2 °C above pre-industrial levels (Art. 1a, PCA), it requires:

- Each party to prepare its NDC, which will include its climate action commitment (Art. 4.2, PCA);
- All parties to take domestic mitigation measures (Art. 4.2, PCA);
- Developed countries to provide support to developing country parties on NDCs (Art. 4.5, PCA);
- All parties to provide necessary information to the CMA in communicating their NDCs (Art. 4.8, PCA);
- Each party to communicate its NDC every five years from 2020 (Art. 4.9, PCA);
- All parties to account for their NDCs (Art. 4.13, PCA);
- All parties to improve environmental integrity, transparency, accuracy, completeness, comparability and consistency and ensure the avoidance of double counting, while accounting for anthropogenic emissions and removals corresponding to their NDCs (Art. 4.13, PCA);
- All parties to take into account the concerns of parties with economies most affected by the impacts of response measures, particularly developing country parties, in their implementation of the Agreement (Art. 4.15, PCA);

- All parties to notify the secretariat the emission level allocated to each party within the relevant time period (Art. 4.16, PCA); and to be responsible for its emission level as set out (Arts. 4.17, 4.18, PCA).

That is, the Agreement creates a high level of commitment basis through ambitious NDCs, which should reflect "highest possible ambition" (Art. 4.3, PCA), requiring maintenance and continuation of ambition to achieve "a progression ahead of the current NDC" (Art. 4.3, PCA). The legal obligations for the Parties with respect to NDCs imply that the failure by a party to comply with its NDC would put that party in breach of its legally binding obligations under the Paris Agreement.

However, it is observed that while the Paris Agreement contains a set of procedural obligations on NDCs, it is silent on the substance of them, i.e. whether and how the parties' compliance with their climate action commitments under NDCs can be ensured; and, in the case of failure to comply with their commitments, what the consequences will be.

The Agreement's mechanisms to help ensure compliance with those commitments related to the NDCs are structured under three main pillars – transparency (Art. 13), periodic global stocktaking (Art. 14), and the compliance mechanism (Art. 15) – and diverse supplementary means, such as loss and damage (Art. 8); financial support (Art. 9); technology development/transfer (Art. 10); capacity-building (Art. 11); and cooperation (Art. 12).

- *Transparency*

The Agreement establishes a legally binding transparency framework for action and support (including reporting-monitoring provisions) coordinated by the UNFCCC (Art. 13.3, PCA). When the related provisions are examined, its general purpose appears to be 'facilitation'. Indeed, its purpose is defined as "to provide a clear understanding of climate change action, clarity and tracking of progress towards achieving parties' individual NDCs, their adaptation actions, and information on the global stocktake" (Art. 13.5). It also aims "to build mutual trust and confidence and to promote effective implementation" (Art. 13.1); and to render "clarity on support provided and received by relevant individual parties, an overview of the total financial support provided, and information on the global stocktake" (Art. 13.6). Supporting its facilitative nature, its main characteristics are as follows:

- Flexibility-based structure (Arts. 13.1, 13.2)
- Facilitative, non-intrusive, non-punitive manner, respectful of national sovereignty, and avoidance of an undue burden-sharing on parties (Art. 13.3)
- Based on the principle of CBDR-RC (providing transparency-related capacity-building support to the developing parties (Arts. 13.14, 13.15); capacity-building initiative (COP 21, Decision 1 2015: paras. 84–89).

The facilitative nature is saved regarding gathering information as well. In fact, a review of the information that should be provided by the parties – including a national inventory report of anthropogenic emissions by sources and removals by sinks of GHGs (Art. 13.7); information on progress made in implementing and achieving the NDCs (Art. 13.7); information on climate change impacts and adaptation (Art. 13.8); information on financial, technology transfer and capacity-building support provided to developing country parties (Art. 13.9); information on the financial, technology transfer and capacity-building support needed and received by developing country parties (Art. 13.10) – should be made by experts and based on the capabilities and special circumstances of the parties, in accordance with COP 21, Decision 1 (Art. 13.11). It should involve:

- consideration of the parties' support (if they provide any), their respective implementation and achievement of their NDCs (Art. 13.12);
- identification of the areas for improvement for the parties (Art. 13.11);
- a review of the consistency of the information with the related modalities, procedures and guidelines (Art. 13.11);
- assistance in identifying capacity-building needs for the relevant developing country parties (Art. 13.11).

As understood from the wording of the related article, it should be a "facilitative, multilateral consideration" in its nature (Art. 13.12).

On the basis of Art. 13, the CMA 1 is entrusted with adopting related/detailed modalities, procedures and guidelines on the transparency framework (Art. 13.13, PCA; COP 21, Decision 1, paras. 91–98), benefiting from the transparency arrangements used under the Convention – national communications, biennial reports and biennial update reports, international assessment and review and international consultation and analysis (Art. 13.4). Therefore, the work by the APA was undertaken for consideration by the COP 24 in 2018, with a view to forwarding the outcome to the CMA 1 (COP 21, Decision 1, para. 91).

There was also a plan to convene a facilitative dialogue at the COP 24 to identify whether the collective efforts of the parties are sufficient to make good progress in leading to the long-term goals of the Paris Agreement; in the preparation of their NDCs, parties provide the necessary information pursuant to Art. 4.8, and find out ways to improve the ambitious efforts for reaching those goals (COP 21, Decision 1, para. 20).

- *Global Stocktake*

The transparency framework is supplemented by the five-year cycle of global stocktake that will review the aggregate effect of NDCs in reaching the goals on mitigation, finance and adaptation embedded in the Agreement (Art. 14, PCA).

The global stocktake will focus on the aggregate state contributions, and not on individual state NDCs. This is to assess the collective progress towards meeting the Agreement's long-term goals through taking stock of the implementation of this Agreement (PCA, Art. 14; COP 21, Decision 1, paras. 99–101). The first one will be undertaken in 2023; the others will rest upon a five-year cycle thereafter unless otherwise decided by the CMA.

The work related to the global stockage issue is aided by both APA on sources of input and modalities for the global stocktake (COP 21, Decision 1 2015: paras. 99, 101) and SBSTA with advice on how the assessments of the Intergovernmental Panel on Climate Change can inform the global stocktake of the implementation of the Agreement (COP 21, Decision 1, para. 100).

Remarkably, it should be stressed that, like the transparency framework, the process should be conducted in a "facilitative manner"; in this sense, the outcome should inform the parties on updating and enhancing their actions and support, as well as improving international cooperation for climate action.

6.2.4.2   Compliance Mechanism: Possible Options

Supplemented and supported by the transparency system and global stocktake, the Agreement establishes a "mechanism to facilitate implementation and promote compliance" (Art. 15.1, PCA). In order to examine the design options for a CM under the Agreement, it is possible, like Oberthür/Northrop (2018), to follow up different ways of establishing an effective mechanism, e.g. stressing some issues as key ones, such as referral of the parties to the Committee, type of measures to be applied, and functions/Powers of the Committee. However, in this section, in accordance with the general framework of the book, the future options for a compliance mechanism under the Paris Climate Agreement will be further discussed on the basis of the lessons learned and experiences attained from the current systems of the compliance mechanisms of both the KP and the MP to find out of which their features better suit the design of a CM under the Paris Agreement. This analysis will be based on three main components of the CMs: gathering information on the parties' performance, non-compliance procedures, and response measures. It will also ask how better coordination in the mechanism and between different mechanisms can be provided to yield the desired result of enhancing compliance under the mechanism. Benefiting from the analysis of the weaknesses and strengths of the current mechanisms, the aim is to develop a more forward-looking perspective with better prospects for the compliance mechanism under the Paris Agreement.

- *Gathering Information on the Parties' Performance*

Given that the system of the Paris Agreement is founded on a transparency framework for action and support coordinated by the UNFCCC (Art. 13.3, PCA), it is clear that it will be based on transparency arrangements used under the Convention – national communications, biennial reports and biennial update reports, international assessment and review and international consultation and analysis (Art. 13.4).

The parties are expected to provide information on their national inventory reports of anthropogenic emissions by sources and removals by sinks of GHGs

(Art. 13.7), progress made in implementing and achieving the NDCs (Art. 13.7), climate change impacts and adaptation (Art. 13.8), financial, technology transfer and capacity-building support provided to developing country parties (Art. 13.9), and financial, technology transfer and capacity-building support needed and received by developing country parties (Art. 13.10). Under the Kyoto Protocol, six specified greenhouse gases and emission targets that should be reported make it easier to estimate the emissions and review the inventories of these estimates, thus to assess compliance. However, under the Paris Agreement, contrary to the KP, there is no categorization between Annex I and non-Annex I country parties involving general obligations valid for all parties and specific ones for Annex I parties. The Agreement instead requires each party: to prepare its NDC, which will include its climate action commitment (Art. 4.2, PCA); to provide necessary information to the CMA in communicating their NDCs (Art. 4.8, PCA); to communicate its NDC every five years from 2020 (Art. 4.9, PCA); to notify the secretariat of the emission level allocated to each party within the relevant time period (Art. 4.16, PCA); and to be responsible for its specified emission level (Arts. 4.17; 4.18, PCA).

The review of the collected data will be made by experts in a technical but also facilitative, non-intrusive, non-punitive manner (Arts. 13.11, 13.12, 13.1, 13.2, 13.3). It will involve not just a review of the consistency of the information with the related modalities, procedures and guidelines (Art. 13.11), or identification of the areas of improvement for the parties (Art. 13.11) and their respective implementation, achievement of their NDCs (Art. 13.12), but also the consideration of their support (if available). Although a system of self-reporting was used under both the MP and KP, in order to obtain more reliable data, under the Kyoto Protocol it is also possible access the relevant factual and technical information provided by qualified IGOs and NGOs and expert advice (NCP, Section VIII, 4/RoP, 20; NCP, Section VIII, 5/RoP, 21). Another significant difference between the two systems is the establishment of expert review teams (ERTs) under the KP.

Since the purpose of the reporting processes under MEAs is to ensure better compliance, gathering information still remains a useful tool for understanding the compliance level of the parties (Savaşan 2012). Under the current systems of both the MP and the KP,[41] there are already some challenges regarding the gathering of information, such as self-reporting, the complexity of the data that should be reported and the technical character of the process, harmonization problems arising from the different methods and deadlines used by parties, problems in the monitoring and verification processes, the intricacies of capacity-building (introducing new regulations and training new personnel etc.) especially for developing country parties, lack of non-governmental organization (NGO) participation in the process, transparency problems, lack of coordination and proper financing.[42]

---

[41]See Savaşan (2012) for a comparative analysis of reporting under both the MP and KP.

[42]For details of the challenges regarding gathering information under CMs, see Savaşan (2015b).

While all those challenges undermine the efforts of the current mechanisms to ensure compliance, at the same time they indicate the areas where improvements could overcome those challenges and enhance compliance. That is, knowing the weaknesses of the present systems can help create a better functioning system under the Paris Agreement, incorporating ways to mitigate the weaknesses in gathering information.[43]

With respect to the self-reporting principle, it may be preferable to get information from other supplementary sources, such as an independent international institution, other states' reports, international organizations and NGOs, and site visits.[44]

The clarity of the reporting requirements, the standardization of the information reported and the methods used for gathering that information are also significant requirements to ensure effective compliance under the Paris Agreement. As mentioned in various parts, the NDCs are the core elements of the Agreement, so it is essential to provide a systematic basis for their submission by the parties.

According to the Agreement, each party communicates its NDC every five years (Art. 4.9) and the common time frames will be determined under the CMA (Art. 4.10). Regarding the information that should be reported under those NDCs, it is stated that parties have to provide the necessary information in accordance with COP 21, Decision 1 and any relevant decisions of the CMA (Art. 4.8); in accounting for anthropogenic emissions and removals they have to follow up the principles and methods again adopted by the CMA (Art. 4.13); in recognizing and implementing mitigation actions with respect to anthropogenic emissions and removals, they have to take into account present methods and guidance under the Convention (Art 4.14); each successive NDC will represent a progression ahead its current NDC (Art. 4.3). Those NDCs are recorded in a public registry maintained by the secretariat (Art. 4.12), allowing the parties to adjust their NDCs at any time (Art. 4.11). The methods and procedures for the operation of this public registry were further negotiated under the SBI 47 (2017) and will continue to be negotiated until their finalization. In the meantime, communicated NDCs will be available in the interim NDC registry.

However, there is still no specific uniform guideline/template or particular method supported by the secretariat for submitting those NDCs.[45] Therefore, the

---

[43]See Savaşan (2017a) for a general analysis of the weaknesses of the CMs.

[44]Under the MP, if the relevant party invites the ImplCom to gather information in its territory, site visits can already be done. Under the KP they are conducted as part of the in-depth reviews with the approval of the relevant party (COP 1, Decision 2 1995: para. 2c), or it may be stipulated that if the party does not submit its data properly and on time, the competent body of the Agreement can benefit from other sources, or specifically from its own investigation/estimation etc. (for a similar view see Sachariew 1991).

[45]See also the document released on 13 November 2017 concerning parties' positions on the information they need to communicate about their national contributions, showing the differences in the organization, delivery and update systems of parties. Available at: http://unfccc.int/files/meetings/bonn_nov_2017/in-session/application/pdf/apa_3_informal_note_final_version.pdf.

specific principles, methods and procedures of NDCs are currently self-determined by the parties themselves; and qualified reporting requires "a uniform format of reporting with clear and precise requirements as to how and what to report" (Wang/Wiser 2002: 183). That is, this situation makes it challenging "to determine the average depth of the commitments made under the Paris Agreement" and so "to lead to the long-term goals of the Agreement" (Bang et al. 2016).

Another important key feature is that NDCs are based on the principle of the CDRP-CR (Art. 4.3), and it is observed in practice that "more than 50 of the NDCs are fully or partly conditional on international support" (Art. 4.5) (Bäckstrand et al. 2017). In accordance with Articles 9, 10 and 11, PCA, the Agreement already provides the basis for financial support, technological development and transfer and also for capacity-building. The use of 'shall' in some parts makes the application of those provisions stronger and more impressive in practice, e.g. developed country parties 'shall' provide financial resources to assist developing country parties, while other parties voluntarily provide such support (Arts. 9.1–2, PCA); support 'shall' be rendered to developing country parties for technological development and transfer (Art. 10.6); capacity-building activities 'shall' be promoted through proper institutional arrangements to support the implementation of the Agreement (Art. 11.5).

These provisions of the Agreement, relying basically on the principle of CBDR-RC, emerge as supporting tools for the mechanism which will be established under the Agreement. Indeed, through this principle, the developed countries have a legal obligation to provide financial and technical assistance to developing countries, but the developing states also have to take the required measures related to their commitments under the Agreement. That is, if developed states fulfil their commitments to provide financial resources on the basis of "compliance requirement", developing states have to implement their obligations (Boisson de Chazournes 2006). Thus, "a new form of reciprocity" has been established between developed and developing states (Wolfrum 1999: 148). If developed states do not fulfil their commitments effectively, developing states can resort to the competent body of the Agreement to decide on the matter.

The kind of support provided by the Agreement (capacity-building, technological/financial assistance) appears to be an effective way for developing countries to improve the facilities they need to meet their commitments under the Agreement. Compelling/punitive tools can be considered in addition to these supporting tools, e.g. under the MP, if developing country parties do not report their base-year data as required by the MP within one year of the approval of their country programme and their institutional strengthening by the Executive Committee, they can lose their status defined under Art. 5, MP (MOP 6, Decision VI/5, 1994: 15–16, para. 84). These tools can also be considered for developed country parties. However, in accordance with the nature of those mechanisms, the facilitative character of the compliance mechanism should not be forgotten or undermined.

Under the current system of gathering the information required by the CMs, monitoring and verification functions are usually carried out by the CMs' bodies or secretariat, not by independent experts. Although there are suggestions to cope with this challenge, such as the establishment of a "Global Environmental Observing, Monitoring and Assessment Programme (GEOMAP)" (Young 2002), or the creation of a "supervision package" in which the reports are evaluated by a body of experts (Sachariew 1991: 50), there is still no development on this issue. So to promote transparency and openness, there is a need for independent expert reviews and simpler evaluations within those reviews. Therefore, when creating mechanisms for monitoring/verification under the Paris Agreement, these issues should be taken into account.

Under the current CMs, NGO participation in the parties' reporting stage is limited to the consent of the concerned party, while in the preparation of summary reports it is limited to the authority given (generally by Secretariat) to the competent organs of the MEA in question. In both, accepting NGO participation as a rule and rejecting it as an exception could strengthen the role of NGOs in gathering information under CMs. Therefore, under the mechanism of the Paris Agreement, this sort of resolution could be considered to improve the operation of the mechanism. The foundation of the Agreement, including the transparency and the bottom-up approach, and its reference to public participation in the Preamble, can be viewed as a possible indication that parties would accede to the non-compliance procedure being triggered by NGOs; if this happens, it will really make the Agreement a 'breakthrough' (Fournier 2017: 15–16). At least, similar to an *amicus* brief process, NGOs could be entitled to make submissions on the related QoIs.

Better coordination both 'in CMs' and 'between CMs' is also very important for obtaining timely, accurate and coherent essential information and preventing duplication among similar bodies. Good communication and dialogue between all bodies and parties of the mechanism can be ensured in the mechanism. Suggestions like establishing a centralized coordinated information management system (Sachariew 1991), or designing a strategy for centralized information management on MEAs for developing countries (Batagodal et al. 2004), or creating a worldwide supervision agency providing a network on the reporting systems and procedures of the different MEAs' competent organs can also be contemplated as solutions to the problem (Sachariew 1991). Furthermore, transgovernmental compliance networks could be proposed for better coordination (Savaşan 2015a: 98–99, b: 186). This would entail designing appropriate communication mechanisms to ensure all or some participants of the networks have access to all information and all or some proceedings of the CMs. Additionally, the reporting arrangements could be adjusted to oblige the relevant participant of the network to report on the related issues notified by the CMs. All such processes need to be improved. So, for an effective mechanism under the Paris Agreement, the challenges regarding coordination that can arise in the mechanism itself and also with the mechanisms of other MEAs should be foreseen while drafting these measures.

- *Non-Compliance Procedure (NCP)*

As with the CMs of both the KP and MP, when creating a non-compliance procedure under the potential CM of the Paris Agreement, the impartiality and independence of the institution that will be created under the NCP (Committee), the procedural safeguards, the role of NGOs and financial problems will each be discussed, taking into account the shortcomings of the current mechanisms and the suggestions submitted for removing those shortcomings. Additionally, the need to provide stronger coordination between CMs for better compliance will also be scrutinized, with the emphasis on the suggestions submitted for improving the conditions for more coordinated mechanisms.

The Impartiality and Independence of the Committee

The institution created under the NCP, usually called the Implementation Committee or Compliance Committee, is often constituted by the representatives of the parties elected by the MOP under the CMs of many MEAs (e.g. MP NCP, para. 5).

In the current system of the CMs, some crucial problems arise, particularly with respect to: the disuse of experts serving in their personal capacity as members of the Committee; the criteria for electing the parties' representatives; the lack of outside technical/legal experts; the reimbursement of their travel and subsistence costs; the possible privileges and immunities of the members; and the consequences that should be connected in case of violation of the principles of independence and impartiality (for details see Savaşan 2017a).

Therefore, it is expected that all these criticisms will be taken into account to ensure that the Committee established under the Agreement is endowed with the necessary impartiality and independence. Indeed, the Committee attached to this mechanism, like the former ones, is 'expert-based and facilitative in nature and function in a manner that is transparent, non-adversarial and non-punitive' (Art. 15.2, PCA). It is also stated that the Committee should consist of twelve members elected by the CMA, including two members each from the five regional groups of the UN and one member each from the small island developing states and the least developed countries, on the basis of equitable geographical representation, like both the KP and MP systems, but also, as a breakthrough, on the basis of 'gender balance' (COP 21, Decision 1 2015: para. 102). It is also expected that the parties of the Paris Agreement would have competence in relevant scientific, technical, social economic or legal fields. Through these provisions, the Paris Agreement establishes a system similar to that of the KP's NCP. In fact, under the KP NCP, members and their alternates of the Committee should be qualified in climate change and in relevant fields (scientific, technical, socio-economic, legal) and for the EB they should also have legal experience (NCP, Section V(3)). Additionally, it is specified that they should serve in their personal capacities, should act in an independent and impartial manner, and should avoid conflicts of interest (NCP, Section II(6); RoP, 4).

As there is no other detail on the operation of the NCP and the Committee under the Agreement, those all modalities and procedures regarding its operation have been left for the CMA to decide upon. Accordingly, it can be helpful to refer to and rely on the rules of the KP NCP, if necessary, improving them in the light of the criticisms mentioned above.

Procedural Phases and Safeguards

As already well-known, under the KP's CM there are two branches working under the Committee which have no hierarchical relationship between each other (NCP, Section II(7); NCP, Section IX(12); RoP, 23). Of those, the Facilitative Branch (FB) merely aims to provide assistance to promote the parties' compliance with their commitments under the KP (NCP, Section IV, 4), while the Enforcement Branch (EB) decides on questions of implementation relating to Annex I parties' reduction commitments (Art. 3.1, KP), methodological and reporting requirements for greenhouse gas inventories (Art. 5.1.2, Art. 7.1.4, KP) and eligibility require-ments for the Kyoto mechanisms (Arts. 6, 12 and 17, KP) (NCP, Section V(4)). So, it is regarded as a quasi-judicial body with the discretionary authority to impose consequences set out in Section XV, NCP (NCP, Section V(6)). In brief, then, in the KP's CM, the MOP has the power neither to decide on non-compliance nor to adopt responses to it. Decisions by each branch are self-executing, and the con-sequences for non-compliance are applied automatically after the decision given by the EB. The MOP has the power to decide on non-compliance only in the case of an appeal against the EB's decision for denial of due process. Through the appeal procedure, the EB's decision can be referred back to the EB, and it is examined again by a 75% majority vote of the parties present and voting at the meeting (NCP, Section XII(e); NCP, Section XI). That is, unlike the CM under the MP, in which the MOP has the final decision-making authority, the Committee only has the authority to make recommendations to the MOP; there is no requirement to adopt the decisions by the MOP.

Given the fact that the Paris Agreement relies upon NDCs, contrary to the KP relying upon emission reduction targets determined in line with the differentiation between Annex I and non-Annex I parties and enforcement-based aspects, it is expected that those quasi-judicial aspects of the KP's NCP are less likely to be applied under the Paris Agreement. It is also unlikely that its two-branch system can be applied under the Agreement.

The Paris Agreement does not include any provision regarding enforcement; instead, it stresses 'facilitation' through the usage of such words as "mechanism to facilitate implementation" (Art. 15.1) and "...facilitative in nature and function in a manner that is transparent, non-adversarial and non-punitive" (Art. 15.2). Hence, it is more likely that it will establish a mechanism in which the Committee has the power to give recommendations to the CMA (as is provided for under the Agreement), and it will be required to report annually to the CMA (Art. 15.3, PCA), i.e. the CMA has the final decision-making authority, as in the MP's CM system.

The analysis shows that, compared with the NCP of the MP, the KP's NCP is characterized by detailed rules on the procedural structure and necessary procedural safeguards adopted by COP 7, Decision 24 and its RoP prepared for the working of ComplCom by MOP 1, Decision 27 (Savaşan 2017a: 845). So the experience attained under the enforcement branch and its facilitative aspects of the KP can be regarded as a point of reference in the elaboration of the rules, methods and pro-cedures that will be adopted under the Paris Agreement. There are also very detailed rules on the rights of the party concerned and strict neat timetables for distinct phases of the proceedings in the KP's NCP to provide voluntary compliance of the parties.

When designing the mechanism for ensuring and promoting compliance under the Paris Agreement, other significant issues that should be considered and discussed carefully in addition to those safeguards are the inclusion of fixed consequences consisting of just an indicative list of measures (see, for example, the MP system), or both a list of predefined measures and non-compliance situations (see e.g.KP system); the limitations on transparency (particularly the confidentiality of the related information); the possibility of appeal (even if it is not anticipated involving it under the Agreement's NCP, as, unlike the KP's NCP, it will not involve a quasi-judicial perspective and two-branch system); and the limited role of NGOs under CMs (for details, see Savaşan 2017a).

Furthermore, it is very important to address financial challenges because, in line with their facilitative characteristic and unlike traditional means of dispute settlement, several compliance mechanisms include financial support and technical assistance for developing country parties. So lack of proper financing in CMs could adversely affect their functionality, potentially reduce participation in the meetings of the ComplCom and other bodies of the Agreement, and impair their ability to carry out on-site examinations, gather information and expert views, conduct scientific research, train the officials responsible for implementation and compliance, and help developing countries constitute, improve and maintain their technical and administrative capacities. All this could result in a failure to encourage the developed countries to comply, decreasing their compliance costs and ultimately leading to delays and deficiencies in compliance.

Under CMs, while assisting the parties (particularly developing and less-developed parties) in their efforts to meet their commitments, financial resources are generally provided from the general budget (involving binding contributions) prepared and adopted by the COP/MOP for the MEA in question. In addition to the contributions from the budget, voluntary contributions can also be obtained from the parties. However, as the voluntary contributions cannot be accurately estimated, and the binding contributions cannot be gathered in full from the parties with sufficient speed, they both fail to guarantee a regular resource for financing compliance mechanisms (Romanin 2009: 437). Therefore, a "specific budget line" in the regular budget could be introduced to finance the CM, allowing the mechanism "to be independent from the often irregular and unpredictable funding approved by the COP" (Romanin 2009: 422, 437).

While reducing the costs of compliance, financial assistance allows more countries to consider compliance more important and thus "mitigates the free-rider problem" (Kolari 2002: 43). However, adopting these kinds of financial mechanisms can be seen as "rather expensive in comparison to a more traditional sanction-based system" (Kolari 2002: 43). In addition, although these efforts aim to increase the contributions towards financing these mechanisms, in general, the bodies competent for financial issues in compliance mechanisms do not have the power to evaluate "the compliance of parties with their obligations to disburse financial resources" (Romanin 2009: 435). Their authority is limited to "facilitative activities, such as the provision of advice and facilitation of financial assistance" (Romanin 2009: 435).

This is also valid for the UN in practice. Indeed, in Art. 19 of the UN Charter, the suspension of the right to vote is allowed for member states which fail to pay their contributions to the organization for two years. However, in practice, it has been observed that the General Assembly has never used this provision against states which fail to pay their contributions.

The other important consideration is that financial resources need to be used in the most efficient way; simply providing proper financial assistance for the parties is not enough to solve the financial problems within the compliance issue. This is because if the assistance is not be managed properly, it can result in waste of time, service and money. Therefore, it should be "targeted toward the most important needs, monitored to avoid waste, provided conditionally in stages to promote real action, continually assessed and reviewed so that procedures can be improved, and coordinated within and across regimes to achieve potential synergies and to avoid duplication and unintended negative consequences" (Chasek et al. 2006: 216). Regarding this issue, it can also be argued that strengthening the staff, powers (in monitoring, assistance, assessment) and budget of the secretariat which functions to support the other bodies of the MEAs and its parties in their activities could also make a crucial contribution to ensuring the compliance of the parties (Chasek et al. 2006).

Based on these reasons, when designing the mechanism for ensuring and promoting compliance under the Paris Agreement, financing that kind of mechanism also arises as a significant challenge which needs to be skilfully addressed.

- *Non-Compliance Response Measures*

In describing the strength and severity of the consequences and measures of a compliance mechanism, Werksman (2005: 19–20) refers to three components of the KP's CM: 'prescriptiveness,' 'punitiveness' and "legal character." Prescriptiveness here implies establishing which consequences should be applied to which situations; punitiveness means making the costs of non-compliance greater than the costs of compliance; and legal character explains the extent to which a non-compliant party is bound by the consequence imposed on it.

Pursuant to Werksman's (2005) identification, the questions regarding multilateral non-compliance response measures which will be adopted under the Paris Agreement can be discussed in three sub-groups: their prescriptiveness, their punitiveness and their legal character.

With regard to prescriptiveness, the MP includes an indicative list of measures for compliance built into the Protocol in the form of positive measures (such as financial and technical assistance) and negative measures (like the suspension of some rights); yet the NCP does not define the situations of non-compliance with the Protocol, nor should they be inferred from the provisions of the Protocol. The response measures which should be applied to the non-compliant party within the NCP of the KP, on the other hand, are predetermined consequences. Indeed, NCP,

Section XIV obviously defines which consequences should be applied by the FB, and Section XV defines which ones should be applied by the EB in response to three circumstances of non-compliance with different commitments: non-compliance with reporting requirements, non-compliance with eligibility requirements, and non-compliance with emission commitments. So, in short, while the CM under the MP adopts a list of measures of non-compliance, the CM under the KP determines both the measures and the circumstances of non-compliance, so it incorporates a differentiated system of response measures on the basis of the nature of the commitment and extent of non-compliance with that commitment.

As the evaluation phase of non-compliance should not involve much flexibility, it could be argued that, when determining the fixed measures according to the circumstances of non-compliance, less flexibility and fewer political but more integral judicial features could enhance predictability and prevent the misuse of power. Nevertheless, it should not be forgotten that even just adopting a list of measures, as in the CM under the MP, can be seen as "an undue constraining factor" on the mechanism and its flexibility (Handl 1997: 44), as this kind of list draws the possible borders between being in compliance or in non-compliance. Therefore clearly establishing which situations would lead to what kind of measures in a list, as in the CM under the KP, could be regarded as inappropriate, as it could also prevent the choice of appropriate consequences to the circumstances of each individual case at hand (Handl 1997; Ulfstein/Werksman 2005). Given the wording of Art. 18, KP, which emphasizes the development of a list of consequences which take into account the cause, type, degree and frequency of non-compliance, it is also questioned whether the use of predesigned, fixed consequences is consistent with this article (Ulfstein/Werksman 2005).

Despite the ongoing debate on the use of predetermined consequences, apparently this system of "predefined responses to predefined non-compliance situations" has been successfully implemented in the KP's CM so far. So it could be argued that it the same system should be used under the Paris Agreement system. However, even under the MP's CM, although it appears that there is a general agreement on the necessity of determining the causes of non-compliance to arrive at appropriate recommendations and consequences, developing such criteria in a formal list for making "an objective judgement on whether a case of non-compliance was a wilful breach or a result of factors beyond the control of the party concerned" is regarded as something that should be achieved over time (Ad Hoc WG 1998: 5–6, para. 27). A proposal to help the ImplCom match responses to particular types of non-compliance was examined by the WG in 1998. It was decided that the ImplCom and the MOP have the discretion to adapt their response to suit each particular case, so there was no need to change the existing text on this issue (Ad Hoc WG 1998: 7, para. 40). This reduces the likelihood that the CM under the Paris Agreement will rely on NDCs, contrary to the KP relying upon emission reduction targets.

When discussing the punitiveness, the views of Hovi et al. (2007) on the KP's system are beneficial. They argue that, first of all, punitive consequences can result in the withdrawal of parties from the Protocol (Art. 27, KP). In addition, for the punitive consequences to restore compliance "there must be incomplete information" (Hovi et al. 2007: 444). In other words, there should be two conditions: first, the non-compliant party should not be anticipating the application of punitive consequences; second, it should be unable to estimate the cost of them. The first condition cannot be met by the KP's CM, as predefined consequences are determined in the NCP, Section XIV for the FB, and in Section XV for the EB. The second condition is also unlikely to be met, because the consequences are applied only to developed country parties, which are generally able to estimate the cost of these consequences due to their financial and administrative capacities (Hovi et al. 2007). Even if all parties have complete information, if the threat of punitive consequences fails to deter a party from non-compliance, the actual application of those consequence may not automatically restore compliance.

By way of example, according to the KP's CM, if the non-compliant is a large economy and a net seller in the emissions trading market, the suspension of its right to engage in transfers under Art. 17 can cause price levels to rise remarkably in the market. This is because the non-compliant reduces its emissions further, and the demand for fossil fuel and its price fall. In addition, as the supply of emission permits is also reduced, the permit price rises. Under such a circumstance, the decision of whether to punish a non-compliant country with a suspension of its right to engage in transfers can imply significant economic losses or gains for other countries that operate in related markets. So it is usually anticipated that parties that are buyers will vote against that kind of suspension. If the members of the EB take into account these implications, a non-compliant party can easily escape the response measures for its non-compliance. This situation can decrease the deterrent effect of the compliance mechanism (Hagem/Westskog 2005).

Based on their findings, scholars conclude that punitive consequences are rarely able "to make much of a *difference* in a country's decision about whether to return to compliance" (Hovi et al. 2007: 438). Hovi et al. (2007) suggest "external pressure" to accomplish a return to compliance rather than applying punitive consequences.

To prevent parties continuously borrowing from future commitment periods, there could be a ban on borrowing in two sequential commitment periods. Instead, non-compliant parties could be required to pay a financial penalty, increased at a rate of 1.3 in the case of repeated failure to meet emission-reduction targets. Other response measures which were historically suggested as ways to ensure and promote compliance under the KP system were subjecting persistent offenders to a more rigorous international review of the compliance action plan and reconsidering the establishment of a compliance fund[46] (Doelle 2010). Despite its shortcomings, the deduction approach was adopted because no other response was found feasible. A compliance fund established through contributions by parties to the KP was regarded by some "as a potential form of financial penalty," and by others as a

---

[46]See Oberthür/Lefeber (2010: 150) for the view stating that the deduction approach does not form a 'penalty' or suitable way of 'penalizing' the party concerned.

means of setting a "price cap" on the compliance costs of parties (Wang/Wiser 2002: 196).

With regard to the MP system, it should be stressed that as its list forms an indicative list, not a complete list of measures that should be taken against non-compliance, the MOP is free to apply different measures not identified in this list. However, all measures adopted by the MOP should be "advisory and concil-iatory" and aim to assist developing countries, in particular, to comply with the Protocol (MOP 3, Decision III/2 1991). This is because the main aim of the CM is not to give any punitive measures against the non-compliant party, but to bring it back to compliance. In cases of conscious non-compliance, trade sanctions would impress on the parties "that they will face penalties for the violation but also that the costs of the violation will exceed the gains expected from it" (Chasek et al. 2006: 222).

In this respect, for a CM under the Paris Agreement which does not include any remark regarding enforcement, enforcement-based or quasi-judicial aspects, but instead remarkably emphasizes 'facilitation' by its wording (Arts. 15.1–2), it is more likely that the mechanism finally adopted will include measures which are advisory and conciliatory, and, contrary to the KP's CM, in the form of recom-mendations that distinguish between positive and negative measures.

Regarding legal character, in the MP's system, measures adopted by the MOP are considered to be recommendatory rather than mandatory. There have been some discussions on the binding nature of the measures, but there is not yet any con-sensus on this matter. Therefore, it can be argued that non-complying parties do not really feel threatened by these measures. Yet, sometimes, even to be identified as a non-complying party in MOP decisions can be a deterrent factor for that party, can hinder its non-compliance and can be useful for ensuring further compliance in the forthcoming period, through "helping[…] to argue with colleagues at home the need for greater urgency in dealing with the issue" (Brack 2003: 218). Therefore, it should be kept in mind that sometimes measures which have no binding nature can lead to greater changes in behaviour than the likely change which binding rules could produce.

In the KP's system, the implications of the question of whether or not a response measure should be considered binding under Art. 18, KP can be different for each of the measures. In brief, except the measure requiring deductions of assigned amounts at a penalty rate (equal to 1.3 times the amount in tonnes of excess emissions) (NCP, Section XV, 5a),[47] the other measures can be considered to be within the implied powers of the Protocol's organs, that is, within the competence of the EB (Ulfstein/Werksman 2005). However, insistence on the adoption of this measure in legally binding form by an amendment to the Protocol could result in the unwillingness of some parties to ratify and be part of it, and thus directly affect

---

[47]Due to the fact that the parties that have withheld their consent to consequences adopted by decisions of COP/MOP can argue that they are not bound by these deductions (COP 7, Decision 24; MOP 1/Decision 27).

the compliance attitudes of the parties, and hence the efficacy of the whole regime (Hovi et al. 2007).

Given the previous options discussed for a CM under the Paris Agreement, it is recognized that there is no definitive decision on the responses issue. It has been observed by some scholars that the compliance system provided for by the Paris Agreement may be considered a regression with respect to the mechanism envisaged by the Kyoto Protocol. The former is facilitative in nature (like the non-compliance procedures established by the majority of MEAs), while the latter is characterized by a 'sanctioning' approach (Montini 2017: 745–746). However, even in these discussions it is apparent that negative and punitive measures and sanctions are not predominantly suggested or offered as an appropriate option under the compliance mechanism of the PCA.

In fact, in the Geneva Negotiating Text (Section K, paras. 194–201), under the third option it has been recommended that, as with the KP's CM, the Committee should have two branches: the EB, which would review compliance with commitments made by developed country parties and developing country parties with emission reduction commitments, biennial reports, and technical expert teams' related reports; and the FB, which would review and facilitate implementation by developing country parties. The response measure which would be applied against the parties by both branches would be 'recommendations.' Similarly, in the Bonn Draft Agreement (Article 11), one of the options regarding measures involves only two kinds of measure: a declaration of non-compliance and a request for the development of a compliance action plan. Another just refers to measures ranging from advice and assistance to the issuance of a statement of concern. In the Lima Call for Climate Action (Section L, para. 88), while one of the options proposes the application of facilitative measures; another supports the application of not only measures but also sanctions; and the third suggests facilitative measures for non-Annex I parties and sanctions for Annex I parties. The last one considers the establishment of expert groups to support developing countries and ensure compliance with their commitments. So, even under the proposals that were discussed and negotiated during implementation and compliance in the Pre-Paris period, it was not decided whether the mechanism would or should involve negative measures and sanctions. However, it is already arguable whether the qualifications of the response measures and consequences (positive or negative, punitive or facilitative, binding or non-binding) significantly promote their effectiveness. This is particularly because, in a CM based on a cooperative approach, in the processes of finding solutions to the problems, relying on consensus by the parties rather than a decision by an institution is more important (Ehrmann 2002). In addition, the consequences of non-compliance are not the only factors that move a party towards compliant behaviour. There are several other factors: the party's characteristics (social-economic conditions), the legitimacy of international norms, their implementation within national systems, etc. So because of the existence of other factors which trigger the party towards non-compliance, the non-compliant party can resist being in compliance despite the existence of negative, punitive, legally binding response measures. However, if parties see that the process and its consequences are fair, they can probably comply with the requirements voluntarily, even though measures are positive, facilitative and non-binding. Indeed, "[m]aking a provision legally binding may provide a greater signal of commitment and greater assurance of compliance; [b]ut transparency, accountability and precision can also make a significant difference, and legal bindingness can be a double-edged sword, if it leads States not to participate or to make less ambitious commitments" (Bodanksy 2016b: 150). The utility of binding measures should therefore be evaluated on a case-by-case basis; the comparative advantages or disadvantages of different solutions should be taken into account when assessing their effectiveness on the basis of their diverse qualifications. That is, under the CM of the Paris Agreement, positive measures alone could be effective in ensuring and promoting compliance based on the features of the case in which they are applied.

- *Need for Stronger Coordination in the Mechanism and Between Different Mechanisms*

In order to achieve better compliance with the MEAs, it is often argued that there is an obvious need to create synergies and interactions between treaty regimes and across respective monitoring bodies (Beyerlin et al. 2006; Carazo/Klein 2017; Chambers 2008; Chasek et al. 2006; Levy et al. 1995; Mrema 2006; Oberthür 2006, 2002, 2001; Oberthür/Gebring 2006; Pitea 2009; Wolfrum/Matz 2003). Therefore providing better coordination in the single CM (between different bodies and parties of the CM) and between different CMs (of different MEAs) emerges as a crucial issue that should be dealt with when considering the design of an effective CM under the Paris Agreement.

Within the single compliance mechanism, effective collaboration may be ensured, encouraged and supported by good communication and dialogue, and by communicative platforms between different components and bodies established by MEAs. As the CMs of both the MP and KP are among the most developed on this issue, as on many others, they are good models for the CM for the Paris Agreement.

For different CMs, particularly those within the same cluster, the first step would be to strengthen the existing informal methods of cooperation and coordination between the parties to the mechanisms (Andresen 2001; Pitea 2009; Raustiala 2001). As a further step, formal ways of coordination can be introduced. But, those ways should be able to enable various compliance mechanisms to regard themselves not as isolated, but rather as parts of a global effort to protect the environment for the benefit of present and future generations.

There are several suggestions to create formal ways to improve the coordination, such as developing a sort of general code for CMs (Epiney 2006), entrusting a single body with addressing compliance issues (Beyerlin et al. 2006; Ehrmann

2002; Pitea 2009), and, on the basis of the clustering issue of MEAs[48]; creating a permanent location for a number of COPs or co-locating MEA secretariats (Oberthür 2002), and arranging regular meetings of representatives from different mechanisms.

Of these suggestions, creating a sort of general code – i.e. categorising different possible procedures – could become a generally accepted part of the process. However, because the tailor-made structure of MEAs means that every MEA has its specific features and specific compliance mechanisms involving specific institutional structures and procedures in accordance with those features, formulating and implementing a general code could prove problematic (Epiney 2006). Therefore, for each MEA, this accepted code would need to be adapted accordingly. If any deviation from the general code was required, the reasons for this would need to be stated openly to demonstrate that the deviation was justifiable and ensure transparency. However, confusion could arise from this kind of categorization if the parties at a general convention were parties to different MEAs with different variants of the general code.

Similar problems emerge in establishing a single body for addressing compliance issues. Here again, as each MEA has its own features, each requires specific expertise, technical information and capability concerning the environmental problem in question (Beyerlin et al. 2006). An example of this kind of body is the central ImplCom, a concept which did not emerge until the Protocols of the 1979 Convention on Long Range Transboundary Air Pollution (Ehrmann 2002). Regarding this issue, the possibility of creating a universal CM is generally evaluated as "an unrealistic and unsuitable perspective" (Pitea 2009: 441).

The clustering issue involves proposals for clustering the organisational elements and for clustering the common functions of MEAs:

- Clustering the bodies of MEAs would involve creating a permanent location for a number of COPs and their subsidiary bodies, establishing an umbrella institution through an MEA secretariat (Fodella 2009) and also co-locating the MEA secretariats (Oberthür 2002). Of those, the importance of MEA secretariats in compliance issues should be stressed. In fact, through their functions, particularly that of providing direct access to information, they play a special part in ensuring compliance within the CMs. They also have an important role in providing coordination and cooperation among and between compliance mechanisms. Yet they are generally hosted by an IGO. On the one hand, this is good for improving the functioning of the CM, as it provides easy access to

---

[48]On the issue of clustering, the Second Consultative Meeting of MEAs on IEG agreed that the clustering of MEAs for promoting collaboration and coordination should be carried out at the sectoral level (e.g. biodiversity-related conventions, land conventions, chemicals and hazardous wastes conventions, atmosphere conventions and regional seas conventions and related agreements), the functional level (e.g. trade-related MEAs, conventions with prior-informed consent procedures, and conventions with customs procedures) and the regional level (e.g. capacity-building, enforcement and compliance etc.) (UNEP 2001: para. 12). See also Oberthür (2002) for the details of clustering.

"technical expertise, inter-institutional networking and resource availability" (Fodella 2009: 366). However, on the other hand, it raises the problem of power-balance between the host organization and the secretariat. In fact, although the IGO hosting the secretariat has no power over the COP or subsidiary bodies, it does have power over the staff's employment conditions, appointment etc. In addition, as host organizations are located in different places from the related IGO's secretariat, the problem of coordination also arises (Chambers 2008).

One option would be to co-locate MEA secretariats instead of applying the present system, which involves rotating meeting locations. However, the possibility of the "fierce competition" which can emerge between host countries would make this difficult in practice (Oberthür 2002: 322). Moreover, it is also argued that the present system has some advantages. For instance, "[a]rranging for separate meetings limits the overall administrative and organisational burden for host countries, [r]otating meeting places also serves the purpose of heightening awareness of the respective environmental problem in the host country/region" (Oberthür 2002: 322). Abandoning this system would mean abandoning its advantages (Oberthür 2002).

- The clustering of MEAs' common functions, such as the implementation review and compliance (including dispute settlement), would have potential advantages relating to reducing the costs, enhancing coherence and making the overall functioning more efficient. Nevertheless, due to the diverse functional needs of CMs in MEAs, successful harmonization would require substantial similarities in their functional needs (Oberthür 2002).

Regarding organizing meetings of representatives from different mechanisms, the meetings of the national ozone focal points in particular regions, funded by the Montreal Protocol Multilateral Fund, can be used as an example. These regional meetings can support the creation and the expansion of networks between government officials for gathering and sharing information, and can also help put pressure (politic or public) on non-compliant actors to bring them back to compliance (Oberthür 2001; Raustiala 2001). Also, the efforts under the CM of the Kyoto Protocol to establish a dialogue with compliance bodies under other treaties to exchange information on compliance-related matters can be expressed as positive attempts towards further dialogue between different bodies studying compliance-related issues (MOP 8_Compliance 2012: para. 6).

Finally, as an additional note, it should be stressed that when considering the interlinkages between different MEAs and their CMs and using these interlinkages to function more effectively, it is certainly necessary to reflect carefully on their own characteristics in a more nuanced way.

By way of example, given the interlinkages between the Montreal and Kyoto Protocols, which have been chosen as case studies in this book, it is clear that both Protocols involve systems created to deal with atmospheric problems, stabilize "greenhouse gas concentrations in the atmosphere" (Art. 2, UNFCCC) and "prevent depletion of the ozone layer" (Art. 2.1, VC). Indeed, by overseeing the planned phase-out of ODSs (which are also GHGs), regulating and preventing the growth of them, the Montreal Protocol has made a major contribution to the combat of climate change and the reduction of its effects on Earth (Newman 2009; Zaelke/Grabiel 2009). Since its creation, the Kyoto Protocol has also contributed to the success of the Montreal Protocol. Nevertheless, each can also contribute to the failure of the other if a properly coordinated relationship between the relevant bodies of the two Protocols cannot be established. To illustrate, HFCs, PFCs and SF_6 were included in the group of GHGs, yet the problem is that as HFCs were promoted as substitutes for CFCs and other ODSs, this could result in compliance problems which delay ODS phase-out under the Montreal Protocol (Oberthür 2001).

Through the Kigali (Rwanda) Amendment to the Protocol, adopted at the MOP 28 (14 October 2016), hydrofluorocarbons (HFCs) are also addressed under the Protocol, in addition to chlorofluorocarbons (CFCs) and hydrochlorofluorocarbons (HCFCs) (MOP 28, Annex I, Kigali Amendment). Thus, it has been agreed to phase out the emissions of HFCs, which have an extremely powerful global-warming potential. That is, the Kigali Amendment is critical, in that it confers obligations to address HFCs. This could cut the global temperature, contributing to the efforts to control anthropogenic climate change, together with the Protocol's basic aim of protecting the ozone layer.

In addition to these suggestions, opportunities for better coordination can be expanded through the transgovernmental compliance networks, which include flexible and informal structures that can deliver better coordination both 'in CMs' and 'between CMs.' Because they have no (or little) relationship with CMs, in addressing the coordination issue between the networks and CMs, new proposals on how they can work together are required, such as participation in the CMs' meetings as observers. In the long term, a global network on compliance issues, e.g. INECE, could undertake the task of coordination between different CMs (Savaşan 2015a: 98–99, b: 186).

Overall, when refining the details of a compliance mechanism under the Paris Agreement, the possible challenges regarding coordination that can arise in the mechanism itself and also with the mechanisms of other relevant MEAs should be taken into consideration and evaluated alongside other potential advantages and disadvantages.

6.2.4.3   Paris Climate Agreement and Its Potential for Better Compliance

Harris/Taedong (2017) focus their attention on the role of consumption behaviours and the policy implications thereof within each country in ensuring compliance with the KP's emissions targets. They argue that the implementation is least likely to be realized in countries with the highest levels of consumption; and the Paris Agreement with its self-determined NDCs seems unlikely to provide the implementation of stronger emissions controls. This is because, despite its provisions on research and development for alternative forms of energy and related technologies, restrictions on consumption are still required for better compliance.

Bang et al. (2016) focus on three factors with respect to compliance issues: the coordination or collaboration problem; the depth of the parties' commitments (shallow or deep); and the presence of enforcement measures. They argue that, in an agreement dealing with merely coordination problem, the parties do not tend to be in non-compliance, contrary to the one dealing with collaboration problems. A shallow commitment does not entail substantial costs, while a deep commitment can do, and even if non-existence of enforcement measures does not influence compliance with international environmental agreements in a negative manner, it is not certain that it is valid for deep climate agreements as well. This is mainly because the climate change problem is a collaboration problem; a climate agreement includes deep commitment, as economic activity mostly requires GHG emissions, and these deep commitments are more costly to implement than in other international environmental agreements. On this basis, they criticize the Paris Agreement as it relies disproportionately on norms, but does not provide for states' incentives to deter free-riding, which is much stronger in the climate change field than in other international environmental cooperation fields.

Nevertheless, due to the fact that there are many different factors that can have an impact on it, like the object and scope of the MEA, the characteristics of the accord, the characteristics of the country or party to the accord, and other factors in the international environment (the role of NGOs, actions of other states and the role of IGOs) (Jacobson/Weiss 2001), compliance under an MEA is not a simple issue to capture with its all aspects.

Hence, no single logic can fully capture the complex nature of the compliance issue and compliance mechanisms. Understanding compliance and preventing non-compliance entails focusing on both the logic of appropriateness and the logic of consequence. According to the appropriateness of the actors' behaviours, to prevent non-compliance, it is recommended to promote cooperation between the actors; and within the logic of consequence, it is advised to enforce, deter and punish based on the calculation of costs and benefits of behaviours (Mitchell 2007). Therefore, it is necessary to understand both of them.

Additionally, neither the rationalist nor the normative theories can fully explain MEAs and their compliance mechanisms, while all contribute to an understanding of some aspects of them. In this respect, the management approach of Abram and Antonia Chayes and the enforcement model championed by critics of the Chayes approach have arisen as two basic explanatory approaches to compliance issues and also compliance mechanisms created under MEAs (see Chap. 3 for details of the theories and these approaches).

The two approaches have both similarities and basic differences. When asked which approach is the most effective at promoting compliance, it is difficult to give a definite response. In their comprehensive study, "Analyzing International Environmental Regimes: From Case Study to Database", Breitmeier et al. (2006) found that international environmental regimes do not rely heavily on enforcement measures to elicit compliance, but instead rely on capacity-building measures and the management approach. They also found that neither the enforcement model nor the management approach is adequate to describe the compliance issues under environmental regimes. So there is a need for a "composite perspective" (Breitmeier et al. 2006: 110) that integrates "incentive mechanisms, juridification, participation of transnational NGOs in the rule-making process, and a responsive approach to the development of compliance mechanisms over time" (Breitmeier et al. 2006: 112). In order to determine whether non-compliance is voluntary (cheating/consideration of the rule as wrong) or involuntary (ambiguities of a prescription/lack of capacity to comply with it) and whether it amounts to a substantial challenge to the concerned rule (consideration of the rule as wrong/lack of capacity) or just involves a technical challenge (cheating/ambiguities), they differentiate four perspectives: 'incentives', 'legalization', 'legitimacy', and 'responsiveness'. In their final analysis, they argue that, though these perspectives have certains methods of responding to compliance problems, nevertheless, "t[h]ey are not full-fledged theories of compliance" (2006: 72). This is particularly because these perspectives have crucial linkages to the broader theories of IR, e.g. the incentives perspective is connected to rational institutionalism; the legalization perspective and the legitimacy perspective are influenced mostly by liberalism and social constructivism; the responsiveness perspective draws on theories which emphasize discourses and communication.

Accordingly, this arguably implies that neither the enforcement model nor the management approach, nor any perspective (even a newly created novel one which has not emerged from existing theories) can provide a full understanding of the patterns of compliance with the commitments created under MEAs. In fact, a compliance mechanism should incorporate elements of both approaches, "not as alternatives but rather as a menu of possible responses to different possible types of non-compliance" (Kolari 2003: 217), "so as to be capable of responding to different types of compliance problems states may face" (Kolari 2003: 227). This stems from a feature of compliance mentioned at the outset: complexity. Therefore, it can be argued that, depending on the problem and its conditions, each relevant approach can explain different aspects of compliance processes. Basically, therefore, two fundamental approaches of compliance (management and enforcement) or possibly another relevant one can be used separately or together depending on the circumstances. So, in some cases, instead of abandoning the one for the sake of the other, it appears more appropriate to place emphasis on interaction between both of these approaches and try to synthesize them. If neither of them is adequate to explain the situation, society, legitimacy, knowledge system and habits or standard operating procedures need to be taken into account, in line with the views of Young's social practice perspective.

Regarding the present CMs, Breitmeier et al. (2006) conclude that enforcement mechanisms increase compliance (Breitmeier et al. 2006: 78). However, they also find that the management approach to compliance is far more dominant in international environmental regimes than the enforcement approach (regarding goal attainment, in 94.1 per cent of the cases; regarding problem-solving in 89.4 per cent of the cases) (Breitmeier et al. 2006: 182, 189–236).

Indeed, the management approach appears to be more dominant than the enforcement approach in the CM of both of the case studies featured in this book, the CM of the KP and the MP.

Under the CM of the MP, there has been no submission by parties against another party to the Protocol, so this type of triggering has not arisen to date. In addition, the Committee does not have competence on the determination of non-compliance, but the MOP involving the state parties has this authority. Even if the debate is still under way on this issue, decisions of the MOP are accepted as recommendatory rather than mandatory (Ad Hoc WG 1989: 4, para. 9i). Though they are not mandatory in nature, except for the application of trade restrictions and conditional assistance on some parties (like Russia, Belarus and Ukraine), negative measures have not been applied against non-compliant parties.

Given those features of the mechanism, even though there seems a combination of two approaches (the enforcement approach with punitive measures like restricting trade with non-parties to the MP, and the management approach with facilitative, positive measures, like granting developing parties financial assistance or a ten-year grace period, and strengthening their capacity-building), the 'facilitation-orientated' management approach appears to be the more dominant approach in the CM of the MP. In practice, it is observed that, despite some shortcomings in the current system – for instance, problems with the compliance of the parties – the system and its functioning in practice are indeed still successful at ensuring and promoting the compliance of the parties. Therefore, it is defined as "a flexible and sophisticated system that continued to function successfully" (MOP 27, para. 97). This is particularly because it has a compliance rate of over 98 per cent. This implies that the Protocol has achieved the phase-out of some 98 per cent (about. 1.8 million ODP-tonnes or 2.5 million metric tonnes) of the production and consumption of 96 ozone-depleting substances globally – 2 per cent is mainly hydrochlorofluorocarbons (HCFCs) (about 32,000 ODP-tonnes or 500,000 metric tonnes) – and its many parties, both developed and developing, have met their phase-out targets (Ozone Secretariat 2015). With over twenty years' experience of its CM, its successful history in compliance matters has been described as "a compliance regime that was widely respected and regarded as a model to be emulated" (MOP 28, para. 97). Indeed, with its features and consequences, it has the potential to be an exemplar for other treaties' compliance mechanisms for ensuring and strengthening parties' compliance with their commitments.

Under the CM of the KP, on the other hand, apart from the submission made to the FB against the 15 Annex I parties by South Africa, all submissions have been initiated by the ERTs. Although the NCP of the Protocol involves various response measures (positive and negative, except financial measures or loss of credits), most of them have been very rarely applied in practice, if at all, and have been very limited in scope. In all eight cases of non-compliance, the EB has not suspended the relevant party's participation in the flexibility mechanisms. To date, the subjects of the QoIs that come before the EB have not been directly related to whether the concerned parties were in compliance with their 2012 emissions target. This is because consideration of compliance for the first period did not begin until July

2015 or later, due to the true-up period. For the second commitment period, it is not expected that this situation will change or that there will be QoIs related to the emission reduction targets of the parties, as even the Doha Amendment, in which the second commitment period with legally binding emissions targets for the years 2013–20 is enshrined, agreed in 2012. For the time being, as of 9 August 2017, 80 countries have ratified the Doha Amendment,[49] but that is not enough for the Amendment to take legal effect under IL; it needs 144 of the 192 parties to the KP pursuant to articles 20 and 21 of the KP (Arts. 20.4; 21.7, KP). Given the fact that it is still not in force, it is not expected that this situation will be different for the second commitment period.

Therefore, even though in the KP there seems to be a system of bringing together the facilitation and enforcement functions under the same mechanism, the management approach (which applies not only LoC but also LoA and is equipped with facilitative-positive measures) emerges as the more dominant approach adopted in both the KP's and the MP's CM. However, compared with the MP, although the KP has pre-determined emission targets, more detailed rules on the procedural structure and necessary procedural safeguards in the process of related proceedings (COP 7, Decision 24; MOP 1, Decision 27), and, more importantly, both positive and negative response measures and an enforcement branch, it is generally seen as a failure with regard to participation in the Protocol and its consequences for compliance. In fact, countries with the highest share of global carbon-dioxide emissions, such as China (30% share of global emissions)[50] and India (7%), did not sign it, while the USA (15%) signed but did not ratify it,[51] and Canada legally withdrew from the Protocol in 2011 (with effect from 15 December 2012). On the other hand, according to an analysis made by Shishlov et al. (2016), only nine of the 36 countries that fully participated in the KP – Austria, Denmark, Iceland, Japan, Liechtenstein, Luxembourg, Norway, Spain and Switzerland – emitted higher levels of GHGs than specified in their commitments on the domestic level, while all Annex B Parties are in compliance on the international level through the usage of the flexibility mechanisms ET, CDM and JI. In addition, most of those countries reduced their domestic emissions more than required by the KP, and overall compliance would have been achieved even without hot air. Furthermore, according to the factsheet prepared by the Climate Institute (2015), all countries except Canada and the USA will meet their reduction targets for the first period. For the second period, only Canada is not expected to meet its target. Indeed, the EU-28 countries are highly expected to meet their targets; while Japan is also possibly

---

[49]Status of approval, acceptance, accession and ratification of the Doha Amendment, see: https://treaties.un.org/Pages/ViewDetails.aspx?src=IND&mtdsg_no=XXVII-7-c&chapter=27&clang=_en.

[50]In recent years, China has started crucial projects to improve environmental protection in the country and has also taken a leading role in the process of the PCA. See Doğan (2017: 193–194) on China's improvement in environmental issues.

[51]For emissions by country, see: https://www.epa.gov/ghgemissions/global-greenhouse-gas-emissions-data.

expected to be on track due to decreasing its target. The USA, Australia and Norway are currently not on track because of the domestic policies applied by the USA for meeting the target and the possibility of Australia and Norway using international carbon credits in the future. Given the numbers achieved by the countries, it is possible to argue that the KP is not a failure at all, but a real success. Nevertheless, it is possible to interpret these emission numbers differently. If the following facts are taken into account, what sounds like a success turns into failure: 0.3 $GtCO_2$ per year of the reductions has been achieved by buying carbon credits; the reduction in emissions in developed countries (1 to 2 $GtCO_2$ per year) occurred because of the financial crisis, not because of the countries' efforts to reduce them; some reductions are because of the transfer of GHGs from countries applying strict emission controls to countries with less stringent regulations (targets were not rigid at the outset) (Page 2016).

In the final analysis, even if those numbers are interpreted as a success, the KP has so far not achieved as much as the MP. This suggests that, even if the Paris Climate Agreement has an enforcement mechanism/branch or strict response measures, there are likely to be two outcomes in practice:

1. Parties will not want to force other parties to comply in an adversarial or confrontational manner.
2. Even if the responses to non-compliance are applied to non-compliant parties, the loopholes mean that related parties can still resist being in non-compliance with their commitments under the Agreement, despite the existence of enforcement tools and the response measures which such parties may encounter.

That is, the ultimate decision to comply with their commitments largely depends on the parties themselves, because there is no enforcement mechanism similar to those of the national systems to force them to comply with their commitments under international law. Under a CM well-founded on a cooperative-facilitative perspective in line with its main objectives, especially one "maintaining the threat of a credible sanction to be applied when a treaty violation proves to be significant and persistent", it is first of all essential to ensure that parties are willing to comply with their commitments (Kolari 2003: 230). How can willingness be ensured? The parties should be convinced that the system is fair to all parties, not just some of them (this is likely to require a careful dialogue between all parties) so that they trust each other to implement their commitments. They should believe that they will be supported when they lack the capability to comply. They should also be sure that when they encounter response measures, they will also receive assistance to return to compliance. They should believe that they will all benefit if full compliance is ensured within the system. Finally, as a prerequisite for achieving all those elements mentioned above, it is undoubtedly necessary to provide further coordination based on coherent attitudes through good communication and dialogues between all parties.

These all are significant in the sense that they align with the core logic of the establishment of compliance mechanisms instead of traditional dispute settlement means, and do not conflict with the reasons for laying their foundations. Indeed, unlike traditional methods, the purpose of establishing CMs is not to find the offender and penalize it, so breach of a rule and direct damage stemming from this breach is not necessary; failure to comply or potential non-compliance are enough to initiate the procedure. This is due to the non-confrontational and non-adversarial nature of the mechanisms. Therefore, the instruments and procedures used in these mechanisms should not be based on the bilateral relationship between the non-compliant party and the directly injured party, but on a collective approach in line with the objective of the CMs.[52]

The purpose of these mechanisms is to identify the reasons for non-compliance and suggest the ways by which the non-compliant parties can re-establish and facilitate better compliance. Hence response measures applied to non-compliant parties basically aim to facilitate rather than penalise those parties, and to support them to return to compliance with financial, technological and capacity-based assistance.[53]

The fundamental criticism levelled against the compliance/implementation issue regulated under the Paris Agreement is that, although it has ambitious goals, it does not have a foundation to actualise them in practice. That is, if the current NDCs are not fully implemented and the parties' commitments (self-determined NDCs) fall well short of the goal set out in the Agreement – the self-determination of the form and substantive content of those NDCs is another significant challenge to be dealt with (Bang et al. 2016)[54] – through which tools will the Agreement tackle the challenges of non-compliance?

As previously mentioned, what is anticipated for ensuring compliance under the Agreement is the establishment of a facilitative, non-adversarial and non-punitive mechanism (Art. 15.1, Art. 15.2, PCA). The response measures which will be adopted and applied should therefore have the same features and be decided under an expert-based and transparent framework (Art. 15.2, PCA). The mechanism should thus work to help and support parties identify what is required, and make recommendations on how to achieve compliance, primarily focusing on the elements in the Agreement that do not establish legally binding obligations and that contribute to the enhanced transparency framework of the Agreement (Voigt 2016a, b). That means that, although detailed rules and methods will be prepared under the CMA, as the Paris Agreement does not, in principle, stipulate any strict measures or consequences to be implemented against a country that fails to fulfil its NDC, it is not anticipated that the CM of the Agreement will use enforcement tools or adversarial punitive measures to deter free-riding. To what extent can a CM of the Paris Agreement based on a facilitative and transparency system in pursuance of the

---

[52]See *supra*, Chap. 4 for details.

[53]See *supra*, Chap. 4 for details.

[54]See ELI (2015) for a "Model Law Implementing the Nationally Determined Contributions Submitted Under the Paris Agreement", designed to help countries construct a law-based system for meeting their NDCs under the Paris Agreement.

objectives of CMs successfully cope with the challenges of compliance?[55] Even
assuming that it will include a mechanism involving enforcement tools, it is easy to
guess that parties will be inclined to withdraw from the Agreement, pursuant to Art.
28, PCA, instead of remaining as a party and suffering the consequences of punitive
measures for non-compliance. The experience of Canada's withdrawal from the KP
already shows that, even without stringent measures, just the possibility of being
declared a non-compliant party is sufficient for parties to withdraw from
Agreements' obligations. Indeed, the effectiveness of the measures adopted under
the compliance mechanism in particular and the effectiveness of CM in general to a
large extent depends on the non-compliant party and its willingness to comply with
them. Therefore, the question of how a non-compliant country can be forced to
comply with strict consequences or sanctions adopted within the CM of the
Agreement remains a thorny issue.

Moreover, if the mechanism is equipped with enforcement tools, how will it
differ from traditional means of dispute settlement? Given the fact that CMs are
already established because traditional means do not work effectively in practice,[56]
does this mean a return to the starting point?

That does not seem feasible in the near future, as much of the existing system
and its operation in practice remains – and in the short term will continue to remain
– mostly under the authority of individual states and their cooperation with each
other; and it is not often possible to persuade states to commit to less flexible
mechanisms involving stricter obligations and response measures. Therefore, CMs
will probably be the most important means of ensuring compliance in the near
future, due to their non-confrontational and non-adversarial features.

Finally, there have been proposals under both the Geneva Negotiating Text and
the Bonn Draft Agreement to establish an international climate justice tribunal to
supervise and sanction non-compliance. There are already earlier proposals for the
creation of a World Environment Court (Pauwelyn 2005; Rest 2000). If these are
adopted to solve the challenges of compliance, why the need for CMs?

To achieve more wide-ranging coordinated efforts for more effective compliance, it is necessary to establish a regime (and hence related mechanisms) "acceptable to most of the countries", in which both developed and developing
countries are active participants, allowing them to make necessary substantive
changes to the operation of the regime, thus making them more committed to
meeting their obligations (Hunter et al. 2002). Accordingly, when designing a
compliance mechanism for the Paris Agreement, if the target is potential change for
better compliance, it is essential to go beyond simply basing the PCA CM on the
systems of existing CMs, or making adjustments to mitigate their shortcomings
according to the features of the Paris Agreement; it is also essential to design a

---

[55]See Mehling (2012) for the view that the inclusion of weaker commitments in terms of their
enforceability results in "a certain risk for the future of international climate cooperation" (Mehling
2012: 214).
[56]See *supra*, Chap. 4, at Sect. 4.2.2.1.

system in which different components of the mechanism can interact more efficiently and effectively to achieve full compliance and reach the ultimate goal of moderating climate change. In fact, with better coordination in the mechanism and, as a further step, better coordination with other compliance mechanisms of different relevant MEAs, all the efforts to address the problems of the mechanisms can help speed up the progress towards better compliance; deliver the most effective use of resources; provide coherence and consistency; and avoid duplication among similar bodies.

Annexes

G. COPs to the UNFCCC and MOPs to the Kyoto Protocol.

References

Andresen, S. (2001). Global Environmental Governance: UN Fragmentation and Co-ordination. Stokke, O.S. and Thommessen, Ø.B. (Eds.), *Yearbook of International Co-operation on Environment and Development 2001/2002* (19–26). London: Earthscan Publications.

Andresen, S. and Gulbrandsen, L.H. (2005). The Role of Green NGOs in Promoting Climate Compliance. Stokke, O.S., Hovi J. and Ulfstein, G. (Eds.), *Implementing the Climate Regime, International Compliance* (169–186). Sterling, VA: The Fridtjof Nansen Institute.

Bäckstrand, K., Kuyper, J. W., Linnér, B. and Lövbrand, E. (2017). Non-state Actors in Global Climate Governance: From Copenhagen to Paris and Beyond, *Environmental Politics*, 26, 4, 561–579, https://doi.org/10.1080/09644016.2017.1327485.

Bang, G., Hovi, J. and Skodvin, T. (2016). The Paris Agreement: Short-Term and Long-Term Effectiveness, *Politics and Governance*, 4, 3, 209–218, https://doi.org/10.17645/pag.v4i3.640. Available at: http://www.cogitatiopress.com/politicsandgovernance/article/view/640/640.

Barrett, S. (2003). *Environment and Statecraft: The Strategy of Environmental Treaty-making*. Oxford: Oxford University Press.

Batagodal, B.M.S., Perera, B.R.L. and De Alwis, J.M.D.D.J. (2004). A New Strategy for Centralized Information Management on (MEAs) in Developing Countries. International Environmental Governance Conference, Paris, 15 & 16 March 2004.

Beyerlin, U., Stoll, P.T. and Wolfrum, R. (2006). Conclusions from MEA Compliance. Beyerlin, U., Stoll, P.T., Wolfrum, R. (Eds.), *Ensuring Compliance with Multilateral Environmental Agreements: Academic Analysis and Views from Practice* (359–369). The Netherlands: Koninklijke Brill NV.

Bodansky, D. (2015). Reflections of the Paris Conference. Available at: http://opiniojuris.org/2015/12/15/reflections-on-the-paris-conference/.

Bodansky, D. (2016a). The Paris Climate Change Agreement: A New Hope? *American Journal of International Law (AJIL)*, 110, 2, 288–319.

Bodansky, D. (2016b). The Legal Character of the Paris Agreement, *Review of European Community & International Environmental Law (RECIEL)*, 25, 2, 142–150. https://doi.org/10.1111/reel.12154.

Bodle, R. and Oberthür, S. (2017). Legal Form of the Paris Agreement and Nature of Its Obligations. Klein, D., Carazo, M.P., Doelle, M., Andrew, J.B. (Eds). *The Paris Agreement on Climate Change: Analysis and Commentary* (91–106). Oxford: Oxford University Press.

Boisson de Chazournes, L. (2006). Technical and Financial Assistance and Compliance: The Interplay. Beyerlin, U., Stoll, P.T. and Wolfrum, R. (Eds.), *Ensuring Compliance with Multilateral Environmental Agreements*, A Dialogue between Practitioners and Academia (273–300). The Netherlands: *Koninklijke Brill NV.*

Brack D. (2003). Monitoring the Montreal Protocol. *Verification Yearbook*, 209–226. Available at: http://www.vertic.org/media/Archived_Publications/Yearbooks/2003/VY03_Brack.pdf.

Breitmeier, H., Young, O.R. and Zürn, M. (2006). *Analyzing International Environmental Regimes: From Case Study to Database.* Cambridge, MA: MIT Press.

Brunnée, J. (2012). Promoting Compliance with Multilateral Environmental Agreements, Brunnée, J., Doelle, M. and Rajamani, L. (Eds). *Promoting Compliance in an Evolving Climate Regime* (38–54). Cambridge, New York: Cambridge University Press.

Brunnée, J. (2003). The Kyoto Protocol: Testing Ground for Compliance Theories? *Zeitschrift für ausländisches öffentliches Recht und Völkerrecht (Heidelberg Journal of International Law)*, 63, 2, 255–280. Available at: http://www.hjil.de/63_2003/63_2003_2_a_255_280.pdf.

Carazo, M. and Klein, D. (2017). Implications for Public International Law: Initial Considerations, Klein, D., Carazo, M.P., Doelle, M., Andrew, J.B. (Eds). *The Paris Agreement on Climate Change: Analysis and Commentary* (389–412). Oxford: Oxford University Press.

Chambers, B.W. (2008). *Interlinkages and the Effectiveness of MEAs.* Tokyo, New York: UN University Press.

Chasek, P.S., Downie, D.L. and Brown, J.W. (2006). *Global Environmental Politics.* Boulder, CO: Westview Press.

Climate Institute (2015). Are Countries Currently on Track To Meet Emissions Targets? *The Climate Institute Level* 15/179. 01 August 2015. Available at: http://www.climateinstitute.org.au/verve/_resources/TCI_On_track_factsheet.pdf.

Crossen, T.E. (2004). The Kyoto Protocol Compliance Regime: Origins, Outcomes and the Amendment Dilemma. *Official Journal of the Resource Management Law Association of New Zealand Inc.*, *I,* XII, 1–6.

Dagnet, Y. and Levin, K. (2017). Transparency (Article 13), Klein, D., Carazo, M.P., Doelle, M., Andrew, J.B. (Eds) *The Paris Agreement on Climate Change: Analysis and Commentary,* (301–318). Oxford: Oxford University Press.

Dagnet, Y. and Northrop, E. (2017). Facilitating Implementation and Promoting Compliance (Article 15), Klein, D., Carazo, M.P., Doelle, M., Andrew, J.B. (Eds,), *The Paris Agreement on Climate Change: Analysis and Commentary,* (338–351). Oxford: Oxford University Press.

Dannenmaier, E. (2012). The Role of Non-State Actors in Climate Compliance, Brunnée, J., Doelle, M. and Rajamani, L. (Eds). *Promoting Compliance in an Evolving Climate Regime* (149–176). Cambridge, New York: Cambridge University Press.

Dimitrov, R.S. (2010). Inside UN Climate Change Negotiations: The Copenhagen Conference, *Review of Policy Research,* 27, 6, 795–821.

Doelle, M. (2017). Assessment of Strengths and Weaknesses, Klein, D., Carazo, M.P., Doelle, M., Andrew, J.B. (Eds) *The Paris Agreement on Climate Change: Analysis and Commentary,* (375–388). Oxford: Oxford University Press.

Doelle, M. (2012). Experience with the Facilitative and Enforcement Branches of the Kyoto Compliance System, Brunnée, J., Doelle, M. and Rajamani, L. (Eds). *Promoting Compliance in an Evolving Climate Regime* (102–121). Cambridge, New York: Cambridge University Press.

Doelle, M. (2010). Early Experience with the Kyoto Compliance System: Possible Lessons for MEA Compliance System Design. *Climate Law,* 1, 237–260, IOS Press. https://doi.org/10.3233/cl-2010-012.

Doğan, F. (2017). *Çin'in Hegemonik Yükselişi: Pax Americana'dan Pax Sinica'ya,* Ankara: Orion Kitabevi.

Ehrmann, M. (2002). Procedures of Compliance Control in International Environmental Treaties. *Colorado Journal of International Environmental Law and Policy,* 13, 2, 377–444.

Environmental Law Institute (ELI) (2015). Model Law Implementing the Nationally Determined Contributions Submitted Under the Paris Agreement. Available at: http://www.eli.org/sites/default/files/modellaw/EN%20Model%20Implementing%20Legislation%20for%20Climate%20NDCs%20Under%20Paris%20Agreement.pdf.

Epiney, A. (2006). The Role of NGOs in the Process of Ensuring compliance with MEAs. Beyerlin, U., Stoll, P.T. and Wolfrum, R. (Eds.), *Ensuring Compliance with Multilateral Environmental Agreements, A Dialogue between Practitioners and Academia* (273–300). Leiden: *Koninklijke Brill NV.*

Eritja, M.C., Pons, X.F. and Sancho, L.H. (2004). Compliance Mechanisms in the Framework Convention on Climate Change and the Kyoto Protocol. *Revue Generale de Droit,* 34, 51–105.

Faure, M.G. and Lefevere, J. (1999). Compliance with International Environmental Agreements. Vig, N.J. and Axelrod, R.S. (Eds.), *The Global Environment: Institutions, Law and Policy* (138–156). Washington: CQ Press.

Fodella, A. (2009). Structural and Institutional Aspects of NCMs. Treves, T., Tanzi, A., Pineschi, L., Pitea, C., Ragni, C. (Eds.), *Non-Compliance Procedures and Mechanisms and the Effectiveness of International Environmental Agreements* (355–372). The Hague: T.M.C. Asser Press.

Fournier, L. (2017). Compliance Mechanisms under the Kyoto Protocol: Lessons for Paris. https://doi.org/10.13140/rg.2.2.13842.66240. Available at: https://www.researchgate.net/publication/316635610.

G20 Leaders' Declaration (2017). G20 Leaders' Declaration, Shaping an Interconnected World. Hamburg, 7–8 July 2017. Available at: http://www.g20.utoronto.ca/2017/2017-G20-leaders-declaration.pdf.

Goldberg, D.M., Wiser, G., Porter S.J. and Lacosta, N. (1998). Building a Compliance Regime under the Kyoto Protocol. CIEL/EuroNatura, 1998) [CC98-3]. Available at: http://ciel.org/Publications/buildingacomplianceregimeunderKP.pdf.

Grantham Research Institute on Climate Change and the Environment and Sabin Center for Climate Change Law (2018). Climate Change Laws of the World Database. Available at: http://www.lse.ac.uk/GranthamInstitute/climate-change-laws-of-the-world/.

Hagem, C. and Westskog, H. (2005). Effective Enforcement and Double-edged Deterrents: How the Impacts of Sanctions also Affect Complying Parties. Stokke, O.S., Hovi J. and Ulfstein, G. (Eds.), *Implementing the Climate Regime, International Compliance* (107–128). Sterling, VA: The Fridtjof Nansen Institute.

Haita, C. (2012). The State of Compliance in the Kyoto Protocol. *International Center for Governance (ICCG) Reflection*, No. 12/2012. Available at: http://www.iccgov.org/wp-content/uploads/2015/05/12_Reflection_December_2012.pdf

Handl, G. (1997). Compliance Control Mechanisms and International Environmental Obligations. *Tulane Journal of International and Comparative Law,* 5, 29–51.

Harris, P.G. and Taedong, L. (2017). Compliance with Climate Change Agreements: The Constraints of Consumption. *International Environmental Agreements.* Springer Science +Business Media Dordrecht. https://doi.org/10.1007/s10784-017-9365-x.

Herold, A. (2012). Experiences with Articles 5, 7, and 8 defining the Monitoring, Reporting and Verification System under the Kyoto Protocol. Brunnée, J., Doelle, M. and Rajamani, L. (Eds). *Promoting Compliance in an Evolving Climate Regime* (122–146). Cambridge, New York: Cambridge University Press.

Hovi, J., Froyn, C.B. and Bang, G. (2007). Enforcing the Kyoto Protocol: Can Punitive Consequences Restore Compliance? *Review of International Studies,* 33, 435–449.

Hovi J., Stokke, O.S. and Ulfstein, G. (2005). Introduction and Main Findings. Stokke, O.S., Hovi J. and Ulfstein, G. (Eds.), *Implementing the Climate Regime, International Compliance* (1–16). Sterling, VA: The Fridtjof Nansen Institute.

Hovi, J. (2005). The Pros and Cons of External Enforcement. Stokke, O.S., Hovi J. and Ulfstein, G. (Eds.), *Implementing the Climate Regime, International Compliance* (129–146). Sterling, VA: The Fridtjof Nansen Institute.

Hunter, D., Salzman, J. and Zaelke, D. (2002). *International Environmental Law and Policy.* New York: Foundation Press.

International Law Association - Washington Conference (2014). Legal Principles Relating to Climate Change. Available at: https://papers.ssrn.com/sol3/papers.cfm?abstract_id=2461556.

Ivanova, M. (2015). COP21 proved the Power of Peer Pressure. Available at: https://www.bostonglobe.com/opinion/2015/12/16/cop-proved-power-peer-pressure/YDP0scVWHYnum8qFKXigCJ/story.html.

Jacobson, H.K. and Weiss, E.B. (2001). Strengthening Compliance with International Environmental Accords. Diehl, P.F. (Eds.), *The Politics of Global Governance, International Organizations in an Interdependent World* (2nd Ed.) (406–435). Boulder, CO: Lynne Rienner Publishers.

Kazokoğlu, C. (2015). 10 grafikte BM İklim Değişikliği Konferansi ve Türkiye. Available at: http://www.bbc.com/turkce/ekonomi/2015/11/151130_cop21_turkiye_cuneyt_kazokoglu.

Kolari, T. (2003). Constructing Non-Compliance Systems into International Environmental Agreements – A Rise of Enforcement Doctrine with Credible Sanctions Needed? *Finnish Yearbook of International Law*, Vol. XIV, 205–232. Leiden/Boston: Martinus Nijhoff Publishers.

Kolari, T. (2002). Promoting Compliance with International Environmental Agreements – An Interdisciplinary Approach with Special Focus on Sanctions. (Pro Gradu Thesis, Faculty of Social Sciences, Department of Law, University of Joensuu, 2002). Available at: http://www.peacepalacelibrary.nl/ebooks/files/C08-0029-Kolari-Promoting.pdf.

Lefeber, R. and Oberthür, S. (2012). Key Features of the Kyoto Protocol's Compliance System, Brunnée, J., Doelle, M. and Rajamani, L. (Eds). *Promoting Compliance in an Evolving Climate Regime* (77–101). Cambridge, New York: Cambridge University Press.

Levy, M.A., Young O.R. and Zürn M. (1995). The Study of International Regimes. *European Journal of International Relations*, 1, 3, 267–330.

Machado-Filho, H. (2012). Financial Mechanisms under the Climate Regime, Brunnée, J., Doelle, M. and Rajamani, L. (Eds). *Promoting Compliance in an Evolving Climate Regime* (216–239). Cambridge, New York: Cambridge University Press.

Mehling, M. (2012). Enforcing Compliance in an Evolving Climate Regime, Brunnée, J., Doelle, M. and Rajamani, L. (Eds). *Promoting Compliance in an Evolving Climate Regime* (194–215). Cambridge, New York: Cambridge University Press.

Mitchell, R.B. (2007). Compliance Theory, Effectiveness, and Behaviour Change in International Environmental Law. Bodansky, B., Brunnée, J. and Hey, E. (Eds.), *The Oxford Handbook of International Environmental Law* (893–921). New York: Oxford University Press.

Mitchell, R.B. (2005). Flexibility, Compliance and Norm Development in the Climate Regime. Stokke, O.S., Hovi J. and Ulfstein, G. (Eds.), *Implementing the Climate Regime, International Compliance* (65–84). Sterling, VA: The Fridtjof Nansen Institute.

Moosmann, L., Neier, H., Mandl, N. and Radunsky, K. (2017). *Implementing the Paris Agreement – New Challenges in view of the COP 23 Climate Change Conference,* Study for the ENVI Committee, European Parliament, Policy Department for Economic and Scientific Policy, Brussels. Available at: http://www.europarl.europa.eu/RegData/etudes/STUD/2017/607353/IPOL_STU(2017)607353_EN.pdf.

Montini, M. (2017). Riflessioni Critiche Sull'accordo Di Parigi Sui Cambiamenti Climatici (Some Critical Reflections on the Paris Agreement on Climatic Changes), *Rivista Di Diritto Internazionale,* Vol. C, Fasc. 3, 719–755.

Montini, M. (2011). Reshaping Climate Governance for Post-2012. *European Journal of Legal Studies*, 4, 1 (Summer 2011), 7–24. Available at: https://ejls.eui.eu/wp-content/uploads/sites/32/pdfs/Spring_Summer2011/COMMENT_%20RE-SHAPING_CLIMATE_GOVERNANCE_FOR_POST-2012_.pdf.

Montini, M. (2009). Procedural Guarantees in NCMs. Treves, T., Tanzi, A., Pineschi, L., Pitea, C., Ragni, C. (Eds.), *Non-Compliance Procedures and Mechanisms and the Effectiveness of International Environmental Agreements* (389–406). The Hague: T.M.C. Asser Press.

Morel, R. and Shishlov, I. (2014). Ex-Post Evaluation of The Kyoto Protocol: Four Key Lessons For The 2015 Paris Agreement, No. 44, May 2014. Available at: http://www.cdcclimat.com/IMG/pdf/14-05_climate_report_no44_-_analysis_of_the_kp-2.pdf.

Morgan, J. (2012). The Emerging Post-Cancun Climate Regime. Brunnée, J., Doelle, M. and Rajamani, L. (Eds). *Promoting Compliance in an Evolving Climate Regime* (17–37). Cambridge, New York: Cambridge University Press.

Mrema, E.M. (2006). Cross-Cutting Issues Related to Ensuring Compliance with MEAs. Beyerlin, U., Stoll, P.T. and Wolfrum, R. (Eds.), *Ensuring Compliance with Multilateral Environmental Agreements, A Dialogue between Practitioners and Academia* (201–228). Leiden: Koninklijke Brill NV.

Nanou, M. (2015). *Greek Financial Crisis and Environment, Can Crisis Be an Opportunity?* MSc Thesis. May, 2015. Available at: http://edepot.wur.nl/357781.

Newman, P.A. (2009). What If There Had Been No Montreal Protocol? Building on the Montreal Protocol's Success and Facing the Challenges Ahead. *Ozone Action, Special Issue.* UNEP. Available at: http://www.unep.fr/ozonaction/information/mmcfiles/3139-e-OASI09_2010andThen.pdf.

Oberthür, S. and E. Northrop (2018). The Mechanism to Facilitate Implementation and Promote Compliance with the Paris Agreement: Design Options. *Working Paper. Washington, DC; Project for Advancing Climate Action Transparency (PACT).* Available at: www.wri.org/publication/pact-compliance-mechanism .

Oberthür, S. (2014). Options for a Compliance Mechanism in a 2015 Climate Agreement, *Working Paper*, September 2014, Institute for European Studies, Vrije Universiteit, Brussels.

Oberthür, S. and Lefeber, R. (2010). Holding Countries to Account: The Kyoto Protocol's Compliance System Revisited After Four Years of Experience. *Climate Law,* 1, 1, 133–158.

Oberthür, S. and Gebring, T. (2006). Conceptual Foundations of Institutional Interaction. Oberthür, S. and Gebring, T. (Eds.), *Institutional Interaction in GEG, Synergy and Conflict among International and EU Policies* (19–52). Cambridge, MA: MIT Press.

Oberthür, S. (2006). The Climate Change Regime: Interactions with ICAO, IMO and the EU Burden-Sharing Agreement. Oberthür, S. and Gebring, T. (Eds.), *Institutional Interaction in GEG, Synergy and Conflict among International and EU Policies* (53–72). Cambridge, Mass.: MIT Press.

Oberthür, S. (2002). Clustering of Multilateral Environmental Agreements: Potentials and Limitations. *International Environmental Agreements: Politics, Law and Economics, 2,* 317–340. Dordrecht: Kluwer Academic Publishers.

Oberthür S. (2001). Linkages between the Montreal and Kyoto Protocols, Enhancing Synergies between Protecting the Ozone Layer and the Global Climate. *International Environmental Agreements: Politics, Law and Economics*, 1, 3, 357–377.

Oberthür, S. and Ott, H. (1999). *The Kyoto Protocol, International Climate Policy for the 21st Century.* New York, Berlin, Heidelberg: Springer.

Oslo Principles on Global Obligations to Reduce Climate Change (2015). Available at: https://law.yale.edu/system/files/area/center/schell/oslo_principles.pdf.

Ozone Secretariat (2015). Montreal Protocol-Achievements to Date and Challenges Ahead.

Page, M.L. (2016). Was Kyoto Climate Deal a Success? Figures Reveal Mixed Results. 14 June 2016. Available at: https://www.newscientist.com/article/2093579-was-kyoto-climate-deal-a-success-figures-reveal-mixed-results.

Pitea, C. (2009). Multiplication and Overlap of NCPs and Mechanisms: Towards Better Coordination? Treves, T., Tanzi, A., Pineschi, L., Pitea, C., Ragni, C. (Eds.), *Non-Compliance Procedures and Mechanisms and the Effectiveness of International Environmental Agreements* (439–452). The Hague: T.M.C. Asser Press.

Rajamani, L. (2016) Ambition and Differentiation in the 2015 Paris Agreement: Interpretative Possibilities and Underlying Politics, *International and Comparative Law Quarterly (ICLQ),* 65 (April 2016), 493–514. https://doi.org/10.1017/s0020589316000130.

Raustiala, K. (2001). Reporting and Review Institutions in 10 Selected Multilateral Environmental Agreements. Nairobi: UNEP. Available at: https://www.peacepalacelibrary.nl/ebooks/files/C08-0025-Raustiala-Reporting.pdf.

Romanin Jacur, F. (2009). Controlling and Assisting Compliance: Financial Aspects. Treves, T., Tanzi, A., Pineschi, L., Pitea, C., Ragni, C. (Ed.), *Non-Compliance Procedures and Mechanisms and the Effectiveness of International Environmental Agreements* (419–438). The Hague: T.M.C. Asser Press.

Sachariew, K. (1991). Promoting Compliance with International Environmental Standards: Reflections on Monitoring and Reporting Mechanisms. *Yearbook of International Environmental Law*, 2, 1, 31–52. https://doi.org/10.1093/yiel/2.1.31.

Sands, P. (2016). Climate Change and the Rule of Law: Adjudicating the Future in International Law. *Journal of Environmental Law,* 28, 19–35. https://doi.org/10.1093/jel/eqw005. Available at: https://academic.oup.com/jel/article-abstract/28/1/19/1748465/Climate-Change-and-the-Rule-of-Law-Adjudicating.

Savaşan, Z. (2012). Climate Change, Compliance and the Role of Reporting: A Comparative Analysis of Reporting under the Montreal and Kyoto Protocols. *World Congress on Water, Climate and Energy 2012, Conference Proceedings*, 13–18 May 2012. Dublin, Ireland. Available at: http://keynote.conference-services.net/resources/444/2653/pdf/IWAWCE2012_0737.pdf.

Savaşan, Z. (2015a). The Role of Networks in Ensuring Compliance And Strengthening Coordination: A Comparative Analysis on INECE, ECENA, RENA and REC Turkey. Faure, M., Smedt, P., Stas, A. (Eds). *Environmental Enforcement Networks: Concepts, Implementation and Effectiveness* (68–104). Cheltenham: Edward Elgar Publishing Limited.

Savaşan, Z. (2015b). Gathering Information under Compliance Mechanisms: Potential New Ways for Current Challenges. De Bree, M. and Ruessink, H. (Eds.), *Innovating Environmental Compliance Assurance* (171–194). The Netherlands: INECE.

Savaşan, Z. (2015c). The Role of Compliance Mechanism under the Kyoto Protocol in Coping with Climate Change. *Energy and Diplomacy Journal*, 1, 1, 2 (Summer 2015), 88–115. Available at: http://enerjivediplomasi.com/makale.aspx?makaleNo=35.

Savaşan, Z. (2017a). Coping with Global Warming: Compliance Issue Compliance Mechanisms Under MEAs. Zhang, XinRong, Dincer, Ibrahim (Eds.), *Energy Solutions to Combat Global Warming*, Lecture Notes on Energy, 33, Switzerland: Springer International Publishing.

Savaşan, Z. (2017b). A Brief Assessment on the Paris Climate Agreement and Compliance Issue. *Uluslararasi İlişkiler*, 14, 54, 107–125.

Shishlov, I., Morel, R. and Bellassen, V. (2016). Compliance of the Parties to the Kyoto Protocol in the First Commitment Period, *Climate Policy,* 6, 6, 768–782. Available at: http://dx.doi.org/10.1080/14693062.2016.1164658.

Ulfstein, G. and Werksman, J. (2005). The Kyoto Compliance System: Towards Hard Enforcement. Stokke, O.S., Hovi J. and Ulfstein, G. (Eds.), *Implementing the Climate Regime, International Compliance* (39–64). Sterling, VA: The Fridtjof Nansen Institute.

Urbinati, S. (2009). Procedures and Mechanisms relating to Compliance under the 1997 Kyoto Protocol to the 1992 UN Framework Convention on Climate Change. Treves, T., Tanzi, A., Pineschi, L., Pitea, C., Ragni, C. (Eds.), *Non-Compliance Procedures and Mechanisms and the Effectiveness of International Environmental Agreements* (63–84). The Hague: T.M.C. Asser Press.

UNEP (2001). Proposal for a Systematic Approach to Coordination of Multilateral Environmental Agreements. The Third Consultative Meeting of the MEA Secretariats on International Environmental Governance. Teleconference, 4 July 2001 Document No. 4/Rev.1, 27 June 2001. Available at: http://archive.unu.edu/inter-linkages/eminent/papers/WG2/unep01.pdf.

Vinuales, J. (2016). The Paris Climate Agreement: An Initial Examination. *Blog of the European Journal of Environmental Law.* Available at: https://www.ejiltalk.org/the-paris-climate-agreement-an-initial-examination-part-i-of-ii/.

Voigt, C. and Ferreira, F. (2016). 'Dynamic Differentiation': The Principles of CBDR-RC, Progression and Highest Possible Ambition in the Paris Agreement. *Transnational Environmental Law*, 5, 2, 285–303.

Voigt, C. (2016a). The Compliance and Implementation Mechanism of the Paris Agreement, *Review of European Community & International Environmental Law (RECIEL)*, 25, 2. https://doi.org/10.1111/reel.12155.

Voigt, C. (2016b). Operationalizing the Implementation and Compliance Mechanism in the Paris Agreement, *Conversations with Climate Experts Series*.

Wang, X. and Wiser, G. (2002). The Implementation and Compliance Regimes under the Climate Change Convention and Its Kyoto Protocol. *Review of European Community and International Environmental Law*, 11, 2, 181–198.

Werksman, J.D. (2005). The Negotiation of a Kyoto Compliance System. Stokke, O.S., Hovi J. and Ulfstein, G. (Eds.), *Implementing the Climate Regime, International Compliance* (17–38). Sterling, VA: The Fridtjof Nansen Institute.

Werksman, J.D. (1996). Compliance and Transition: Russia's Non-Compliance Tests the Ozone Regime. *Zeitschrift für ausländisches öffentliches Recht und Völkerrecht (ZaoRV)*, 56, 3, 750–773. Available at: http://www.zaoerv.de/56_1996/56_1996_3_a_750_773.pdf.

Wolfrum, R. and Friedrich, J. (2006). The Framework Convention on Climate Change and the Kyoto Protocol. Beyerlin, U., Stoll, P.T., Wolfrum, R. (Eds.), *Ensuring Compliance with Multilateral Environmental Agreements: Academic Analysis and Views from Practice* (53–68). Leiden: *Koninklijke Brill NV*.

Wolfrum, R. and Matz, N. (2003). *Conflicts in International Environmental Law*. Berlin, Heidelberg: Springer.

Wolfrum, R. (1999). *Recueil des cours: Collected Courses of the Hague Academy of International Law*, Vol. 272 (1998). The Hague, Boston, London: Martinus Nijhoff Publishers.

Young, O.R. (2002). Matching Institutions and Ecosystems: The Problem of Fit. *Les seminaires de l'Iddri*, No. 2, Iddri.

Zaelke, D. and Grabiel, P.M. (2009). New Strategies to Leverage the Montreal Protocol to Protect the Climate. Building on the Montreal Protocol's Success and Facing the Challenges Ahead. *Ozone Action, Special Issue*. UNEP. Available at: http://www.unep.fr/ozonaction/information/mmcfiles/3139-e OASI09_2010andThen.pdf.

Official Documents Related to CM Under Kyoto Protocol

Consolidated Rules of Procedure of the Compliance Committee of the Kyoto Protocol (RoP). Decision 4/MOP 2; Decision 4/MOP 4 (version of 3 February 2014). Available at: http://unfccc.int/files/kyoto_protocol/compliance/application/pdf/consolidated_rop_with_cmp_4&cmp9_amend_2014feb03.pdf.

Kyoto Protocol to the United Nations Framework Convention on Climate Change (1998). United Nations. Available at: http://unfccc.int/resource/docs/convkp/kpeng.pdf.

Outcome of the work of the Ad Hoc Working Group on Further Commitments for Annex I Parties under the Kyoto Protocol. Amendment to the Kyoto Protocol pursuant to its Article 3, Paragraph 9. (Draft Decision-/MOP 8). (FCCC/KP/CMP/2012/L.9). Available at:http://unfccc.int/documentation/documents/advanced_search/items/6911.php?priref=600007290.

Procedures and Mechanisms relating to Compliance under the Kyoto Protocol (NCP). Decision 27/MOP 1. FCCC/KP/CMP/2005/8/Add.3. Available at: http://unfccc.int/resource/docs/2005/cmp1/eng/08a03.pdf#page=92.

United Nations Framework Convention on Climate Change (UNFCCC) (1992). United Nations. FCCC/INFORMAL/84. Available at: http://unfccc.int/resource/docs/convkp/conveng.pdf.

Reports of the Conference of the Parties (COP)

COP 1 (1995). Report of the COP on its First Session. Part Two: Action taken by the COP at its
   First Session. Berlin, 28 March–7 April 1995. FCCC/CP/1995/7/Add.1. Available at: http://
   unfccc.int/resource/docs/cop1/07a01.pdf.
COP 2 (1996). Report of the COP on its Second Session. Part Two: Action taken by the COP at its
   Second Session. Geneva, 8–19 July 1996. FCCC/CP/1996/15/Add.1. Available at: http://
   unfccc.int/resource/docs/cop2/15a01.pdf.
COP 3 (1997). Report of the COP on its Third Session. Part Two: Action taken by the COP at its
   Third Session. Kyoto, 1–11 December 1997. FCCC/CP/1997/7/Add.1. Available at: http://
   unfccc.int/resource/docs/cop3/07a01.pdf.
COP 4 (1998). Report of the COP on Its Fourth Session. Part Two: Action taken by the COP at its
   Fourth Session. Buenos Aires, 2–14 November 1998. FCCC/CP/1998/16/Add.1. Available at:
   http://unfccc.int/resource/docs/cop4/16a01.pdf.   .
COP 5 (1999). Report of the COP on Its Fifth Session. Part Two: Action taken by the COP at its
   Fifth Session. Bonn, 25 October–5 November 1999. Available at: http://unfccc.int/resource/
   docs/cop5/06a01.pdf.
COP 6 (2000). Report of the COP on the First Part of Its Sixth Session. Part Two: Action taken by
   the COP at its Sixth Session. The Hague, 13–25 November 2000. FCCC/CP/2000/5/Add.2.
   Available at: http://unfccc.int/resource/docs/cop6/05a02.pdf.
COP 6 (2001). Report of the COP on the Second Part of Its Sixth Session. Part II. Bonn, 16–27
   July 2001. FCCC/CP/2001/5. Available at: http://unfccc.int/resource/docs/cop6secpart/05.pdf.
COP 7 (2001). Report of the COP on its Seventh Session. Part Two: Action taken by the COP at its
   Seventh Session. Volume III. Marrakesh, 29 October–10 November 2001. FCCC/CP/2001/13/
   Add.3. Available at: http://unfccc.int/resource/docs/cop7/13a03.pdf.
COP 9 (2003). Report of the COP on its Ninth Session. Part Two: Action taken by the COP at its
   Ninth Session. Milan, 1–12 December 2003. FCCC/CP/2003/6/Add.2. Available at: http://
   unfccc.int/resource/docs/cop9/06a02.pdf.
COP 10 (2004). Report of the COP on its Tenth Session. Part Two: Action taken by the COP at its
   Tenth Session. Buenos Aires, 6–18 December 2004. FCCC/CP/2004/10/Add.2. Available at:
   http://unfccc.int/resource/docs/cop10/10a02.pdf.
COP 11 (2005). Report of the COP on its Eleventh Session. Part Two: Action taken by the COP at
   its Eleventh Session. Montreal, 28 November–10 December 2005. FCCC/CP/2005/5/Add.2.
   Available at: http://unfccc.int/resource/docs/2005/cop11/eng/05a02.pdf.
COP 12 (2006). Report of the COP on its Twelfth Session. Part Two: Action taken by the
   Conference of the Parties at its Twelfth Session. Nairobi, 6–17 November 2006. FCCC/CP/
   2006/5/Add.1. Available at: http://unfccc.int/resource/docs/2006/cop12/eng/05a01.pdf.
COP 13 (2007). Bali Action Plan. Decision 1/COP 13. Report of the Conference of the Parties on
   its thirteenth session. Bali, 3–15 December 2007. Addendum Part Two: Action taken by the
   Conference of the Parties at its thirteenth session. FCCC/CP/2007/6/Add.1. Available at: http://
   unfccc.int/resource/docs/2007/cop13/eng/06a01.pdf#page=3.
COP 15 (2009). Copenhagen Accord. Decision 2/COP 15. Report of the Conference of the Parties
   on its fifteenth session, Copenhagen, 7–19 December 2009. Addendum Part Two: Action taken
   by the Conference of the Parties at its fifteenth session. FCCC/CP/2009/11/Add.1. Available at:
   http://unfccc.int/resource/docs/2009/cop15/eng/11a01.pdf#page=4.
COP 16 (2010). Cancun Agreements: Outcome of the work of the Ad Hoc Working Group on
   Long-term Cooperative Action under the Convention. Decision 1/COP 16, Report of the
   Conference of the Parties on its sixteenth session, Cancun, 29 November–10 December 2010.
   Addendum Part Two: Action taken by the Conference of the Parties at its sixteenth session. FCCC/
   CP/2010/7/Add.1. Available at: http://unfccc.int/resource/docs/2010/cop16/eng/07a01.pdf.

COP 17 (2011). Decision 1/COP 17. Report of the Conference of the Parties on its seventeenth session, Durban, 28 November–11 December 2011. Addendum Part Two: Action taken by the Conference of the Parties at its seventeenth session. FCCC/CP/2011/9/Add.1. Available at: http://unfccc.int/resource/docs/2011/cop17/eng/09a01.pdf#page=2.

COP 18 (2012). Decision 1/COP 18. Agreed outcome pursuant to the Bali Action Plan. Report of the Conference of the Parties on its eighteenth session, Doha, 26 November–8 December 2012. Addendum Part Two: Action taken by the Conference of the Parties at its eighteenth session. FCCC/CP/2012/8/Add.1. Available at: http://unfccc.int/resource/docs/2012/cop18/eng/08a01.pdf#page=3.

COP 19 (2013). Decision 1/COP 19. Further advancing the Durban Platform. Report of the Conference of the Parties on its nineteenth session, Warsaw, 11–23 November 2013. Addendum Part Two: Action taken by the Conference of the Parties at its nineteenth session. FCCC/CP/2013/10/Add.1. Available at: http://unfccc.int/resource/docs/2013/cop19/eng/10a01.pdf#page=3.

COP 20 (2014). Lima Call for Climate Action, Decision 1/COP 20, Report of the Conference of the Parties on its twentieth session, Lima, 1–14 December 2014. Addendum Part Two: Action taken by the Conference of the Parties at its twentieth session. FCCC/CP/2014/10/Add.1. Available at: http://unfccc.int/resource/docs/2014/cop20/eng/10a01.pdf.

COP 21 (2015). Decision 1/COP 21, Adoption of the Paris Agreement. Report of the Conference of the Parties on its twenty-first session, Paris, 30 November–13 December 2015. Addendum Part Two: Action taken by the Conference of the Parties at its twenty-first session. FCCC/CP/2015/10/Add.1. Available at: http://unfccc.int/resource/docs/2015/cop21/eng/10a01.pdf#page=2.

COP 23 (2017). Provisional agenda. Conference of the Parties Twenty-third session, Bonn, Germany, 6–17 November 2017. Available at: http://unfccc.int/files/meetings/bonn_nov_2017/application/pdf/cop23_pa_web.pdf.

Reports of the Meeting of the Parties (MOP)

MOP 1 (2005a). Report of the COP serving as the MOP to the Kyoto Protocol on its First Session. Part Two: Action taken by the COP serving as the MOP at its First Session. Montreal, 28 November–10 December 2005. FCCC/KP/CMP/2005/8/Add.3. Available at: https://unfccc.int/resource/docs/2005/cmp1/eng/08a03.pdf.

MOP 1 (2005b). Report of the COP serving as the MOP to the Kyoto Protocol on its First Session. Part Two: Action taken by the COP serving as the MOP at its First Session. Montreal, 28 November–10 December 2005. FCCC/KP/CMP/2005/8/Add.2. Available at: http://unfccc.int/resource/docs/2005/cmp1/eng/08a02.pdf.

MOP 1 (2005c). Report of the COP serving as the MOP to the Kyoto Protocol on its First Session. Part Two: Action taken by the COP serving as the MOP at its First Session. Montreal, 28 November–10 December 2005. FCCC/KP/CMP/2005/8/Add.1. Available at: http://unfccc.int/resource/docs/2005/cmp1/eng/08a01.pdf.

MOP 1 (2005d). Report of the COP serving as the MOP to the Kyoto Protocol on its First Session. Part One: Proceedings. Montreal, 28 November–10 December 2005. FCCC/KP/CMP/2005/8. Available at: http://unfccc.int/resource/docs/2005/cmp1/eng/08.pdf.

MOP 6 (2010). Report of the COP serving as the MOP to the Kyoto Protocol on its Sixth Session. Part One: Proceedings. Cancun, 29 November–10 December 2010. FCCC/KP/CMP/2010/12. Available at: http://unfccc.int/resource/docs/2010/cmp6/eng/12.pdf.

MOP 7 (2011). Report of the COP serving as the MOP to the Kyoto Protocol on its Seventh Session. Part Two: Action taken by the COP serving as the MOP to the Kyoto Protocol on its Seventh Session. Durban, 28 November–11 December 2011. FCCC/KP/CMP/2011/10/Add.2. Available at: http://unfccc.int/resource/docs/2011/cmp7/eng/10a01.pdf.

MOP 8_Compliance (2012). Draft decision-/CMP 8. Available at: https://unfccc.int/files/bodies/application/pdf/awgkp_vice_chair_revised_29112012.pdf.

MOP 8 (2012). Doha Amendment. Decision 1/MOP 8. Amendment to the Kyoto Protocol
    pursuant to its Article 3, paragraph 9. Available at: http://unfccc.int/resource/docs/2012/cmp8/
    eng/13a01.pdf.
MOP 10 (2014). Report of the COP serving as the MOP to the Kyoto Protocol on its Tenth
    Session. Lima, 1–14 December 2014. Part two: Action taken by the COP serving as the MOP
    to the Kyoto Protocol at its Tenth Session. FCCC/KP/CMP/2014/9/Add.1. Available at: http://
    unfccc.int/resource/docs/2014/cmp10/eng/09a01.pdf#page=13.

Reports of the Compliance Committee (ComplCom)

ComplCom 2 (2007). Annual Report of the ComplCom to the COP serving as the MOP to the
    Kyoto Protocol. Bali, 3–14 December 2007. FCCC/KP/CMP/2007/6. Available at: http://
    unfccc.int/resource/docs/2007/cmp3/eng/06.pdf.
ComplCom 3 (2008). Annual Report of the ComplCom to the COP serving as the MOP to the
    Kyoto Protocol. Poznan, 1–12 December 2008. FCCC/KP/CMP/2008/5. Available at: http://
    unfccc.int/resource/docs/2008/cmp4/eng/05.pdf.
ComplCom 4 (2009). Annual Report of the ComplCom to the COP serving as the MOP to the
    Kyoto Protocol. Copenhagen, 7–18 December 2009 FCCC/KP/CMP/2009/17. Available at:
    http://unfccc.int/resource/docs/2009/cmp5/eng/17.pdf.
ComplCom 8 (2013). Annual Report of the ComplCom to the COP serving as the MOP to the
    Kyoto Protocol. Warsaw, 11–22 November 2013. FCCC/KP/CMP/2013/3. Available at: http://
    unfccc.int/resource/docs/2013/cmp9/eng/03.pdf.

Reports of the Subsidiary Body for Implementation (SBI)

SBI 24 (2006). Report of the SBI on its Twenty-Fourth Session. Bonn, 18–25 May 2006. FCCC/
    SBI/2006/11. Available at: https://unfccc.int/resource/docs/2006/sbi/eng/11.pdf.
SBI 25 (2006). Report of the SBI on its Twenty-Fifth Session. Nairobi, 6–14 November 2006.
    FCCC/SBI/2006/28. Available at: https://unfccc.int/resource/docs/2006/sbi/eng/28.pdf.
SBI 26 (2007). Report of the SBI on its Twenty-Sixth Session. Bonn, 7–18 May 2007. FCCC/SBI/
    2007/15. Available at: https://unfccc.int/resource/docs/2007/sbi/eng/15.pdf.
SBI 27 (2007). Report of the SBI on its Twenty-Seventh Session. Bali, 3–11 December 2007.
    FCCC/SBI/2007/34. Available at: https://unfccc.int/resource/docs/2007/sbi/eng/34.pdf.
SBI 28 (2008). Report of the SBI on its Twenty-Eighth Session. Bonn, 4–13 June 2008. FCCC/
    SBI/2008/8. Available at: https://unfccc.int/resource/docs/2008/sbi/eng/08.pdf.
SBI 29 (2008). Report of the SBI on its Twenty-Ninth Session. Poznan, 1–10 December 2008.
    FCCC/SBI/2008/19. Available at: https://unfccc.int/resource/docs/2008/sbi/eng/19.pdf.
SBI 30 (2009). Report of the SBI on its Thirtieth Session. Bonn, 1–10 June 2009, FCCC/SBI/
    2009/8. Available at: https://unfccc.int/resource/docs/2009/sbi/eng/08.pdf.
SBI 32 (2010). Report of the SBI on its Thirty-Second Session. Bonn, 31 May–9 June 2010.
    FCCC/SBI/2010/10. Available at: https://unfccc.int/resource/docs/2010/sbi/eng/10.pdf.
SBI 33 (2010). Report of the SBI on its Thirty-Third Session. Cancun, 30 November-4 December
    2010. FCCC/SBI/2010/27. Available at: https://unfccc.int/resource/docs/2010/sbi/eng/27.pdf.
SBI 37 (2012). Amendment of the Kyoto Protocol in respect of Procedures and Mechanisms
    relating to Compliance. Doha, 26 November–1 December 2012. FCCC/SBI/2012/L.40.
    Available at: https://unfccc.int/sites/default/files/resource/docs/2012/sbi/eng/l40.pdf.

Documents Related to Cases Under CM of Kyoto Protocol

CC-2007-1/Greece/EB. Question of Implementation-Greece. Available at: http://unfccc.int/kyoto_ protocol/compliance/enforcement_branch/items/5455.php.

CC-2008-1/Canada/EB. Question of Implementation-Canada. Available at: http://unfccc.int/ kyoto_protocol/compliance/enforcement_branch/items/5298.php.

CC-2009-1/Croatia/EB. Questions of Implementation- Croatia. Available at: http://unfccc.int/ kyoto_protocol/compliance/enforcement_branch/items/5456.php.

CC-2010-1/Bulgaria/EB. Question of Implementation-Bulgaria. Available at: http://unfccc.int/ kyoto_protocol/compliance/questions_of_implementation/items/5538.php.

CC-2011-1/Romania/EB. Question of Implementation-Romania. Available at: http://unfccc.int/ kyoto_protocol/compliance/questions_of_implementation/items/6030.php.

CC-2011-2/Ukraine/EB. Question of Implementation-Ukraine I. Available at: http://unfccc.int/ kyoto_protocol/compliance/questions_of_implementation/items/6077.php.

CC-2016-1-1/Ukraine/EB. Question of Implementation-Ukraine II. Available at: http://unfccc.int/ kyoto_protocol/compliance/questions_of_implementation/items/9575.php.

CC-2011-3/Lithuania/EB. Question of Implementation-Lithuania. Available at: http://unfccc.int/ kyoto_protocol/compliance/questions_of_implementation/items/6195.php.

CC-2012-1/Slovakia/EB. Question of Implementation-Slovakia. Available at: http://unfccc.int/ kyoto_protocol/compliance/questions_of_implementation/items/6920.php.

Reports on the Enforcement Branch (EB) Meetings

EB 27 (2015). Enforcement Branch. Twenty-seventh meeting. Report on the meeting. Bonn, Germany, 6 September 2015. CC/EB/27/2015/2. Available at: http://unfccc.int/files/kyoto_ protocol/compliance/enforcement_branch/application/pdf/cc-eb-27-2015-2_report_on_the_ meeting.pdf.

EB 21 (2012). Enforcement Branch. Twenty-First Meeting. Report on the Meeting. Bonn, Germany, 22–24 October 2012. CC/EB/21/2012/2. Available at: https://unfccc.int/files/kyoto_ protocol/compliance/enforcement_branch/application/pdf/cc-eb-21-2012_2_report_on_the_ meeting.pdf.

EB 20 (2012). Enforcement Branch. Twentieth Meeting. Report on the Meeting. Bonn, Germany, 9–14 July 2012. CC/EB/20/2012/2. Available at: https://unfccc.int/files/kyoto_protocol/ compliance/enforcement_branch/application/pdf/cc-eb-20-2012-2_report_on_the_meeting.pdf.

EB 19 (2012). Enforcement Branch. Nineteenth Meeting. Report on the Meeting. Bonn, Germany, 8–9 March 2012. CC/EB/19/2012/2. Available at: https://unfccc.int/files/kyoto_protocol/ compliance/enforcement_branch/application/pdf/cc-eb-19-2012-2_report_on_the_meeting.pdf.

EB 18 (2012). Enforcement Branch. Eighteenth Meeting. Report on the Meeting. Bonn, Germany, 7, 8–10 February 2012. CC/EB/18/2012/3. Available at: https://unfccc.int/files/kyoto_protocol/ compliance/enforcement_branch/application/pdf/cc-eb-18-2012-3_report_on_the_meeting.pdf.

EB 16 (2011). Enforcement Branch. Sixteenth Meeting. Report on the Meeting. Bonn, Germany, 14–18 November 2011. CC/EB/16/2011/2. Available at: https://unfccc.int/files/kyoto_protocol/compliance/ enforcement_branch/application/pdf/cc-eb-16-2011-2_report_on_the_meeting.pdf.

EB 15 (2011). Enforcement Branch. Fifteenth Meeting. Report on the Meeting. Bonn, Germany, 11–12 October 2011. CC/EB/15/2011/2. Available at: https://unfccc.int/files/kyoto_protocol/ compliance/enforcement_branch/application/pdf/cc-eb-15-2011-2_report_on_the_meeting.pdf.

EB 13 (2011). Enforcement Branch. Thirteenth Meeting. Report on the Meeting. Bonn, Germany, 6–8 July 2011. CC/EB/13/2011/2. Available at: https://unfccc.int/files/kyoto_mechanisms/ compliance/enforcement_branch/application/pdf/cc-eb-13-2011-2_report_on_the_meeting.pdf.

EB 12 (2011). Enforcement Branch. Twelfth Meeting. Report on the Meeting. Bonn, Germany, 3–4 February 2011. CC/EB/12/2011/2. Available at: https://unfccc.int/files/essential_background/convention/application/pdf/cc_eb_12_2011_2_report_on_the_meeting.pdf.

EB 11 (2010). Enforcement Branch. Eleventh Meeting. Report on the Meeting. Bonn, Germany, 16 September 2010. CC/EB/11/2010/2. Available at: https://unfccc.int/files/kyoto_protocol/compliance/enforcement_branch/application/pdf/cc-eb-11-2010-2_report_on_the_meeting.pdf.

EB 10 (2010). Enforcement Branch. Tenth Meeting. Report on the Meeting. Bonn, Germany, 28 June 2010. CC/EB/10/2010/2. Available at: https://unfccc.int/files/kyoto_protocol/compliance/enforcement_branch/application/pdf/cc-eb-10-2010-2_report_on_the_meeting.pdf.

EB 9 (2010). Enforcement Branch. Ninth Meeting. Report on the Meeting. Bonn, Germany, 10–12 May 2010. CC/EB/9/2010/2. Available at: https://unfccc.int/files/kyoto_mechanisms/compliance/enforcement_branch/application/pdf/cc-eb-9-2010-2_report_on_the_meeting.pdf.

EB 8 (2009). Enforcement Branch. Eighth Meeting. Report on the Meeting. Bonn, Germany, 23–24 November 2009. CC/EB/8/2009/2. Available at: https://unfccc.int/files/kyoto_protocol/compliance/enforcement_branch/application/pdf/cc-eb-8-2009-2_report_on_the_8th_meeting_of_the_eb.pdf.

EB 7 (2009). Enforcement Branch. Seventh Meeting. Report on the Meeting. Bonn, Germany, 11–13 October 2009. CC/EB/7/2009/2. Available at: https://unfccc.int/files/kyoto_protocol/compliance/enforcement_branch/application/pdf/cc-eb-7-2009-2_report_on_the7th_meeting_of_the_eb.pdf.

EB 6 (2008). Enforcement Branch. Sixth Meeting. Report on the Meeting. Bonn, Germany, 6–7 October 2008. CC/EB/6/2008/3. Available at: https://unfccc.int/files/kyoto_protocol/compliance/enforcement_branch/application/pdf/cc-eb-6-2008-3_report_on_the_6th_mtg_of_the_eb.pdf.

EB 4 (2008). Enforcement Branch. Fourth Meeting. Report on the Meeting. Bonn, Germany, 16–17 April 2008. CC/EB/4/2003/2. Available at: https://unfccc.int/files/kyoto_protocol/compliance/enforcement_branch/application/pdf/cc-eb-4-2008-2_report_on_the_4th_meeting_of_the_eb.pdf.

Secretariat (2014). Canada's withdrawal from the Kyoto Protocol and its effects on Canada's reporting obligations under the Protocol. Note by the Secretariat, 20 August 2014. CC/EB/25/2014/2. Available at: https://unfccc.int/files/kyoto_protocol/compliance/enforcement_branch/application/pdf/cc-eb-25-2014-2_canada_withdrawal_from_kp.pdf.

Facilitative Branch (FB) QoI Documents

FB Reports to the Compliance Committee on the Deliberations in the Facilitative Branch relating to the Submission entitled "Compliance with Article 3.1 of the Kyoto Protocol." (Parties concerned: Austria; CC-2006-1-2/FB, Bulgaria, CC-2006-2-3/FB; Canada, CC-2006-3-3/FB; France, CC-2006-4-3/FB; Germany, CC-2006-5-2/FB; Ireland, CC-2006-6-2/FB; Italy, CC-2006-7-2/FB; Latvia, CC-2006-8-4/FB, Liechtenstein, CC-2006-9-2/FB; Luxembourg, CC-2006-10-2/FB; Poland, CC-2006-11-3/FB; Portugal, CC-2006-12-3/FB; Russian Federation, CC-2006-13-2/FB; Slovenia, CC-2006-14-3/FB; Ukraine, CC-2006-15-2/FB). Available at: https://unfccc.int/process/bodies/constituted-bodies/compliance-committee-cc/questions-of-implementation/question-of-implementation-numerous-annex-1-party.

CC-2006-8-3/Latvia/FB. FB Decision not to Proceed against Latvia. Available at: https://unfccc.int/process/bodies/constituted-bodies/compliance-committee-cc/questions-of-implementation/question-of-implementation-numerous-annex-1-party.

CC-2006-14-2/Slovenia/FB. FB Decision not to Proceed against Slovenia. Available at: https://unfccc.int/process/bodies/constituted-bodies/compliance-committee-cc/questions-of-implementation/question-of-implementation-numerous-annex-1-party.

FB 12 (2012). Facilitative Branch. Twelfth meeting. Report on the meeting. Bonn, Germany, 22–23 October 2012. CC/FB/12/2012/3. Available at: http://unfccc.int/files/kyoto_protocol/compliance/facilitative_branch/application/pdf/cc-fb-12-2012-3_report_on_the_meeting.pdf.

FB 16 (2014). Facilitative Branch. Sixteenth meeting. Report on the meeting. Bonn, Germany, 4 September 2014. CC/FB/16/2014/2. Available at: http://unfccc.int/files/kyoto_protocol/compliance/facilitative_branch/application/pdf/cc-fb-16-2014-2_report_on_the_meeting.pdf.

Reports of the Individual Review of the Annual Submission (ARRs)

FCCC/ARR/2011/ROU. Report of the Individual Review of the Annual Submission of Romania submitted in 2011. Available at: https://unfccc.int/sites/default/files/resource/docs/2012/arr/rou. pdf.

FCCC/ARR/2010/ROU. Report of the Individual Review of the Annual Submission of Romania submitted in 2010. Available at: http://unfccc.int/resource/docs/2011/arr/rou.pdf.

FCCC/ARR/2011/SVK. Report of the Individual Review of the Annual Submission of Slovakia submitted in 2011. Available at: https://unfccc.int/sites/default/files/resource/docs/2012/arr/ svk.pdf.

FCCC/ARR/2011/UKR. Report of the Individual Review of the Annual Submission of Ukraine submitted in 2011. Available at: http://unfccc.int/resource/docs/2012/arr/ukr.pdf.

FCCC/ARR/2011/UKR/Corr.1. Report of the Individual Review of the Annual Submission of Ukraine submitted in 2011. Corrigendum. Available at: http://unfccc.int/resource/docs/2012/ arr/ukrc01.pdf.

FCCC/ARR/2010/UKR. Report of the Individual Review of the Annual Submission of Ukraine submitted in 2010. Available at: http://unfccc.int/resource/docs/2011/arr/ukr.pdf.

FCCC/ARR/2010/LTU. Report of the Individual Review of the Annual Submission of Lithuania submitted in 2010. Available at: http://unfccc.int/resource/docs/2011/arr/ltu.pdf.

FCCC/ARR/2009/BGR. Report of the Individual Review of the Annual Submission of Bulgaria submitted in 2009. Available at: http://unfccc.int/resource/docs/2010/arr/bgr.pdf.

FCCC/ARR/2008/GRC. Report of the Individual Review of the Greenhouse Gas Inventories of Greece submitted in 2007 and 2008. Available at: http://unfccc.int/resource/docs/2008/arr/grc2. pdf.

Reports of the Individual Review of the Annual Submissions Submitted and Forwarded to the Compliance Committee

CC/ERT/ARR/2012/17. Report of the Individual Review of the Annual Submission of Slovakia submitted in 2011. Available at: https://unfccc.int/files/kyoto_protocol/compliance/plenary/ application/pdf/cc-ert-arr-2012-17_arr_2011_of_slovakia.pdf.

CC/ERT/ARR/2012/4. Report of the Individual Review of the Annual Submission of Romania submitted in 2011. Available at: https://unfccc.int/files/kyoto_protocol/compliance/plenary/ application/pdf/cc-ert-arr-2012-4_arr_2011_of_romania.pdf.

CC/ERT/ARR/2012/2. Report of the Individual Review of the Annual Submission of Ukraine submitted in 2011. Available at: https://unfccc.int/files/kyoto_protocol/compliance/plenary/ application/pdf/cc-ert-arr-2011-2_of_ukraine.pdf.

CC/ERT/ARR/2012/2/Corr.1. Report of the Individual Review of the Annual Submission of Ukraine submitted in 2011. Corrigendum. Available at: https://unfccc.int/files/kyoto_protocol/ compliance/plenary/application/pdf/cc-ert-arr-2012-2-corr.1_2011_arr_of_ukraine.pdf.

CC/ERT/ARR/2011/33. Report of the Individual Review of the Annual Submission of Lithuania submitted in 2010. Available at: https://unfccc.int/files/kyoto_protocol/compliance/plenary/ application/pdf/cc-ert-arr-2011-33_arr_2010_of_lithuania.pdf.

CC/ERT/ARR/2011/28. Report of the Individual Review of the Annual Submission of Ukraine submitted in 2010. Available at: https://unfccc.int/files/kyoto_protocol/compliance/plenary/ application/pdf/cc-ert-arr-2010-28_arr_2010_of_ukraine.pdf.

CC/ERT/ARR/2011/21. Report of the Individual Review of the Annual Submission of Romania submitted in 2010. Available at: https://unfccc.int/files/kyoto_protocol/compliance/plenary/ application/pdf/cc-ert-arr-2011-21_arr_2010_of_romania.pdf.

CC/ERT/ARR/2010/18. Report of the Individual Review of the Annual Submission of Bulgaria submitted in 2009. Available at: https://unfccc.int/files/kyoto_protocol/compliance/plenary/ application/pdf/cc-ert-arr-2010-18_rep_of_the_ind_rev_of_annual_subm_of_bulgaria_2009. pdf.

Reports of the Review of the Initial Reports (IRRs)

FCCC/IRR/2008/HRV. Report of the Review of the Initial Report of Croatia. Available at: http://unfccc.int/resource/docs/2009/irr/hrv.pdf.

FCCC/IRR/2007/GRC. Report of the Review of the Initial Report of Greece. Available at: http://unfccc.int/resource/docs/2007/irr/grc.pdf.

FCCC/IRR/2007/CAN. Report of the Review of the Initial Report of Canada. Available at: http://unfccc.int/resource/docs/2008/irr/can.pdf.

Official Documents Related to CM Under Montreal Protocol

Ad Hoc WG (1989). Report of the 1st Meeting of the Ad Hoc WG. Geneva, 14 July 1989. UNEP/OzL.Pro.LG.1/3.

Ad Hoc WG (1998). Report on the Work of the Ad Hoc WG of Legal and Technical Experts on Non-compliance with the Montreal Protocol. Geneva, 3–4 July 1998 and Cairo, 17–18 November 1998. UNEP/OzL.Pro/WG.4/1.

Montreal Protocol on Substances that Deplete the Ozone Layer (1987). Available at: https://treaties.un.org/doc/publication/unts/volume%201522/volume-1522-i-26369-english.pdf.

MOP 3 (1991). Nairobi, 19–21 June 1991. UNEP/OzL.Pro.3/11. Available at: https://www.informea.org/en/event/third-meeting-parties-montreal-protocol-substances-deplete-ozone-layer.

MOP 4 (1992). Copenhagen, 23–25 November 1992. UNEP/OzL.Pro.4/15. Available at: https://www.informea.org/en/event/fourth-meeting-parties-montreal-protocol-substances-deplete-ozone-layer.

MOP 10 (1998). Cairo, 23–24 November 1998. UNEP/OzL.Pro.10/9. Available at: https://www.informea.org/en/event/tenth-meeting-parties-montreal-protocol-substances-deplete-ozone-layer.

MOP 28 (2016). Kigali, 10–15 October 2016. UNEP/OzL.Pro.28/12. Available at: http://conf.montreal-protocol.org/meeting/mop/mop-28/final-report/English/MOP-28-12E.pdf.

Procedures and Mechanisms relating to Compliance under the Montreal Protocol (NCP) (2006). *Handbook for the Montreal Protocol on Substances that Deplete the Ozone Layer* (419–421). Nairobi: UNEP.

Official Documents Related to the Paris Climate Agreement

ADP 2.8 (2015a). Ad Hoc Working Group on the Durban Platform for Enhanced Action (2015a). Negotiating Text, Second Session, Part Eight. Geneva, 8–13 February 2015. Agenda item 3 Implementation of all the Elements of Decision 1/COP 17. FCCC/ADP/2015/1. Available at: http://unfccc.int/resource/docs/2015/adp2/eng/01.pdf.

ADP 2.11 (2015b). Ad Hoc Working Group on the Durban Platform for Enhanced Action. Draft Agreement and Draft Decision on Workstreams 1 and 2 of the Ad Hoc Working Group on the Durban Platform for Enhanced Action. Second session, Part Eleven, Bonn, 19–23 October 2015. Available at: http://unfccc.int/files/bodies/application/pdf/ws1and2@2330.pdf.

APA 1.1 (2016). Report of the Ad Hoc Working Group on the Paris Agreement on the first part of its first session, Bonn, 16–26 May 2016. FCCC/APA/2016/2. Available at: http://unfccc.int/resource/docs/2016/apa/eng/02.pdf.

APA 1.2 (2016). Report of the Ad Hoc Working Group on the Paris Agreement (APA) on the second part of its first session. Marrakech, 7–14 November 2016. FCCC/APA/2016/4. Available at: http://unfccc.int/resource/docs/2016/apa/eng/04.pdf.

APA 1.3 (2017). Report of the Ad Hoc Working Group on the Paris Agreement (APA) on the third part of its first session, Bonn, 8–18 May 2017. FCCC/APA/2017/2. Available at: http://unfccc. int/resource/docs/2017/apa/eng/02.pdf.

APA 1.4 (2017). Report of the Ad Hoc Working Group on the Paris Agreement (APA) on the fourth part of its first session, Bonn, 7–18 November 2017. FCCC/APA/2017/4. Available at: https://unfccc.int/sites/default/files/resource/APA_2017_4.pdf.

COP 23 (2017). Draft Decision 1, Preparations for the implementation of the Paris Agreement and the first session of the Conference of the Parties serving as the meeting of the Parties to the Paris Agreement, Proposal by the President. FCCC/CP/2017/L.13, 18 November 2017. Available at: http://unfccc.int/resource/docs/2017/cop23/eng/l13.pdf.

Paris Climate Agreement (2016). Available at: http://unfccc.int/files/home/application/pdf/paris_ agreement.pdf.

SBI 47 (2017). Subsidiary Body for Implementation. Forty-seventh Session. Bonn, 6–15 November 2017. Provisional agenda. Available at: http://unfccc.int/files/meetings/bonn_nov_ 2017/application/pdf/sbi_47_provisional_agenda.pdf.

Submission by the Plurinational State of Bolivia to the Ad-Hoc Working Group on Long-Term Cooperative Action (AWG-LCA) (2010). Available at: http://unfccc.int/files/meetings/ad_hoc_ working_groups/lca/application/pdf/bolivia_awglca10.pdf.

Other Official Documents

UN Charter (1945). Available at: https://treaties.un.org/doc/publication/ctc/uncharter.pdf.

Vienna Convention on the Law of Treaties (VCLT) (1969). Available at: https://treaties.un.org/ doc/Publication/UNTS/Volume%201155/volume-1155-I-18232-English.pdf.

Chapter 7
Conclusion

> *However much you know, what you say is as much as what is*
> *understood of it.*
>
> Mevlana Jalaluddin Rumi (1207–1273).

In line with Mevlana's words, in order to make the book's argument more clear and understandable, this section presents a brief summary of the book's previous chapters, highlighting the fundamental points that should be gleaned from them and also basic findings which suggest that there is no reason to dismiss the current structure of CMs and their possible contribution to the supply of better compliance and better GEG, although they have some weaknesses and need to be supported by strengthened coordination.

7.1 Summary

The book started by outlining a general framework which clarifies its purpose, the research design (consisting of research question, sub-questions, methodology and the rationale for case selection) and its main argument. Here, the aim was to provide a brief and clear introduction to the research subject: compliance and CMs under MEAs.

In this respect, it stated that its main purpose is not to focus on the need for a new alternative structure for the CMs or to question their effectiveness, but just to investigate the features of the current system and to discuss the ways of enhancing compliance with CMs under MEAs, particularly for a better system for improving compliance under the PCA, benefiting from the illustrations from the two case studies in the book: the CM under the Kyoto Protocol (KP) and the CM under the Montreal Protocol (MP).

The design of the research was based on the main research question asking whether CMs created under MEAs can contribute to ensure and improve the compliance of the parties with their commitments under MEAs. On the basis of this question, sub-questions which have formed the different sections of the book were also asked, e.g. questions on the meaning of the concepts of compliance, CM and

© Springer Nature Switzerland AG 2019

Z. Savaşan, *Paris Climate Agreement: A Deal for Better Compliance?*
The Anthropocene: Politik—Economics—Society—Science 11,
https://doi.org/10.1007/978-3-030-14313-8_7

MEA (Chap. 2), on the related theories and the two basic explanatory models (approaches) of compliance (Chap. 3), on the development of CMs, the reasons of their emergence, the preparatory and triggering reasons behind their development, also the questions on the limitations of traditional means and on the main characteristics of CMs (Chap. 4), the case study of the compliance mechanism regarding ozone depletion-CM under the Montreal Protocol (its main components – gathering information, NCPs and response measures – and their operation in practice) (Chapter 5), the case study of the compliance mechanism regarding climate change-CM under the Kyoto Protocol (its main components – gathering information, NCPs and response measures – and their operation in practice) (Chap. 6), and finally, benefiting from the lessons learned by the CMs of the Montreal Protocol and the Kyoto Protocol, questions about ways to improve compliance under the PCA by designing a compliance mechanism which addresses the weaknesses of the current compliance mechanisms under MEA (Chap. 6).

The research method applied is a comparative case analysis based on both international relations and international environmental law literature.

With respect to the cases, it should be underlined that they were chosen particularly because it was considered that the Protocols of both Kyoto and Montreal can render beneficial lessons for the themes under discussion in the face of the great number of practices experienced, and that outcomes stemming from them can be employed to find out the possible ways of enhancing compliance. The PCA was chosen as a field study because it provides a different system from the UNFCCC system, and this requires the Agreement to be examined in terms of its implications for compliance issues and compliance mechanisms.

Overall, the introductory part of the book, identifying its outline and the way forward, highlighted the major argument of the study: to improve the current system of CMs, eliminate their shortcomings and strengthen the coordination between different mechanisms in order to ensure better compliance with the commitments under MEAs in the short term and perhaps better environmental governance in the long term.

In the second chapter, in order to enable and facilitate the discussion on compliance and compliance mechanisms under MEAs, the focus was on clarifying what the concepts of compliance and compliance mechanisms and MEA mean both in general and in the context of this study in particular. Regarding the concept of compliance, it was stated that, in this study, compliance would be used to explain exclusively whether the behaviour of the states which are party to the related MEA conform to the agreement-based requirements (procedural-substantive) because of MEAs' compliance mechanisms, rather than conform to the broader categories of other non-explicit rules, like principles and norms and upholding the spirit of the treaty. Thus, according to this definition, non-compliance occurs when the states' behaviour departs from the related MEAs' requirements.

Before proceeding further, the concept of the compliance mechanism (CM) within multilateral environmental agreements (MEA) was scrutinized to identify what exactly it refers to and how it can be defined within the context of this study. For this purpose, firstly the terms 'MEA' and 'environmental regime' were defined, due to the fact that IEL has been developed mostly through environmental regimes created under these MEAs. Then, after outlining diverse definitions of CMs made by different scholars, the definition of CMs under MEAs which is employed in this book was put forward. According to this definition, CMs were identified as international internal mechanisms of MEAs involving three components: information-gathering, NCPs, and response measures.

The third chapter firstly gives brief explanations of the theoretical perspectives on compliance from both IL and IR. Then it focuses on two basic explanatory models of compliance mechanisms: the management approach, which has been very effective in the design of the CMs under MEAs; and the enforcement approach, which has basically been established by critics of the management approach. It was noted that both have the potential to provide often complementary and beneficial insights into the compliance debate.

The examination made of these two models revealed that both approaches have similarities, but at the same time, basic differences, particularly regarding the reasons for non-compliance and the forms of sanctions (either in the form of coercive economic or military, or in the form of social enforcement). There are also some missing elements regarding other forms of free-riding, such as non-participation, in both approaches. In addition, depending on the problem and its conditions, each can explain different aspects of compliance processes, and under some circumstances, neither is sufficient to explain the existing situation. Therefore, different approaches may be required for different countries, depending on their features with respect to compliance. Regarding the question of which is better at promoting compliance and improving environmental quality, it was found that, in the current system, combining the two approaches (management approach-enforcement approach), dominantly leaning on the management approach in the short term, has emerged as the best way, instead of merely leaning on more severe mechanisms in line with the enforcement approach.

The third chapter involved mainly three sub-sections: development of CMs, fundamental bodies of the CMs, and the components of the CMs.

Taking these sub-sections in order, the development of compliance mechanisms and the reasons behind their development were first discussed briefly along with preparatory and triggering reasons.

With regard to the preparatory reasons, the changes in the international system and the concept of sovereignty were analysed, as were the features of IEL which make it distinct from other fields of international law – and hence require the application of a primarily preventive approach – as well as MEAs, which, unlike bilateral agreements, do not have reciprocal obligations. With regard to the triggering reasons, the limitations of traditional means and the features which make CMs more attractive and preferable to other methods were studied.

After giving a brief explanation of the operation of the CMs via their main bodies, three main components of the CM which complement and support each other in line with the definition given in the first chapter were focused on: gathering information to review the parties' performance, institutionalised multilateral NCPs (including the institutions created under NCPs: committees, procedural phases and safeguards), and multilateral non-compliance response measures (including positive measures, negative measures and their binding effect). In that part, the aim was simply to give a general overview of the system of the CMs in order to lay a foundation for the case studies in subsequent chapters of the book.

In the fifth and sixth chapters of the book, building on the background information on compliance mechanisms provided in the previous chapters, a comparative analysis was made between two case studies: ozone depletion and climate change.

In the fifth chapter, all aspects of the compliance mechanism under the Montreal Protocol (MP) were discussed. During this analysis, the history of their development up to the present was studied, taking the evolution of the NCPs – which have formed the building block of the entire mechanism – as a basis for the examination. Then, their three components and the application of the mechanisms in practice were explored. Thus, the aim was to make the analysis on the basis of four basic dimensions: their three main components – gathering information, NCPs (involving committees, procedures and phases), and response measures – and the way they function in practice. Finally, based on an overall assessment, some tentative conclusions were drawn to find out whether it is necessary to improve compliance by making improvements which address the weaknesses of the current mechanism.

In the first part of the sixth chapter, the compliance mechanism under the Kyoto Protocol (KP) was studied. In line with the format of the previous chapter, the study was concerned with the development of the CM up to the present – the components of the mechanism (gathering information, procedures/institutional structure, measures), the functioning of the mechanism in practice, and the lessons learned from the operation of the current system.

In the second section of the sixth chapter, the PCA was illustrated in detail to discuss the future options for a compliance mechanism under the PCA and its potential for better compliance. Benefiting from the lessons learned by the CMs of the Montreal Protocol and Kyoto Protocol, the weaknesses of the current compliance mechanisms under MEAs that figured throughout the book and the ways/options/proposals for a better system for improving compliance under the PCA were evaluated.

7.2   Basic Findings

After submitting this brief summary of the previous chapters, from now on, the basic findings of the book will be indicated on the basis of the two case studies – ozone depletion (CM under the MP) and climate change (CM under the KP) – and their four dimensions – gathering information, procedures/institutional structure, responses and their operation in practice – plus the PCA.

Regarding the first dimension, gathering information on the parties' performance, it should be stated that nearly all the parties of both the Montreal and Kyoto Protocols have tended to comply with reporting obligations from the beginning. But there are still some challenges for both CMs which should be dealt with, such as self-reporting, the complexity of the substances that should be reported, the challenge of capacity-building, the quality and reliability of data, the need for harmonization, the lack of third-party monitoring and verification, and the insufficiency of public and NGO participation.

With respect to non-compliance procedures, the main findings can be categorized in four sub-groups: institutional structure, procedural safeguards, the role of NGOs in CMs, and financial challenges.

On the institutional structure, it is concluded that each mechanism under the KP and MP gives competences to different organs to decide on non-compliance and also to adopt responses to it. Under the CM of the MP, the recommendations of the ImplCom can have a strong impact on the decisions made by the MOP, as the MOP usually evaluates the circumstances of the matter on the basis of the ImplCom's report containing these recommendations. The CM under the KP seems more promising than that of the MP, as it grants the power to decide on non-compliance to the Committee and its two branches, FB and EB, whose members should be qualified in climate change and relevant fields (scientific, technical, socio-economic or legal), independent of the political organ of the mechanism.

Regarding the Committees' impartiality and independence, it is underlined that under the CM of the MP, ImplCom consists of parties elected by the MOP. This is questionable as experts serving in their personal capacity as members of the Committee are expected to be more independent than representatives of the parties. Additionally, in terms of the obligations of parties' representatives, there are no criteria in the NCP, MP. This is probably because the dominant view among the parties is that each party should have the right to decide on the background of its representative. Under the Kyoto Protocol, on the other hand, members of the EB should also have legal experience and should serve in their personal capacities. There are crucial provisions supporting the independence and impartiality of the Compliance Committee, such as suspension of the membership of any member who is found to have materially violated the independence and impartiality of the Committee.

The research also indicates that the Kyoto Protocol's NCP generally has more detailed rules on the procedural structure and necessary procedural safeguards in the process of related proceedings. Indeed, in contrast to the Kyoto Protocol's NCP, in the MP's NCP, it is not possible to find provisions concerning the rights of the party in question. The mechanism under the MP includes strict timelines for different processes of the proceedings. However, contrary to the Kyoto Protocol's CM, there are not so many and a clearly defined timetable is not provided for each step in the proceedings of the ImplCom. There is also no expedited procedure in the CM, MP. The CM, KP, through its strict predetermined timetables and the expedited procedure, emerges as having more potential to prevent time-consuming delays. With respect to transparency, the NCP Section VIII(6), KP allows information to be kept from the public until the conclusion of the proceedings at the request of the party being investigated and at the discretion of the EB. However, this has not been used to date, and transparency has been supported both by the ComplCom and its two branches. The NCP, Montreal Protocol, para.16, opens the way for making the ImplCom's reports available to anyone upon request, thus providing its publicity. On request, it also makes available to any party all information exchanged by or with the ImplCom relating to any of its recommendations to the MOP. However, it restricts this opportunity to reports which do not contain any confidential information, and with the obligation to protect the confidentiality of information that the party has received in confidence. Moreover, it obliges the members of the ImplCom and the parties involved in its deliberations to guarantee the confidentiality of any information they receive in confidence. Here, questions can arise about determining whether information is confidential or not, and about the possible results of not complying with these rules of the NCP. Finally, under the CM, KP, as a rule, there is no opportunity to appeal against the decisions of the branches of the ComplCom. As an exception, it is possible for the parties to appeal to the MOP against final decisions of the EB relating to emissions targets concerning whether the rules of the due process are applied to the party concerned. Under the CM, MP, even this kind of appeal opportunity is not possible. There is also no support for creating such a mechanism, particularly because it would be against the decisions of the MOP, as distinct from the appeal applied in the Kyoto Protocol's CM.

With regard to strengthening the role of NGOs in CMs, it is found that under the MP, NGOs can participate as observers in the meetings of the MEA's organs if there is no objection from the parties. However, in the assessment stage of the reporting, their role is restricted further, as they cannot usually attend the meeting to assess the report unless their presence has been specifically requested by at least one party. If they are admitted, they do not have the right of veto (e.g. VC, Art. 6(5); MP Art. 11(5); RoP 7(2)). Under the KP mechanism, they are merely allowed to submit relevant factual and technical information to either branch of the ComplCom. As in the MP, in the CM of the KP there is no formal way for NGOs to participate in ComplCom proceedings. Under both CMs, the IGOs and NGOs are also not allowed to trigger the NCP; they can only submit relevant factual and technical information, and NGOs' participation in monitoring activities remains at a low level, as it is restricted to only some areas of monitoring.

For both, lack of proper financial resources also arises as a significant problem, especially since both mechanisms have to support financial and technical assistance towards developing country parties under the principle of common but differentiated responsibility.

With respect to the third dimension of the analysis, it is found that while the CM under the MP adopts a list of measures of non-compliance, yet no list of possible situations of non-compliance, the KP includes a system of predefined non-compliance situations with predefined responses. This results in a debate over which system is better for the compliance mechanisms, leading to the emergence of different views on the situation.

Indeed, it is possible to argue that, as the evaluation phase of non-compliance should not involve much flexibility but instead less flexibility and less political but more integral judicial features, determining the fixed measures according to the circumstances of non-compliance can enhance the predictability. Nevertheless, even just adopting a list of measures, as in the CM under the MP, can be seen as a restrictive factor on the mechanism and its flexibility, even though there is not a complete list of measures that should be taken against non-compliance, and the MOP is still free to apply different measures for different circumstances. This is because this kind of list draws the possible borders of being in compliance and in non-compliance. So the system in the CM under the KP can be seen as improper, as it can also prevent the choice of consequences which are appropriate for the circumstances of each individual case at hand. However, because this system has so far been successfully implemented in practice, it is proposed for the Montreal Protocol system as well. Yet, under the CM of the MP, developing such criteria in a formal list is regarded as something that should be achieved over time.

It is noted that for both, there have been some discussions on the binding status of the decisions. Although it is observed that facilitative response measures rather than punitive ones have been preferred under both Protocols in practice, there has still been no consensus on this matter. However, all measures applied by the MOP under the MP are necessarily accepted as recommendatory, rather than mandatory. In the Kyoto Protocol's system, apart from the measure requiring deductions of assigned amounts at a penalty rate, the other measures can be considered to be within the implied powers of the Protocol's organs, that is, within the competence of the EB. It should be kept in mind that as the cooperative-facilitative approach is dominant under both mechanisms, the application of response measures can be effective even without a formally binding status when strong cooperation and coordination between parties can be ensured within the system.

Finally, on the last dimension – the functioning of the mechanisms in practice – the findings stemming from the analysis of the different non-compliance cases brought before the CMs of both the MP and the KP are as follows:

Under the CM of the MP, the first type of triggering by the party against another party has not arisen to date, while the second type arose by CEIT in the 1990s. But the most used one is the third type: triggering by the secretariat. Many parties, both developed and developing, have met their phase-out targets in line with their schedules. Thus, a high level of compliance rate on the commitments of the parties, contributing to a decline in the production and consumption of ODSs, has been achieved. None of the parties has been deprived of international assistance or encountered the measures consistent with item C of the indicative list of measures. Except for the application of trade restrictions and conditional assistance (even it was usually considered favourably in order to provide funding for projects to implement its national programme) on some parties (like Russia, Belarus and Ukraine), negative measures have not been applied against non-compliant parties. Even though there seems to be a combination of two approaches (the enforcement approach with punitive measures like restricting trade with non-parties to the MP, and the management approach with facilitative-positive measures, like granting developing parties financial assistance or a ten-year grace period, and strengthening their capacity-building), the management approach appears to be the more dominant approach in the CM of the MP.

Under the CM of the KP, likewise under the CM of the MP, despite the existence of opportunities for the parties to trigger the mechanism, there has been no submission to the EB by the parties. The submissions made, up to now, have been triggered by the ERTs. FB has been applied only in the case of QoIs raised against the 15 Annex I parties by South Africa. However, no decision could be taken about these submissions, except that in the Latvia and Slovenia cases the FB decided not to proceed, but could not provide reasons for its decision. So, it is usually argued that the FB needs to be further developed to address potential non-compliance through its early warning function. To date, the subjects of the QoIs that come before the EB have generally been about national system requirements, but not have been related to whether the concerned parties are in compliance with their 2012 emissions target, because non-compliance with emissions targets is a subject that could not be brought before the EB until after the end of the first commitment period in 2012. Although the NCP of the Protocol involves various response measures (positive-negative, except financial measures or loss of credit), most of them have been very rarely applied in practice, if applied at all, and have been very limited in scope. In parallel with the MP's CM, there even seems to be a system of bringing together the facilitation and enforcement functions under the same mechanism, hence a management approach applying not only LoC but also LoA and equipped with facilitative-positive measures, emerges as the more dominant approach adopted in the KP's CM as well.

Overall, keeping all these findings of this study in mind, the following conclusions can be drawn about the two case studies: although the CM under the KP is grounded on a shorter length of experience than the CM under the MP, both CMs under the MP and the KP offer a reasonably strong basis for ensuring and improving compliance with the related MEAs's requirements. As the CM under the KP is better structured than that of MP, with a more detailed and comprehensive institutional structure (with two branches), formalized procedures based on a quasi-judicial character, precise time limitations on the duration of proceedings and specific deadlines for intermediate steps in the process, it appears to function as well as that of the MP and, in some cases, more effectively than that of the MP.

It can therefore be concluded that the assessment of existing compliance mechanisms made so far in the book indicates that CMs under the MP and the KP have already played a more effective role in ensuring compliance than traditional means, such as the Law of Treaties, the responsibility of states, and dispute settlement procedures (DSPs) , so even their current system reasonably offers a strong basis for ensuring and improving compliance with the related MEAs's requirements.

When weighing up their strengths and weaknesses, it also becomes clear that they have the potential to seriously contribute to better compliance. Indeed, their primary potentials – for reducing the workload, costs, duplication and conflicts of efforts and for creating consistency to prevent diverging assessments on the similar issues, thereby reducing the potential tensions between the mechanisms – can eventually speed up the progress towards better compliance in the short run and better environmental governance through CMs in the long run. Even if the level of such a contribution may appear modest and remains uncertain at present, on balance it appears to be positive rather than negative, if the improvement of their components and functioning in practice and the coordination between them can be achieved properly.

Nevertheless, these conclusions should not be accepted as valid for all CMs, as it is naturally not possible to draw firm conclusions about all CMs' overall functioning capacity and effects on compliance on the basis of just the two cases analysed within this book, even if they are generally viewed as the most successful. Nevertheless, the lessons learnt from the Montreal and Kyoto Protocol CMs experiences can foster understanding of the weak sides of these mechanisms, which involve both common deficiencies related to their each component and also special compliance challenges resulting from their specific characteristics. Thus, they can also contribute to the establishment of more efficient compliance mechanisms under different MEAs.

In line with the lessons learned from both cases, CMs under the MP and the KP, the discussion on the design of a compliance mechanism under the Paris Climate Agreement provides some basic findings:

First of all, the Paris Climate Agreement does not include any elaborated formulation on the mechanism, but just refers to the establishment of a mechanism consisting of a Committee which is expert-based and facilitative in nature to facilitate implementation of and promote compliance with the provisions of the Agreement. Therefore, there are still major issues that need to be further clarified in the next CMAs, such as the name, structure of the mechanism, phases of the non-compliance procedure, triggering process, deadlines for both the submission of the parties and the decisions to be taken by the relevant bodies, the rights granted to the parties under scrutiny and the rights provided to the submitting parties, the rights of confidentiality and transparency and publicity within the procedures, the impartiality and independence of the Committee, the existence of response measures and their types etc.

Given the current conditions, to make a discussion on possible future options for a compliance mechanism under the Paris Climate Agreement, the NDCs emerge as key commitments shaping and forming the main structure/bone of the Agreement's system. That is, the Paris Agreement does not put a categorization between Annex I and non-Annex I country parties, but on the basis of its basic objective – holding the global average temperature increase well below 2 °C above pre-industrial levels (Art. 1a, PCA), each party to prepare its NDC which will include its climate action commitment (Art. 4.2, PCA). That is, in the case that a party fails to comply with its NDC, this would put that party in breach of its obligations made legally binding by the Agreement.

However, the questions of whether and how the parties' compliance with their climate action commitments under NDCs can be ensured and what the consequences of non-compliance will be, are still not put forward with tangible words under the Agreement.

The Agreement's mechanisms to ensure compliance with those commitments are structured under three main pillars: transparency system (Art. 13), global stocktake

(Art. 14) and compliance mechanism (Art. 15). There are also diverse supplementary means, such as loss and damage (Art. 8); financial support (Art. 9); technology development/transfer (Art. 10); capacity-building (Art. 11); and cooperation (Art. 12).

Of those, the future options for a compliance mechanism under the Paris Climate Agreement are discussed as proper to the general framework of the book, relying on three main components of the CMs – gathering information on the parties' performance, non-compliance procedures, and response measures. During this discussion, the weaknesses and strengths of the current mechanisms, and the need for stronger coordination in CMs and between CMs, are used as the basis for consideration of a compliance mechanism under the Paris Agreement.

Accordingly, it is found that the quasi-judicial/enforcement-based aspects of the KP's NCP and its two-branch system are less likely to be applied under the Paris Agreement, as the Paris Agreement relies upon NDCs, in contrast to the KP relying upon emission reduction targets differing between Annex I and non-Annex I parties. The Agreement indeed does not include any remark regarding enforcement, but instead stresses 'facilitation' (Art. 15.1–2). So then it is more likely that it will be a mechanism in which the Committee has the power to give recommendations to the CMA (Art. 15.3, PCA); and the CMA has the final decision-making authority, likewise being under the CM of the MP. With respect to the inclusion of the response measures and their types, it is again more likely that it will be a mechanism involving measures which are necessarily advisory and conciliatory, in the form of recommendations, in contrast to the KP's CM, dividing between positive and negative measures. Nonetheless, the lessons learned under the enforcement branch and the facilitative aspects of the KP can be considered to elaborate the CM under the Paris Agreement, due to the fact the KP's CM has very detailed rules on the rights of the party concerned and strict, neat timetables for distinct phases of the proceedings.

The question to what extent the CM of the Paris Agreement based on a facilitative system can be successful arises here as a significant concern in dealing with the challenges of compliance.

The analysis stresses that the effectiveness of the CM to a large extent depends on the non-compliant party's tendency to comply with its commitments. So, under a CM, it should be primarily essential to ensure the conditions of willingness of parties to comply with their commitments. In order to achieve this, there is a need for wide-ranging coordinated efforts between the parties, and thus in the mechanism itself, and also between different similar compliance mechanisms. In short, while designing a compliance mechanism for the Paris Agreement, if the target is better compliance, it should be of great importance to have a mechanism in which its different components can interact, cooperate and coordinate in a more effective manner; in addition, in which the weaknesses or shortcomings of the current mechanisms are improved according to the Paris Agreement's principles.

Based on the findings, last but not least, it should be stated that in such a decentralized system of environmental agreements and institutions, benefiting from the potential contributions of the current system through further improving its functioning and creating conditions for stronger coordination between different mechanisms – even if it should not be advocated as a panacea for compliance issues – emerges as an important means of enhancing CMs of the MEAs. Therefore, in this book, it is argued that if the weaknesses of the compliance mechanisms are improved and the coordination between them is ensured, the current system of compliance mechanisms can be responsive for better compliance in the short term, and for better environmental governance in the long term. So, the CM under the Paris Climate Agreement should be constructed on the basis of two key principles: improvement of the challenges of the current CMs and their proper application to the Paris Agreement's features, and providing strong coordination between parties/bodies of the CM (and in the long term between different CMs) to strengthen voluntary compliance.

The focus of the Conference of the Parties (COP 23) convened in Bonn from 6 to 17 November 2017 was the technical implementation of the Paris Agreement on various topics, including guidance and modalities for the Committee to promote compliance. At the time of writing, its work programme, called the Paris Rulebook, is scheduled to be finalised and adopted by COP 24 in December 2018, after the convention of the APA, together with the SBSTA and SBI, in Bonn from 30 April to 10 May 2018 to improve this work programme. Therefore, it is hopefully expected that although not all questions regarding the Paris Climate Agreement and its implementation can be solved, most of them will be dealt with during 2018. However, with regard to compliance issues, more time is needed to understand and measure its practical consequences, even if the structure has been fully constructed.

Annexes

Annex A: The List of COPs to the Vienna Convention and MOPs to the Montreal Protocol

COPs To the vienna convention			MOPs to the montreal protocol		
Session no.	Date	Location	Session no.	Date	Location
COP 1	26–28 April 1989	Helsinki, Finland	MOP 1	2–5 May 1989	Helsinki, Finland
			MOP 2	27–29 June 1990	London, England
COP 2	17–19 June 1991	Nairobi, Kenya	MOP 3	19–21 June 1991	Nairobi, Kenya
			MOP 4	23–25 Nov 1992	Copenhagen, Denmark
COP 3	23 Nov 1993	Bangkok, Thailand	MOP 5	17–19 Nov 1993	Bangkok, Thailand
			MOP 6	6–7 Oct 1994	Nairobi, Kenya
			MOP 7	5–7 Dec 1995	Vienna, Austria
COP 4	25–27 Nov 1996	San José, USA	MOP 8	25–27 Nov 1996	San José, USA
			MOP 9	15–17 Sept 1997	Montreal, Canada
			MOP 10	23–24 Nov 1998	Cairo, Egypt

(continued)

© Springer Nature Switzerland AG 2019
Z. Savaşan, *Paris Climate Agreement: A Deal for Better Compliance?*
The Anthropocene: Politik—Economics—Society—Science 11,
https://doi.org/10.1007/978-3-030-14313-8

(continued)

COPs To the vienna convention			MOPs to the montreal protocol		
Session no.	Date	Location	Session no.	Date	Location
COP 5	29 Nov–3 Dec 1999	Beijing, China	MOP 11	29 Nov–3 Dec 1999	Beijing, China
			MOP 12	11–14 Dec 2000	Ouagadougou, Burkina Faso
			MOP 13	16–19 Oct 2001	Colombo, Sri Lanka
COP 6	25–29 Nov 2002	Rome, Italy	MOP 14	25–29 Nov 2002	Rome, Italy
			MOP 15	10–14 Nov 2003	Nairobi, Kenya
			MOP 16	22–26 Nov 2004	Prague, Czechia
COP 7	12–16 Dec 2005	Dakar, Senegal	MOP 17	12–16 Dec 2005	Dakar, Senegal
			MOP 18	30 Oct–3 Nov 2006	New Delhi, India
			MOP 19	17–21 Sept 2007	Montreal, Canada
COP 8	16–20 Nov 2008	Doha, Qatar	MOP 20	16–20 Nov 2008	Doha, Qatar
			MOP 21	4–8 Nov 2009	Port Ghalib, Egypt
			MOP 22	8–12 Nov 2010	Bangkok, Thailand
COP 9	21–25 Nov 2011	Bali, Indonesia	MOP 23	21–25 Nov 2011	Bali, Indonesia
			MOP 24	12–16 Nov 2012	Geneva, Switzerland
			MOP 25	21–25 Oct 2013	Bangkok, Thailand
COP 10	17–21 Nov 2014	Paris, France	MOP 26	17–21 Nov 2014	Paris, France
			MOP 27	1–5 Nov 2015	Dubai, United Arab Emirates
			MOP 28	10–15 Oct 2016	Kigali, Ruanda
COP 11		Montreal, Canada	MOP 29	20–24 Nov 2017	Montreal, Canada

Source Prepared by the author on the basis of information/documentation on the official website of UNEP/Ozone Secretariat. See: http://conf.montreal-protocol.org/SitePages/Home.aspx

Annex B: The List of MOPs and Their Decisions Related to Compliance

MOP 7 (5–7 December 1995) Vienna	**Decisions VII/15–19 (respectively)**: Compliance with the Montreal Protocol by Poland, Bulgaria, Belarus, the Russian Federation, Ukraine
MOP 8 (25–27 November 1996) San Jose	**Decisions VIII/22, 23, 25**: Compliance by Latvia, Lithuania and the Russian Federation
MOP 9 (15–17 September 1997) Montreal	**Decisions IX/29–31**: Compliance by Latvia, Lithuania and the Russian Federation **Decision IX/32**: Compliance by the Czech Republic with the freeze in consumption of methyl bromide in 1995
MOP 10 (23–24 November 1998) Cairo	**Decisions X/20–28**: Compliance by Azerbaijan, Belarus, the Czech Republic, Estonia, Latvia, Lithuania, the Russian Federation, Ukraine, Uzbekistan
MOP 11 (11–14 December 1999) Beijing	**Decisions XI/24–25**: Compliance by Bulgaria, Turkmenistan
MOP 13 (16–19 October 2001) Colombo	**Decision XIII/16**: Potential non-compliance with the freeze on CFC consumption in article 5 parties in the control period 1999–2000 **Decisions XIII/17–25**: Compliance by the Russian Federation, Armenia, Kazakhstan, Tajikistan, Argentina, Belize, Cameroon, Ethiopia, Peru
MOP 14 (25–29 November 2002) Rome	**Decision XIV/17**: Potential non-compliance with the freeze on CFC consumption by parties operating under article 5 for the control period July 2000 to June 2001 **Decision XIV/28**: Non-compliance with consumption phase-out by parties not operating under article 5 in 2000 (on Belarus and Latvia) **Decisions XIV/18–26, 35**: Compliance by Albania, Bahamas, Bolivia, Bosnia and Herzegovina, Namibia, Nepal, Saint Vincent and the Grenadines, Libyan Arab Jamahiriya, Maldives, the Russian Federation
MOP 15 (10–14 November 2003) Nairobi	**Decision XV/21**: Potential non-compliance with consumption of Annex A, group I, ozone-depleting substances (CFCs) by Article 5 parties for the control period 1 July 2001–31 December 2002 **Decision XV/22**: Potential non-compliance with consumption of Annex A, group II, ozone-depleting substances (halons) by article 5 parties in 2002 **Decision XV/23**: Potential non-compliance with consumption of the Annex C, group II, ODS (hydrobromofluorocarbons) by Morocco in 2002 **Decision XV/24**: Potential non-compliance with consumption of the controlled substance in Annex E (methyl bromide) by non-Article 5 Parties in 2002 (on Latvia and Israel) **Decision XV/2**: Potential non-compliance with consumption of the ODS in Annex E (methyl bromide) by article 5 parties in 2002 **Decisions XV/26–45**: Compliance by Albania, Armenia, Azerbaijan, Bolivia, Bosnia and Herzegovina, Botswana,

(continued)

(continued)

	Cameroon, the Democratic Republic of the Congo, Guatemala, Honduras, the Libyan Arab Jamahiriya, Maldives, Namibia, Nepal, Papua New Guinea, Qatar, Saint Vincent and the Grenadines, Uganda, Uruguay, Vietnam
MOP 16 (22–26 November 2004) Prague	**Decision XVI/19**: Potential non-compliance with consumption of Annex A, group II, ODS (halons) by Somalia in 2002 and 2003 **Decision XVI/20**: Potential non-compliance in 2003 with consumption of the controlled substance in Annex B, group III (methyl chloroform) by article 5 parties **Decisions XVI/21–30**: Compliance by Azerbaijan, Chile, Fiji, Guinea-Bissau, Lesotho, Libyan Arab Jamahiriya, Nepal, Oman, Pakistan, Saint Vincent and the Grenadines
MOP 17 (12–16 December 2005) Dakar	**Decision XVII/30**: Potential non-compliance in 2004 with consumption of the controlled substances in Annex B group I (other fully halogenated chlorofluorocarbons) by China **Decision XVII/35**: Potential non-compliance in 2004 with the controlled substances in Annex A, group I (CFCs) by Kazakhstan **Decisions XVII/25–29, 31–33, 36–38**: Compliance by Armenia, Azerbaijan, Bangladesh, Bosnia and Herzegovina, Chile, Ecuador, Federated States of Micronesia, Fiji, Kyrgyzstan, the Libyan Arab Jamahiriya and Sierra Leone
MOP 18 (30 Oct–3 Nov 2006) New Delhi	**Decision XVIII/24**: Potential non-compliance in 2005 with the control measures of the MP governing consumption of the controlled substances in Annex A, group I, (CFCs) by Eritrea **Decision XVIII/27**: Potential non-compliance in 2005 with the control measures of the MP governing consumption of the controlled substance in Annex B group II, (carbon tetrachloride) by the Islamic Republic of Iran **Decisions XVIII/20–23, 25, 28, 30–33**: Compliance by Armenia, the Democratic Republic of the Congo, Dominica, Ecuador, Greece, Kenya, Mexico, Pakistan, Paraguay, Serbia
MOP 19 (17–21 September 2007) Montreal	**Decision XIX/23**: Potential non-compliance in 2005 with the provisions of the MP governing consumption of the controlled substance in Annex E (methyl bromide) by Saudi Arabia **Decisions XIX/21–22, 27**: Compliance by Greece, Paraguay, the Islamic Republic of Iran
MOP 20 (16–20 November 2008) Doha	**Decision XX/18**: Potential non-compliance in 2006 with the provisions of the MP in respect of consumption of the controlled substances in Annex A, group I, (chlorofluorocarbons) by Solomon Islands **Decisions XX/16, 19**: Compliance by Ecuador, Somalia
MOP 21 (4–8 November 2009) Port Ghalib	**Decisions XXI/17–23, 25–26**: Compliance by Bangladesh, Bosnia and Herzegovina, the Federated States of Micronesia, Mexico, Saudi Arabia, Solomon Islands, Somalia, Turkmenistan, Vanuatu
MOP 22 (8–12 November 2010) Bangkok	**Decisions XXII/13, 15, 16, 18**: Compliance by Singapore, Saudi Arabia, the Republic of Korea and Vanuatu

(continued)

(continued)

MOP 23 (21–25 November 2011) Bali	**Decision XXIII/23**: Potential non-compliance in 2009 with the provisions on consumption of the controlled substances in Annex A, group II (halons), by Libya **Decisions XXIII/26, 27**: Compliance by the European Union, the Russian Federation
MOP 24 (12–16 November 2012) Geneva	**Decision XXIV/18**: Non-compliance with the MP by Ukraine
MOP 25 (21–25 October 2013) Bangkok	**Decision XXV/10**: Non-compliance with the Montreal Protocol by Azerbaijan **Decision XXV/11**: Non-compliance with the Montreal Protocol by France **Decision XXV/12**: Non-compliance with the Montreal Protocol by Kazakhstan
MOP 26 (17–21 November 2014) Paris	**Decision XXVI/13**: Non-compliance with the Montreal Protocol by Kazakhstan **Decision XXVI/15**: Non-compliance with the Montreal Protocol by the Democratic People's Republic of Korea **Decision XXVI/16**: Non-compliance with the Montreal Protocol by Guatemala
MOP 27 (1–5 November 2015) Dubai	**Decision XXVII/10**: Non-compliance with the Montreal Protocol by Bosnia and Herzegovina **Decision XXVII/11**: Non-compliance with the Montreal Protocol by Libya
MOP 28 (10–15 October 2016) Kigali	**Decision XXVIII/10**: Non-compliance by Israel with its data and information reporting obligations **Decision XXVIII/11**: Non-compliance in 2014 by Guatemala with the provisions of the Montreal Protocol governing consumption of the controlled substances in Annex C, group I (hydrochlorofluorocarbons)
MOP 29 20–24 November 2017 Montreal	**Decision XXIX/14**: Non-compliance in 2015 and 2016 with the provisions of the Montreal Protocol governing consumption of the controlled substance in Annex C, group I (hydrochlorofluorocarbons), by Kazakhstan

Source Prepared on the basis of information/documentation given on the official web site of UNEP/Ozone Secretariat. See: http://conf.montreal-protocol.org/SitePages/Home.aspx

Annex C: Summary on ImplCom Reports on Compliance of the Parties to the Montreal Protocol

ImplCom 3 Geneva 11 April 1992	No cases of non-compliance with the control measures of the Protocol. However, many parties failed to report complete data on time
ImplCom 4 Geneva 14 Sept 1992	Concern was expressed at non-reporting or late reporting by many parties. Although many parties reduced their consumption of controlled substances, some Art. 5 parties increased consumption of controlled substances, particularly halons
ImplCom 5 Geneva 9 March 1993	The status of reporting from the Central and Eastern European Countries (CEECs) was discussed
ImplCom 6 Geneva 26 August 1993	The Committee noted the improvement in the status of reporting and its satisfaction with the reduction of consumption of ODS by many of the parties
ImplCom 7 Bangkok 16–17 Nov 1993	Nearly all non-Art. 5 parties had reported their data for 1991 and most of them had reported their data for 1992. The parties that had consistently failed to provide data concerning their production and consumption of controlled substances as required by the Protocol were also reported
ImplCom 8 Nairobi 4 July 1994	Even though there had been encouraging progress, no data had yet been received from some of the parties (only two parties operating under Art. 5 – Argentina and China – had reported production data)
ImplCom 9 Nairobi 3 October 1994	Due to its problems in implementing a reduction of CFC consumption in the years 1994 and 1995, Poland wished to join the limits for the two years and import the maximum possible in 1994 and 1995 to avoid tensions within Polish industry and the economy. Since Poland had not yet ratified the Copenhagen Amendment, it was not legally bound by Art. 2, para. 3 of the Protocol. So it was decided by the ImplCom that once Poland had ratified the Amendment, it could explain its non-compliance under para. 4 of the NCP
ImplCom 10 Geneva 25 August 1995	ImplCom decided that the statement made by the Russian Federation on behalf of Belarus, Bulgaria, Poland and Ukraine constituted a submission under para. 4 of the NCP
ImplCom 11 Geneva 31 August 1995	Consultations with the other countries (Belarus, Bulgaria, Poland and Ukraine) concerned by the statement of the Russian Federation regarding non-fulfilment of their obligations under the MP
ImplCom 12 27–29 Nov and 1 Dec 1995	Information was submitted by the Russian Federation, Belarus and Ukraine on recycling facilities, statistical data and measures on the phase-out of ozone-depleting substances. The Committee noted the countries from which reports were overdue by more than two years. It stressed that the trend of late reporting should end, particularly in respect of those countries in which institutional strengthening projects have been carried out under the Multilateral Fund

(continued)

(continued)

ImplCom 13 Geneva 18–19 March 1996	The ImplCom noted, on Belarus, the Russian Federation and Ukraine, that while the information available showed a situation of non-compliance for 1996 for these parties, they had taken important steps in complying with relevant MOP Decisions (Decision VII/17, 18, 19) and towards achieving full compliance with the control measures of the Protocol. The secretariat received a letter in December 1995 from the Governments of Estonia, Latvia and Lithuania requesting a longer time frame for phasing out ozone-depleting substances because of the institutional and financial problems facing their countries. The secretariat had advised Estonia that the NCP was only applicable to parties, and Latvia and Lithuania to make a formal submission under para. 4 of the NCP. In response, Lithuania had submitted a plan of action to phase out ozone depleting substances, while no response was received from Latvia
ImplCom 14 Geneva 23 August 1996	• On Latvia and Lithuania: The ImplCom noted that Latvia and Lithuania would be in a situation of non-compliance with the Montreal Protocol in 1996 and that there was a possibility of non-compliance by them also in 1997, despite their efforts to meet its obligations under the Protocol, even in the absence of external financial assistance for investment projects • On the Russian Federation: it also noted that it was in a situation of non-compliance with the Protocol in 1996, despite making considerable progress, so the situation regarding the phase-out of ozone-depleting substances should be kept under review and the disbursement of financial assistance for ODS-phase-out in the country should continue to be contingent on developments with regard to non-compliance and the settlement of any problems with the ImplCom
ImplCom 15 San José 18 November 1996	• On Belarus: Belarus was in a situation of non-compliance with the Protocol in 1996, and it ratified the London Amendment. • On Bulgaria: the information provided by the Government of Bulgaria and its efforts to implement the Protocol fully • On Poland: Its ratification of both the London and Copenhagen Amendments to the Montreal Protocol. • On Ukraine: It was in the process of ratifying the London Amendment. It had established a final phase-out for ODS in 1999, and had drafted provisions to license the import and export of ODS. It had submitted data on consumption and production of ODS up to 1995. It had phased out the production of carbon tetrachloride, the only ODS produced in Ukraine, by 1994 • On Latvia and Lithuania: draft decisions VIII/19 and VIII/20 regarding non-compliance with the Montreal Protocol by Latvia and Lithuania, respectively • On the Russian Federation: draft decision VIII/21 regarding non-compliance with the MP by the Russian Federation • On the Czech Republic: non-compliance by the Czech Republic with the halons reduction schedule for 1994, as production of halons in 1994 was in violation of the control measures established in the amended Montreal Protocol. No further action was necessary regarding non-compliance in 1994, in view of the total phase-out in 1995

(continued)

(continued)

ImplCom 16 San José 20 November 1996	• On the Russian Federation: according to its written submissions and the statements of the representative of the Russian Federation at the ImplCom 13, 14, 15 and 16, the Russian Federation was in a situation of non-compliance with the Montreal Protocol in 1996 • On the Czech Republic: no further action necessary in view of the Czech Republic's complete phase-out of halon consumption according to the data submitted to the secretariat pursuant to Art. 7 of the Montreal Protocol for 1995
ImplCom 17 Geneva 15–16 April 1997	• On Latvia, Lithuania: the situation regarding ODS phase-out in Lithuania should be kept under review • On Poland: Poland's consumption of CFCs in 1996 was below the level of the essential-use exemptions granted it by the MOP 6 • On the Russian Federation: the Russian Federation was in non-compliance with the Protocol for 1996; it had continued to produce ODS during 1996, contrary to the provisions of the Montreal Protocol and decision VII/18; it had exported both new and reclaimed substances to, and also imported ODS from, many arties operating under Art. 5 and those parties not operating under that Art.
ImplCom 18 Nairobi 2–4 June 1997	• On Latvia and Lithuania: both urged to ratify the London Amendment by October 1997 as indicated in their timetable; they are in a situation of non-compliance with the MP in 1997 and there is a possibility of non-compliance in 1998 in light of the commitment reflected in their country programmes and related official communications to the parties; international assistance, particularly by the GEF, should be considered favourably in order to provide funding for projects to implement their programme for phasing out ozone-depleting substances in the countries; the situation with regard to ODS phase-out to be kept under review • On the Russian Federation: the Russian Federation was in a situation of non-compliance with the Protocol in 1996 and there is an expectation of non-compliance in 1997. The situation regarding the phase-out of ozone-depleting substances in the Russian Federation to be kept under review
ImplCom 19 Montreal 8–10 Sept 1997	• On the Russian Federation: international assistance, particularly by the GEF, should continue to be considered favourably in order to provide funding for the Russian Federation for projects to implement the programme for the phase-out of the production and consumption of ozone-depleting substances in the country • On the Czech Republic: although the 1995 imports of methyl bromide exceeded the freeze level of 6.0 ODP-tonnes for the Czech Republic, the average annual consumption for the two years 1995 and 1996 was below that level; no action should be taken on this non-compliance, but the Czech Republic was to be asked to ensure that similar cases did not occur again
ImplCom 20 Geneva 6–7 July 1998	• On Latvia and Lithuania: both had been requested to implement decisions IX/29 and IX/30. Lithuania had ratified both the London and Copenhagen Amendments, but Latvia did not complete the ratification of the London Amendment. Both were in non-compliance in 1996 with the control measures in Art. 2 of the Protocol

(continued)

(continued)

	• The secretariat's report listed a number of parties whose data suggested they were in non-compliance in 1996 with the control measures in Art. 2 of the Protocol (Azerbaijan, Belarus, Czech Republic, Estonia, Russian Federation, Ukraine, Uzbekistan). Their explanations regarding their non-compliance in 1996 and their plans to phase out ozone-depleting substances were outlined by the secretariat, and they were requested to attend the ImplCom 20 to provide further information and discuss the matter in detail
ImplCom 21 Cairo 16 November 1998	• The Committee agreed to recommend decisions (a–i) (Azerbaijan (a), Belarus (b), the Czech Republic (c), Estonia (d), Latvia (e), Lithuania (f), Russian Federation (g), Ukraine (h) and Uzbekistan (i)) contained in Annex I to the present report, for adoption by the MOP. • Annex I: Draft Decisions Submitted by the Implementation Committee For the Consideration of the meeting of the parties
ImplCom 22 Geneva 14 June 1999	• On Bulgaria: the Committee noted that Bulgaria had apparently been in non-compliance with the Protocol in 1996 and 1997, although it appeared to have been in compliance in 1998 • On Turkmenistan: the Committee noted that Turkmenistan had apparently been in non-compliance with the Protocol in 1996 and 1997, although it appeared to have been in compliance in 1998; consideration of the updated data report for 1996, 1997 and 1998 under Art. 7 of the protocol
ImplCom 23 Beijing 27 Nov 1999	Consideration of information relating to any situations of non-compliance by some parties as well as their statements and adoption of any recommendations to the parties at their 11th meeting (Azerbaijan, Bulgaria, Latvia, the Russian Federation, Turkmenistan, Ukraine)
ImplCom 24 Geneva 10 July 2000	Some parties operating under Art. 5 had not reported some or all data for 1995, 1996 and 1997, so the secretariat was unable to determine the baseline for phase-out for Annex A substances. The Committee decided to ask the secretariat to send a letter to Estonia to alert that country to the deviation from the reduction schedule and request clarification on why the benchmark had not been met for Annex A and B substances. As a satisfactory explanation was not given for the deviation from the consumption reduction schedule in the cases of Israel, Kazakhstan and Turkmenistan, the Committee decided to ask the secretariat to send letters to those countries requesting an explanation. Report of the secretariat on compliance and on the follow-up on the recommendations of the previous meetings of the Impl Com Evaluation of Data Reported and Policies Adopted by Art. 5, Parties to Achieve Compliance With the Initial Control Measures of the MP Fund Secretariat Report on CEIT-GEF Secretariat. Progress in establishing licensing systems, regulations and policies – UNEP OzonAction Programme
ImplCom 25 OuagadougoBurkinaFaso 9 Dec 2000	• The observer from Israel informed the Committee that, while it was true that a 3 per cent increase in methyl bromide consumption occurred in 1998, in 1999 the country had achieved a reduction of 30 per cent, as against the commitment of 25 per cent. Israel had also introduced a licensing system and was considering further

(continued)

(continued)

	regulations to control the handling and price of methyl bromide to prevent profiteering. He also assured the Committee of full compliance by Israel in future • The Russian Federation appealed to the parties for understanding after failing to complete the closure of ODS production facilities by June 2000 as agreed. The Committee noted that an agreement had been reached to halt all production of CFCs by 20 December 2000, and looked forward to receiving a report confirming the halt at its next meeting. With respect to those countries with economies in transition, specifically Armenia, Kazakhstan, Kyrgyzstan and Tajikistan, that had not yet submitted their phase-out plans to the parties, the secretariat was asked to request them to submit their plans to be formulated in cooperation with the implementing agencies. The Committee agreed to recommend to the MOP that it should request Bosnia and Herzegovina and Togo, which were classified as Art. 5 countries, to provide an explanation in time for consideration by the Committee at its 26th meeting, as they had passed every deadline for the submission of baseline data
ImplCom 26 Montreal 23 July 2001	• Status of compliance with Decisions X/20 (Azerbaijan); X/21 (Belarus); X/22 (Czech Republic); X/23 (Estonia); X/24 (Latvia); X/25 (Lithuania); X/26 (Russian Federation); X/27 (Ukraine); and X/28 (Uzbekistan). The Committee agreed that it was important to determine the underlying country-specific reasons for any party's inability to achieve compliance and also that it was important to take into account whether a party was persistently in a state of non-compliance, or had only recently agreed phase-out benchmarks with either the parties or the GEF • Other countries with economies in transition: Armenia: To alert Armenia to its situation of potential non-compliance, and request explanatory information about its consumption figures. Bulgaria: To alert Bulgaria to its situation of potential non-compliance, and request explanatory information about its consumption figures • Potential non-compliance of Art. 5 parties: arising out of the data report, as well as the status of compliance with decisions of the parties, and recommendations of the Implementation Committee • Since the first control period for Art. 5 phase-out of CFCs was 1 July 1999–30 June 2000, it was not possible to tell for certain whether any of these parties was actually in non-compliance, but it was reasonable to assume that at least those parties reporting excess consumption in both 1999 and 2000 were potentially in non-compliance
ImplCom 27 Colombo 13 Oct 2001	The secretariat had been requested to write letters to several parties alerting them to their state of potential non-compliance

(continued)

(continued)

ImplCom 28 Montreal 20 July 2002	Review of the status of compliance with decisions of the parties and recommendations of the implementation committee on non-compliance issues: Argentina, Armenia, Bangladesh, Belize, Bulgaria, Cameroon, Chad, Comoros, Dominican Republic, Ethiopia, Honduras, Kazakhstan, Kenya, Mongolia, Morocco, Niger, Nigeria, Oman, Papua New Guinea, Paraguay, Peru, Russian Federation, Samoa, Solomon Islands, Tajikistan, Yemen
ImplCom 29 Rome 23–24 Nov 2002	Review of the status of compliance with specific decisions of the parties by Bangladesh, Chad, Comoros, Honduras, Mongolia, Niger, Nigeria, Oman, Papua New Guinea, Paraguay, Samoa (Decision XIII/16), Russian Federation (Decision XIII/17), Armenia (Decision XIII/18), Kazakhstan (Decision XIII/19), Tajikistan (Decision XIII/20), Belize (Decision XIII/22), Cameroon (Decision XIII/23), Ethiopia (Decision XIII/24), Peru (Decision XIII/25)
ImplCom 30 Montreal 4–7 July 2003	Review of the status of compliance with specific decisions of the Parties on: (a) Non-compliance with data reporting requirements under Art. 7 of the MP by parties temporarily classified as operating under Art. 5 of the Protocol concerning (b) Non-compliance with data reporting requirement for the purpose of establishing baselines under Art. 5, paras. 3 and 8 (d) concerning (c) Non-compliance with specific decisions
ImplCom 31 Nairobi 5–7 Nov 2003	Review of the status of compliance with specific previous decisions of the parties and recommendations by the ImplCom: (a) Non-compliance with data reporting requirements under Art. 7 of the MP by parties temporarily classified as operating under Art. 5 of the Protocol, (b) Non-compliance with data reporting requirements for the purpose of establishing baselines under Art. 5, paras. 3 and 8 (d), (c) Review of the previous recommendations by the Implementation Committee and new information on specific parties
ImplCom 32 Geneva 17–18 July 2004	Review of the status of non-compliance with specific decisions of the parties and recommendations of the Implementation Committee: (a) Non-compliance with data-reporting requirements for one or more of the base years (1986, 1989 or 1991) for one or more groups of controlled substances under Art. 7 of the MP by parties operating under Art. 5 of the Protocol, (b) Non-compliance with data reporting requirements under Art. 7 of the MP by Parties temporarily classified as operating under Art. 5 of the Protocol, (c) Non-compliance with data reporting requirements for the purpose of establishing baselines under Art. 5, paras. 3 and 8 (d), (d) Potential non-compliance with consumption of substances in Annex A, Groups I and II (CFCs and halons), Annex C, Group II (hydrobromofluorocarbons) and Annex E (methyl bromide) (decisions XV/21, XV/22 and XV/25), (e) Review of compliance with specific decisions by individual parties

(continued)

(continued)

ImplCom 33 Prague 17–19 Nov 2004	• Review of the status of compliance with specific decisions of the parties and recommendations of the Implementation Committee on non-compliance: (a) Parties required to limit their consumption of ozone-depleting substances according to the agreed benchmarks applicable for 2003, (b) Follow-up on previous decisions and recommendations for individual parties, (c) Follow-up on recommendations by the Implementation Committee for groups of parties • Consideration of compliance issues arising out of the data report (a) Data reporting; (b) Compliance with control measures
ImplCom 34 Montreal 2 July 2005	Review of the status of compliance with specific decisions of the parties on non-compliance: (a) Non-compliance with data reporting for 2003 (Decision XVI/17), (b) Non-compliance with data reporting for Parties temporarily classified as operating under Art. 5, para. 1, of the MP (Decision XVI/18), (c) Follow-up on previous decisions requesting parties to submit explanations or plans of action for their return to compliance, (d) Follow-up on previous decisions of the parties and recommendations of the Implementation Committee regarding compliance by parties with commitments contained in their approved plans of action for their return to compliance with the MP
From ImplCom 35 Dakar 7–9 Dec 2005 to ImplCom 49 Geneva 8–9 Nov 2012	The ImplCom has usually adopted the following agenda related to the compliance issue, which can change in some degree in some cases • Report by the secretariat on data and information under Arts. 7 and 9 of the MP and on related issues • Information provided by the secretariat for the implementation of the MP on relevant decisions of the Executive Committee of the Fund and on activities carried out by implementing agencies (the UNDP, UNEP, UNIDO and the World Bank) to facilitate compliance by parties • Follow-up on previous decisions of the parties and recommendations of the ImplCom on non-compliance-related issues (a) Data-reporting obligations (b) Existing plans of action to return to compliance (c) Draft plans of action to return to compliance (d) Other recommendations and decisions on compliance (e) Plans of action for the establishment and operation of licensing systems for ozone-depleting substances (for example, at ImplCom 42) (f) Requests for changes to baseline data (for example, at ImplCom 41, ImplCom 49) • Consideration of other non-compliance issues arising out of the data report • Consideration of the report of the secretariat on parties that have established licensing systems (Art. 4B, para. 4, MP) • Information on compliance by parties present at the invitation of the ImplCom

(continued)

(continued)

ImplCom 50
Bangkok
21–22 June 2013
	• Follow-up on previous decisions of the parties and recommendations of the Implementation Committee on non-compliance-related issues: (a) Data-reporting obligations: (i) Mali (Decision XXIV/13); (ii) Sao Tome and Principe (Decision XXIV/13); (b) Existing plans of action to return to compliance: (i) Ecuador (Decision XX/16); (ii) Uruguay (Decision XVII/39)
	• Consideration of other possible non-compliance issues arising from the data report
	  – Recommendation 50/8 regarding Azerbaijan's excess HFC consumption in 2011
	  – Recommendation 50/9 regarding France's excess HFC consumption in 2011
	  – Recommendation 50/10 regarding Israel's reporting of process agent uses
	  – Recommendation 50/11 regarding Kazakhstan's HCFC and methyl bromide consumption reduction
	• Consideration of the report of the secretariat on the establishment of licensing systems: status of establishment of licensing systems under Art. 4B of the Montreal Protocol (Decision XXIV/17)
	  – Recommendation 50/12 on Botswana, Gambia, South Sudan

ImplCom 51
Bangkok
18–19 Oct 2013
	• Follow-up on previous decisions of the parties and recommendations of the Implementation Committee on non-compliance-related issues: (a) Data-reporting obligations: Israel (recommendation 50/10); (b) Existing plans of action to return to compliance: Ecuador (Decision XX/16 and recommendation 50/1); (c) Other recommendations and decisions on compliance: (i) Azerbaijan (Recommendation 50/8); (ii) Kazakhstan (Recommendation 50/11)
	• Consideration of other possible non-compliance issues arising out of the data report
	  – Recommendation 51/5: The Central African Republic, Eritrea, Gabon, Israel, Jordan, Kazakhstan, Kuwait, Latvia, Liechtenstein, Saint Kitts and Nevis, South Sudan, Switzerland, the Syrian Arab Republic, Uzbekistan and Yemen were still in non-compliance with the requirement
	• Consideration of additional information on compliance-related submissions by parties participating in the meeting at the invitation of the Implementation Committee
	  – The Committee considered information provided by the representative of Azerbaijan

ImplCom 52 Paris 9–10 July 2014	• Follow-up on previous decisions of the parties and recommendations of the Implementation Committee on non-compliance-related issues: (a) Data-reporting obligations: (i) Eritrea (Decision XXV/14); (ii) South Sudan (Decision XXV/14); (iii) Yemen (Decision XXV/14); (b) Existing plans of action to return to compliance; (i) Ecuador (Decision XX/16); (ii) Ukraine (Decision XXIV/18); (ii) Uruguay (Decision XVII/39); (c) Other recommendations and decisions on compliance: Kazakhstan (Decision XXV/12) • Possible non-compliance with hydrochlorofluorocarbon phase-out by the Democratic People's Republic of Korea and request for assistance

(continued)

(continued)

	• Consideration of other possible non-compliance issues arising out of the data report/Recommendation 52/4 • Status of establishment of licensing systems under Art. 4B of the Montreal Protocol by Botswana and South Sudan (Decision XXV/15)/Recommendation 52/5 • Consideration of additional information on compliance-related submissions by parties participating in the meeting at the invitation of the Implementation Committee – The Committee considered information provided by the representative of Kazakhstan
ImplCom 53 Paris 14–15 Nov 2014	• Follow-up on previous decisions of the parties and recommendations of the Implementation Committee on non-compliance-related issues: (a) Data-reporting obligations: (i) Ukraine (Decision XXIV/18), (ii) South Sudan (Decision XXV/14 and Recommendation 52/1); (b) Other recommendations and decisions on compliance: Israel (Recommendation 52/4) • Non-compliance with hydrochlorofluorocarbon phase-out by the Democratic People's Republic of Korea and request for assistance • Consideration of other possible non-compliance issues arising out of the data report – Non-reporting of 2013 process agent uses by Israel (Recommendation 53/4) – Non-compliance with control measures under the Protocol: Guatemala (Recommendation 53/5) • Status of establishment of licensing systems under Article 4B of the Montreal Protocol (Decision XXV/15 and Recommendation 52/5 on Botswana and South Sudan) • Consideration of additional information on compliance-related submissions by parties participating in the meeting at the invitation of the Implementation Committee – The Committee considered information provided by the representatives of the Democratic People's Republic of Korea and South Sudan
ImplCom 54 Paris 27–28 July 2015	• Follow-up on previous decisions of the parties and recommendations of the Implementation Committee on non-compliance-related issues: (a) Data- and information-reporting obligations: (i) Central African Republic (Decision XXVI/12); (ii) Israel (Recommendation 53/4); (b) Existing plans of action to return to compliance: (i) Democratic People's Republic of Korea (Decision XXVI/15); (ii) Ecuador (Decision XX/16); (iii) Guatemala (Decision XXVI/16); (iv) Kazakhstan (Decision XXVI/13); (v) Ukraine (Decision XXIV/18) • Possible non-compliance with: (a) Bosnia and Herzegovina: hydrochlorofluorocarbon phase-out; (b) Libya: hydrochlorofluorocarbon phase-out; (c) South Africa: methyl chloroform phase-out • Status of establishment of licensing systems under Article 4B of the Montreal Protocol by Botswana and South Sudan (Decision XXV/15 and Recommendation 53/6)

(continued)

(continued)

ImplCom 55 Dubai, United Arab Emirates 28 October 2015	• Follow-up on previous decisions of the parties and recommendations of the Implementation Committee on non-compliance-related issues: existing plans of action to return to compliance: (a) Kazakhstan (Decision XXVI/13 and Recommendation 54/2); (b) Ukraine (Decision XXIV/18 and Recommendation 54/3) • Possible non-compliance with control measures under the Protocol: (a) Libya: hydrochlorofluorocarbon phase-out (Recommendation 54/5); (b) South Africa: methyl chloroform phase-out (Recommendation 54/6) • Consideration of other possible non-compliance issues arising out of the data report (Recommendations 55/3–55/4) • Status of establishment of a licensing system under Article 4B of the Montreal Protocol by South Sudan (Decision XXV/15 and Recommendation 54/8) • Consideration of additional information on compliance-related submissions by parties participating in the meeting at the invitation of the Implementation Committee – The representative of Libya had been invited to attend but had been unable to do so
ImplCom 56 Vienna 24 July 2016	• Follow-up on previous decisions of the parties and recommendations of the Implementation Committee on non-compliance-related issues: (a) Data-reporting obligations (Decision XXVII/9): (i) Democratic Republic of Congo; (ii) Dominica; (iii) Somalia; (iv) Yemen; (b) Existing plans of action to return to compliance: (i) Democratic People's Republic of Korea (Decision XXVI/15); (ii) Kazakhstan (Decision XXVI/13); (iii) Libya (Decision XXVII/11); (iv) Ukraine (Decision XXIV/18); (c) Non-reporting of process agent uses for 2014: Israel (Recommendation 55/4); (d) Possible non-compliance with the hydrochlorofluorocarbon phase-out (control measures): Guatemala • Consideration of other possible non-compliance issues arising out of the data report: Israel's reporting excess production of 17.3 ODP-tonnes of bromochloromethane in 2014 (Recommendation 56/7) • Status of establishment of a licensing system under Article 4B of the Montreal Protocol by South Sudan (Decision XXV/15 and Recommendation 55/5)
ImplCom 57 Kigali 9 October 2016	• Follow-up on previous decisions of the parties and recommendations of the Implementation Committee on non-compliance-related issues: (a) Yemen: Data-reporting obligations (Decision XXVII/9 and Recommendation 56/1); (b) Existing plans of action to return to compliance: (i) Kazakhstan (Decision XXVI/13 and Recommendation 56/2); (ii) Libya (Decision XXVII/11 and Recommendation 56/3); (iii) Ukraine (Decision XXIV/18 and Recommendation 56/4); (c) Israel: Non-reporting of process agent uses for 2014 (Recommendation 56/5) and excess production of bromochloromethane (Recommendation 56/7) • Consideration of additional information on compliance-related submissions by parties participating in the meeting at the invitation of the Implementation Committee

(continued)

(continued)

	– The representative of Fiji attended part of the meeting to present information regarding his Government's request for a change to its HCFC consumption baseline data
ImplCom 58 Bangkok 9 July 2017	Follow-up on previous decisions of the parties and recommendations of the Implementation Committee on non-compliance-related issues: (a) Data-reporting obligations (Decision XXVIII/9); (b) Existing plans of action to return to compliance: (i) Democratic People's Republic of Korea (Decision XXVI/15); (ii) Kazakhstan (Decision XXVI/13 and Recommendation 57/1); (iii) Libya (Decision XXVII/11); (iv) Ukraine (Decision XXIV/18 and Recommendation 57/2); (c) Israel: data- and information-reporting obligations (Decision XXVIII/10) • Request by Fiji for a change to its baseline data for hydrochlorofluorocarbons (UNEP/OzL.Pro/ImpCom/57/4, para. 56) • Consideration of other possible non-compliance issues arising out of the data report • Consideration of additional information on compliance-related submissions by parties participating in the meeting at the invitation of the Implementation Committee – The representatives of Israel and Kazakhstan attended the meeting at the invitation of the ImplCom to present further information on their compliance issues under agenda items 5(c) and 5(b)(ii) respectively

Source Prepared on the basis of information/documentation given on the official website of UNEP/Ozone Secretariat. See: http://conf.montreal-protocol.org/SitePages/Home.aspx

Annex D: MOP Decisions on the Compliance of All the Related Parties to the Montreal Protocol

Parties categorized as operating under Art. 5, para. 1 of the MP	Related decisions	Parties not categorized as operating under Art. 5, para. 1 of the MP	Related decisions
Albania	Decision XIV/18 Decision XV/26	Azerbaijan	Decision X/20 Decision XV/28 Decision XVI/21 Decision XVII/26 Decision XXV/10
Argentina	Decision XIII/21	Belarus	Decision VII/17 Decision X/21 Decision XIV/28
Armenia[a]	Decision XIII/18 Decision XIV/31 Decision XV/27 Decision XVII/25 Decision XVIII/20	Bulgaria	Decision VII/16 Decision XI/24

(continued)

(continued)

Parties categorized as operating under Art. 5, para. 1 of the MP	Related decisions	Parties not categorized as operating under Art. 5, para. 1 of the MP	Related decisions
Bahamas	Decision XIV/19	Czech Republic	DecisionVIII/24 Decision IX/32 Decision X/22
Bangladesh	DecisionXIV/29 Decision XVII/27 Decision XXI/17	Estonia	Decision X/23
Belize	Decision XIII/22 Decision XIV/33	European Union	Decision XXIII/26
Bolivia	Decision XIV/20 Decision XV/29	Greece	Decision XVIII/25 Decision XIX/21
		Iceland	Decision XXVIII/9
Bosnia and Herzegovina	Decision XIV/21 Decision XV/30 Decision XVII/28 Decision XXI/18 Decision XXVII	Israel	Decision XXIII/7 Decision XXVIII/10
Botswana	Decision XV/31 Decision XXV/15 Decision XXIV/17	Kazakhstan	Decision XIII/19 Decision XVII/35 Decision XXVI/13 Decision XXV/12 Decision XXIX/14
Cameroon	Decision XIII/23 Decision XIV/32 Decision XV/32	Latvia	Decision VIII/22 Decision IX/29 Decision X/24 Decision XIV/28
Central African Republic	Decision XXVI/12	Lithuania	Decision VIII/23 Decision IX/30
Chile	Decision XVI/22 Decision XVII/29		Decision X/25
China	Decision XVII/30	Poland	Decision VII/15
Democratic Republic of the Congo	Decision XV/33 Decision XVIII/21 Decision XXVI/15 Decision XXVII/9	Russian Federation	Decision VII/18 Decision VIII/25 Decision IX/31 Decision X/26 Decision XIII/17 Decision XIV/35 Decision XXIII/27
Dominica	Decision XXVII/9 Decision XVIII/22	Tajikistan	Decision XIII/20
Ecuador	Decision XVII/31 Decision XVIII/23 Decision XX/16	Ukraine	Decision VII/19 Decision X/27 Decision XXIV/18
Ethiopia	Decision XIII/24 Decision XIV/34	Uzbekistan	Decision X/28

(continued)

(continued)

Parties categorized as operating under Art. 5, para. 1 of the MP	Related decisions	Parties not categorized as operating under Art. 5, para. 1 of the MP	Related decisions
Eritrea	Decision XVIII/24		
Federated States of Micronesia	Decision XVII/32 Decision XXI/19		
Fiji	Decision XVI/23 Decision XVII/33		
Guatemala	Decision XV/34 Decision XXVIII/11 Decision XXVI/16		
Guinea-Bissau	Decision XVI/24 Decision XXVI/13		
Honduras	Decision XV/35 Decision XVII/34		
Islamic Rep. of Iran	Decision XIX/27		
Kenya	Decision XVIII/28		
Kyrgyzstan	Decision XVII/36		
Lesotho	Decision XVI/25		
Libya	Decision XIV/25 Decision XV/36 Decision XVI/26 Decision XVII/37 Decision XXIII/23 Decision XXVII/11		
Maldives	Decision XIV/26 Decision XV/37		
Mali	Decision XXIV/13		
Mexico	Decision XVIII/30 Decision XXI/20		
Namibia	Decision XIV/22 Decision XV/38		
Nepal	Decision XIV/23 Decision XV/39 Decision XVI/27		
Nigeria	Decision XIV/30		
Oman	Decision XVI/28		
Pakistan	Decision XVI/29 Decision XVIII/31		
Paraguay	Decision XVIII/32 Decision XIX/22		
Papua New Guinea	Decision XV/40		
Peru	Decision XIII/25		
Republic of Korea	Decision XXII/16		
Sao Tome and Principe	Decision XXIV/13		
Saudi Arabia	Decision XXI/21		

(continued)

(continued)

Parties categorized as operating under Art. 5, para. 1 of the MP	Related decisions	Parties not categorized as operating under Art. 5, para. 1 of the MP	Related decisions
	Decision XXII/15		
Serbia	Decision XVIII/33		
Solomon Islands	Decision XXI/22		
Somalia	Decision XX/19 Decision XXI/23 Decision XXVII/9		
Sudan	Decision XXV/15 Decision XXIV/17		
Qatar	Decision XV/41		
Saint Lucia	Decision XXV/13		
Saint Vincent and the Grenadines	Decision XV/42 Decision XVI/3 Decision XIV/24		
Sierra Leone	Decision XVII/38		
Singapore	Decision XXII/13		
Turkmenistan[b]	Decision XI/25 Decision XXI/25		
Uganda	Decision XV/43		
Uruguay	Decision XV/44 Decision XVII/39		
Vanuatu	Decision XXI/26 Decision XXII/18		
Vietnam	Decision XV/45		
Yemen	Decision XXVII/9 Decision XXVIII/9		

Source Prepared on the basis of information/documentation given on the official website of UNEP/Ozone Secretariat. See: http://conf.montreal-protocol.org/SitePages/Home.aspx
[a]Armenia was reclassified as a developing country operating under Art. 5 through Decision XIV/2, MOP 14 (2002)
[b]Turkmenistan was reclassified as a developing country operating under Art. 5 through Decision XVI/39, MOP 17 (2005)

Annex E: MOP Decisions on the Compliance of Countries with Economies in Transition (CEIT) to the Montreal Protocol

MOP 9 (15–17 Sept 1997)	**Decision IX/32**: Non-compliance by the Czech Republic with the freeze in consumption of methyl bromide in 1995
MOP 10 (23–24 Nov 1998)	**Decisions X/20, 22, 23, 28**: Compliance by Azerbaijan, the Czech Republic, Estonia, Uzbekistan
MOP 13 (16–19 Oct 2001)	**Decision XIII/19**: Compliance by Kazakhstan
MOP 15 (10–14 Nov 2003)	**Decisions XV/28**: Compliance by Azerbaijan
MOP 16 (22–26 Nov 2004)	**Decision XVI/21**: Compliance by Azerbaijan
MOP 17 (12–16 Dec 2005)	**Decision XVII/35**: Potential non-compliance in 2004 with the controlled substances in Annex A, group I (CFCs) by Kazakhstan **Decision XVII/ 26**: Compliance by Azerbaijan
MOP 18 (30 Oct–3 Nov 2006)	**Decision XVIII/34**: Compliance by Uzbekistan
MOP 25 (21–25 Oct 2013)	**Decision XXV/10, 12**: Compliance by Azerbaijan, Kazakhstan
MOP 26 (17–21 Nov 2014)	**Decision XXVI/13**: Compliance by Kazakhstan
MOP 29 (20–24 Nov 2017)	**Decision XXIX/14**: Compliance by Kazakhstan

Source Prepared on the basis of information/documentation given on the official website of UNEP/ Ozone Secretariat. See: http://conf.montreal-protocol.org/SitePages/Home.aspx

Annex F: MOP Decisions on the Compliance of Developing Country Parties to the Montreal Protocol

MOP 13 (16–19 October 2001) Colombo	**Decision XIII/16**: Potential non-compliance with the freeze on CFC consumption in Art. 5 parties in the control period 1999–2000 **Decisions XIII/21–25**: Compliance by Argentina, Belize, Cameroon, Ethiopia, Peru
MOP 14 (25–29 November 2002) Rome	**Decision XIV/17**: Potential non-compliance with the freeze on CFC consumption by parties operating under Art. 5 for the control period July 2000 to June 2001 **Decisions XIV/18–26**: Compliance by Albania, Bahamas, Bolivia, Bosnia and Herzegovina, Namibia, Nepal, Saint Vincent and the Grenadines, Libyan Arab Jamahiriya and Maldives
MOP 15 (10–14 November 2003) Nairobi	**Decision XV/21**: Potential non-compliance with consumption of Annex A, group I, CFCs by Art. 5 parties for the control period 1 July 2001–31 December 2002 **Decision XV/22**: Potential non-compliance with consumption of Annex A, group II, halons by Article 5 parties in 2002 **Decision XV/23**. Potential non-compliance with consumption of the Annex C, group II, hydrobromo fluorocarbons by Morocco in 2002 **Decision XV/25**. Potential non-compliance with consumption of the methyl bromide by Art. 5 parties in 2002 **Decisions XV/26, 27, 29–45**: Compliance by Albania, Armenia,[a] Bolivia, Bosnia and Herzegovina, Botswana, Cameroon, the Democratic Republic of the Congo, Guatemala, Honduras, the Libyan Arab Jamahiriya, Maldives, Namibia, Nepal, Papua New Guinea, Qatar, Saint Vincent and the Grenadines, Uganda, Uruguay, Vietnam
MOP 16 (22–26 November 2004) Prag	**Decision XVI/19**: Potential non-compliance with consumption of Annex A, group II, halons by Somalia in 2002 and 2003 **Decision XVI/20**: Potential non-compliance in 2003 with consumption of the controlled substance in Annex B, group III (methyl chloroform) by parties operating under Art. 5 **Decisions XVI/22–30**: Compliance by Chile, Fiji, Guinea-Bissau, Lesotho, Libyan Arab Jamahiriya, Nepal, Oman, Pakistan, Saint Vincent and the Grenadines
MOP 17 (12–16 December 2005) Dakar	**Decision XVII/30**: Potential non-compliance in 2004 with consumption of the controlled substances in Annex B group I (other halogenated chlorofluorocarbons) by China **DecisionXVII/35**: Potential non-compliance in 2004 with the controlled substances in Annex A, group I (CFCs) by Kazakhstan **Decisions XVII/25, 27, 28, 29, 31–33, 36–38**: Compliance by Armenia, Bangladesh, Bosnia and Herzegovina, Chile, Ecuador, Federated States of Micronesia, Fiji, Kyrgyzstan, the Libyan Arab Jamahiriya and Sierra Leone
MOP 18 (30 October–3 November 2006) New Delhi	**Decision XVIII/24**: Potential non-compliance in 2005 with the control measures of the MP governing consumption of the controlled substances in Annex A, group I (CFCs) by Eritrea

(continued)

(continued)

	Decision XVIII/27: Potential non-compliance in 2005 with the control measures of the MP governing consumption of the controlled substance in Annex B group II (carbon tetrachloride) by the Islamic Republic of Iran Decisions XVIII/20–24, 28, 30–33: Compliance by Armenia, the Democratic Republic of the Congo, Dominica, Ecuador, Eritrea, Kenya, Mexico, Pakistan, Paraguay, Serbia
MOP 19 (17–21 September 2007) Montreal	Decision XIX/23: Potential non-compliance in 2005 with the provisions of the MP governing consumption of the controlled substance in Annex E (methyl bromide) by Saudi Arabia Decisions XIX/22, 27: Compliance by Paraguay, the Islamic Republic of Iran
MOP 20 (16–20 November 2008) Doha	Decision XX/18: Potential non-compliance in 2006 with the provisions of the MP in respect of consumption of the controlled substances in Annex A, group I (chlorofluorocarbons) by Solomon Islands Decisions XX/16, 19: Compliance by Ecuador, Somalia
MOP 21 (4–8 November 2009) Port Ghalib	Decisions XXI/17–23, 25–26: Compliance by Bangladesh, Bosnia and Herzegovina, the Federated States of Micronesia, Mexico, Saudi Arabia, Solomon Islands, Somalia, Turkmenistan,[b] Vanuatu
MOP 22 (8–12 November 2010) Bangkok	Decisions XXII/13, 15, 16, 18: Compliance by Singapore, Saudi Arabia, the Republic of Korea and Vanuatu
MOP 23 (21–25 November 2011) Bali	Decision XXIII/23: Potential non-compliance in 2009 with the provisions on consumption of the controlled substances in Annex A, group II (halons) by Libya
MOP 26 17–21 November 2014 Paris	Decision XXVI/14: Requests for the revision of baseline data by Libya and Mozambique Decision XXVI/15: Non-compliance with the Montreal Protocol by the Democratic People's Republic of Korea Decision XXVI/16: Non-compliance with the Montreal Protocol by Guatemala
MOP 27 1–5 November 2015 Dubai	Decision XXVII/10: Non-compliance with the Montreal Protocol by Bosnia and Herzegovina. Decision XXVII/11: Non-compliance with the Montreal Protocol by Libya
MOP 28 10–15 October 2016 Kigali	Decision XXVIII/11: Non-compliance in 2014 by Guatemala with the provisions of the Montreal Protocol governing consumption of the controlled substances in Annex C, group I (hydrochlorofluorocarbons)

Source Prepared on the basis of information/documentation given on the official website of UNEP/ Ozone Secretariat. See: http://conf.montreal-protocol.org/SitePages/Home.aspx
[a]Armenia was reclassified as a developing country operating under Art. 5 through Decision XIV/2 (MOP 14, 2002)
[b]Turkmenistan was reclassified as a developing country operating under Art. 5 through Decision XVI/39 (MOP 17, 2005)

Annex G: The List of COPs to the UNFCCC and MOPs to the Kyoto Protocol

Session no.	Date		Location
COP 1	28 March–7 April 1995		Berlin, Germany
COP 2	8–19 July 1996		Geneva, Switzerland
COP 3	1–11 December 1997		Kyoto, Japan
COP 4	2–14 November 1998		Buenos Aires, Argentina
COP 5	25 October–5 November 1999		Bonn, Germany
COP 6	Part I	13–25 November 2000	The Hague, The Netherlands
	Part II	16–27 July 2001	Bonn, Germany
COP 7	29 October–10 November 2001		Marrakesh, Morocco
COP 8	23 October–1 November 2002		New Delhi, India
COP 9	1–12 December 2003		Milan, Italy
COP 10	6–18 December 2004		Buenos Aires, Argentina
COP 11 (MOP 1)	28 November–10 December 2005		Montreal, Canada
COP 12 (MOP 2)	6–17 November 2006		Nairobi, Kenya
COP 13 (MOP 3)	3–15 December 2007		Bali, Indonesia
COP 14 (MOP 4)	1–12 December 2008		Poznan, Poland
COP 15 (MOP 5)	7–19 December 2009		Copenhagen, Denmark
COP 16 (MOP 6)	29 November–10 December 2010		Cancun, Mexico
COP 17 (MOP 7)	28 November– 9 December 2011		Durban, South Africa
COP 18 (MOP 8)	26 November–8 December 2012		Doha, Qatar
COP 19 (MOP 9)	11–23 November 2013		Warsaw, Poland
COP 20 (MOP 10)	1–14 December 2014		Lima, Peru
COP 21 (MOP 11)	30 November–13 December 2015		Paris, France
COP 22 (MOP 12)	7–18 November 2016		Marrakech, Morocco
COP 23 (MOP 13)	6–18 November 2017		Bonn, Germany

Source Prepared on the basis of information/documentation given on the official website of the UNFCCC. See: https://unfccc.int/documents

About the Author

Zerrin Savaşan obtained her Ph.D. degree from the Department of International Relations, Middle East Technical University (METU); her Masters degree from the Department of European Studies, METU; and her Bachelors degree from Ankara University's Law Faculty. She is Assistant Professor, Department of International Relations, Sub-department of International Law, Faculty of Economic and Administrative Sciences, Selçuk University.

Postdoc Research: "Enhancing Compliance Through Cooperation Opportunities Between Informal Environmental Compliance Networks And Multilateral Environmental Agreements," conducted with a scholarship granted by The Scientific and Technological Research Council of Turkey (TÜBİTAK-2219) for research at the Institute for Transnational Legal Research (METRO), Law Faculty, Maastricht University, Maastricht, The Netherlands. Ph.D. Research: "Compliance Mechanisms under Multilateral Environmental Agreements", supported by scholarships provided by The Council of Higher Education (YOK) of Turkey at the Max Planck Comparative Public Law and International Law Institute, Heidelberg University, Germany; and by the Academic Staff Training Programme (OYP) of Turkey in the Center for Environmental Studies, Vrijie University, Amsterdam, The Netherlands. Aas an Assistant Professor at Selçuk University she teaches courses on Turkish Law, EU Law, International Law and International Environmental Law. She is also a climate leader under the Climate Reality Project; a research fellow under the Earth System Governance Project; and a correspondent in the International Network of Environmental Compliance and Enforcement (INECE).

Address: Zerrin Savaşan, Selçuk Üniversitesi, Alaaddin Keykubat Kampüsü, İktisadi ve İdari Bilimler Fakültesi, Uluslararası İlişkiler Bölümü, Devletler Hukuku ABD, Oda No: A 112, Selçuklu, Konya, Turkey.

© Springer Nature Switzerland AG 2019
Z. Savaşan, *Paris Climate Agreement: A Deal for Better Compliance?*
The Anthropocene: Politik—Economics—Society—Science 11,
https://doi.org/10.1007/978-3-030-14313-8

Email: szerrin@selcuk.edu.tr
Website: https://www.selcuk.edu.tr/iktisadi_ve_idari_bilimler/uluslararasi_iliskiler/akademik_personel/bilgi/7223/en.
LinkedIn: https://tr.linkedin.com/pub/zerrin-sava%C5%9Fan/36/b96/5a9.
Academic web page: https://selcuk.academia.edu/ZerrinSavasan.
Research gate: https://www.researchgate.net/profile/Zerrin_Savasan.

Index

© Springer Nature Switzerland AG 2019
Z. Savaşan, *Paris Climate Agreement: A Deal for Better Compliance?*
The Anthropocene: Politik—Economics—Society—Science 11,
https://doi.org/10.1007/978-3-030-14313-8